METHODS IN MOLECULAR BIOLOGY™

AF768093

Series Editor
John M. Walker
School of Life Sciences
University of Hertfordshire
Hatfield, Hertfordshire, AL10 9AB, UK

For further volumes:
http://www.springer.com/series/7651

PCR Detection of Microbial Pathogens

Second Edition

Edited by

Mark Wilks

Deptartment of Microbiology, Barts Health Trust, London, United Kingdom

Humana Press

Editor
Mark Wilks
Deptartment of Microbiology
Barts Health Trust
London, UK

ISSN 1064-3745 ISSN 1940-6029 (electronic)
ISBN 978-1-4939-6099-6 ISBN 978-1-60327-353-4 (eBook)
DOI 10.1007/978-1-60327-353-4
Springer New York Heidelberg Dordrecht London

Softcover re-print of the Hardcover 1st edition 2013

Printed on acid-free paper

Humana Press is a brand of Springer
Springer is part of Springer Science+Business Media (www.springer.com)

Preface

PCR methods for the detection of microbial pathogens have made relatively little impact in diagnostic microbiology laboratories. In contrast, diagnostic virology has been revolutionized by the wholesale adoption of PCR and, in particular, real-time PCR methods. Those microbiology laboratories that have introduced PCR methods have often opted to use expensive commercially produced tests, rather than the far cheaper alternative of developing their own tests or introducing tests developed by other workers.

This book makes the case for using "home brew" PCR methods to detect microbial pathogens. There is no *a priori* reason why such tests should not be as good as or better than a commercially produced equivalent. The relative cheapness of thermal cyclers and automated DNA extraction platforms makes this approach increasingly attractive, especially for high-volume tests to detect pathogens described in the chapters on MRSA and *Clostridium difficile*. Alternatively components of a commercial test can be judiciously incorporated into "in-house" methods to develop a test as shown here in the chapter on the identification and resistance detection in mycobacteria.

The weakness of many books of PCR protocols is that they are just that—a collection of unrelated methods which may or may not appeal to a reader, depending on whether they have an interest in the particular pathogen being described. Little attention is paid to the general problems of developing and introducing a PCR test into the lab, save the oft-stated advice that separate rooms should be used for separate aspects of the PCR. Although the majority of the chapters in this book are devoted to the detection of specific pathogens, the first chapters in this book should appeal to anyone working in this field regardless of their particular interests.

The relative ease of PCR and especially real-time PCR has meant that virtually any test devised will give a result of some sort. Numerous tests have been developed and introduced into practice without any knowledge of factors such as the efficiency of the reaction, its detection limits, and the specificity of the assay. In some cases this has severely hampered the acceptance of PCR for the detection of a disease, for example, invasive aspergillosis. These aspects and several others are discussed in the first chapter of this book with particular attention to the detection of *Aspergillus fumigatus*.

Another major problem with the use of PCR is the presence of PCR inhibitors in clinical specimens. The next chapter in the book gives a very comprehensive guide to PCR inhibition in different sample types and the methods used to overcome it. Because PCR of blood poses its own inhibition problems, these are discussed separately in the following chapter.

The last of the general chapters in the book looks in detail at the whole area of quality control of PCR. This has been one of the most important factors holding up the introduction of PCR into the diagnostic lab where robust QC has become an increasingly important issue and laboratories have often not felt completely confident in their results compared to those obtained by more established methods.

The main part of the book is devoted to describing methods for the detection of a wide range of pathogens and from widely different specimens and situations. The list of pathogens detected is not intended to be comprehensive but have been chosen as exemplars to illustrate the wide range of solutions that have been devised to cope with particular problems. The majority of chapters, but not all, use real-time PCR technology rather than gel-based approaches, but these may still have their place where, for example, low sample numbers are processed at a time and expensive probes may degrade before much use has been made of them. One chapter uses an approach which is not strictly a PCR at all. Loop-mediated isothermal amplification (LAMP) is a relatively new technique which has several advantages over PCR, which are well described and are bound to find increasing application in the diagnostic lab. Most chapters concentrate on the detection of a single pathogen but some use multiplex PCR to detect closely related or widely different pathogens.

Whether the reader finds his or her exact pathogen or clinical specimen that he or she is interested in, the sheer variety of approaches used together with the general principles described in the first chapters should enable the reader to design and introduce diagnostic tests in the routine laboratory with confidence.

London, UK ***Mark Wilks***

Contents

Contributors

GUMA M. K. ABDELDAIM • *Department of Clinical Microbiology, Uppsala, University Hospital, Uppsala, Sweden; Department of Clinical Microbiology, National Center for Diseases Control, Benghazi, Libya*
DENNIS BAKKER • *Department of Medical Microbiology, Centre of Infectious Diseases, Leiden University Medical Centre, Leiden, The Netherlands*
FRANCESCA BARLETTA • *Instituto de Medicina Tropical Alexander von Humboldt, Universidad Peruana Cayetano Heredia, Lima, Peru*
MICHEL G. BERGERON • *Département de Microbiologie-infectiologie et Immunologie, Faculté de médecine, Centre de Recherche en Infectiologie, Centre de Recherche du CHUQ, Université Laval, Québec, Canada*
ANNEKE M. C. BERGMANS • *Laboratory of Medical Microbiology, Franciscus Hospital, Roosendaal, The Netherlands; Laboratory of Microbiology and Infection Control, Amphia Hospital, Breda, The Netherlands*
ANTOINE BERRY • *Service de Parasitologie-Mycologie, Hôpital Rangueil, Toulouse, France*
LUC BISSONNETTE • *Département de Microbiologie-infectiologie et Immunologie, Faculté de Médecine, Centre de Recherche en Infectiologie, Centre de Recherche du CHUQ, Université Laval, Québec, Canada*
CATHRIEN A. BRUGGEMAN • *Department of Medical Microbiology, Maastricht University Medical Center, Maastricht, The Netherlands*
STEPHEN A. BUSTIN • *Blizard Institute of Cell and Molecular Science, Barts and the London School of Medicine and Dentistry, Queen Mary University of London, London, UK*
THOMAS G. CLEARY • *Center for Infectious Diseases, University of Texas School of Public Health, Houston, TX, USA*
JENS DREIER • *Institut für Laboratoriums- und Transfusionsmedizin, Herz- und Diabeteszentrum Nordrhein-Westfalen, Bad Oeynhausen, Germany*
JUDITH FILLAUX • *Service de Parasitologie-Mycologie, Hôpital Rangueil, CHU Toulouse, Toulouse, France*
JANET A.M. FYFE • *Mycobacterium Reference Laboratory, Victorian Infectious Diseases Reference Laboratory, North Melbourne, VIC, Australia*
GWENDOLYN L. GILBERT • *Centre for Infectious Diseases and Microbiology, Institute of Clinical Pathology and Medical Research, Western Clinical School, University of Sydney, Sydney, NSW, Australia Westmead Hospital, Westmead, NSW, Australia*
DAVID Y. GRAHAM • *Department of Medicine, Michael E. DeBakey Veterans Affairs Medical Center, Houston, TX, USA; Baylor College of Medicine, Houston, TX, USA*
WENDY L. J. HANSEN • *Department of Medical Microbiology, Maastricht University Medical Center, Maastricht, The Netherlands*
JOHANNES HEDMAN • *Swedish National Laboratory of Forensic Science (SKL), Linköping, Sweden; Division of Applied Microbiology, Faculty of Engineering, Lund University, Lund, Sweden*
BJÖRN HERRMANN • *Department of Clinical Microbiology, Uppsala, University Hospital, Uppsala, Sweden*

Anne Holmes • *Edinburgh Specialist Virology Centre, Directorate of Laboratory Medicine, Royal Infirmary of Edinburgh, Edinburgh, Scotland, UK*

Gemma Johnson • *Blizard Institute of Cell and Molecular Science, Barts and the London School of Medicine and Dentistry, Queen Mary University of London, London, UK*

Knut Kleesiek • *Institut für Laboratoriums- und Transfusionsmedizin, Herz- und Diabeteszentrum Nordrhein-Westfalen, Bad Oeynhausen, Germany*

Fanrong Kong • *Centre for Infectious Diseases and Microbiology, Institute of Clinicals Pathology and Medical Research, Western Clinical School, University of Sydney, Sydney, NSW, AustraliaWestmead Hospital, Westmead, NSW, Australia*

Ed J. Kuijper • *Department of Medical Microbiology, Centre of Infectious Diseases, Leiden University Medical Centre, Leiden, The Netherlands*

Caroline J. Lavender • *Mycobacterium Reference Laboratory, Victorian Infectious Diseases Reference Laboratory, North Melbourne, VIC, Australia*

Clare L. Ling • *London Regional Microbiology Network, Health Protection Agency, London, UK*

William G. MacKay • *Quality Control for Molecular Diagnostics, Glasgow, Scotland, UK*

Andrée F. Maheux • *Département de Microbiologie-infectiologie et Immunologie, Faculté de médecine, Centre de Recherche en Infectiologie, Centre de Recherche du CHUQ, Université Laval, Québec, Canada*

Timothy D. McHugh • *Centre for Clinical Microbiology, Royal Free Campus, University College London, London, UK*

Michelle L. McKechnie • *Centre for Infectious Diseases and Microbiology, Institute of Clinical Pathology and Medical Research, Western Clinical School, University of Sydney, Sydney, NSW, Australia; Westmead Hospital, Westmead, NSW, Australia*

Stephanie L. Mitchell • *Respiratory Diseases Branch, Centers for Disease Control and Prevention, Atlanta, GA, USA*

Tania Nolan • *Sigma-Aldrich, Haverhill, UK*

Theresa J. Ochoa • *Instituto de Medicina Tropical Alexander von Humboldt, Universidad Peruana Cayetano Heredia, Lima, Peru; Center for Infectious Diseases, University of Texas School of Public Health, Houston, TX, USA*

Peter Rådström • *Division of Applied Microbiology, Faculty of Engineering, Lund University, Lund, Sweden*

Lynne Renwick • *Edinburgh Specialist Virology Centre, Directorate of Laboratory Medicine, Royal Infirmary of Edinburgh, Edinburgh, Scotland, UK*

Emiko Rimbara • *Department of Bacteriology II, National Institute of Infectious Diseases, Japan*

John W.A. Rossen • *Laboratory of Medical Microbiology and Immunology, St. Elisabeth Hospital, Tilburg, The Netherlands*

Masanori Sasatsu • *Department of Microbiology, School of Pharmacy, Tokyo University of Pharmacy and Life Sciences, Tokyo, Japan*

Kit-Hang Gilman Siu • *Department of Microbiology, Li Ka Shing Faculty of Medicine, The University of Hong Kong, Hong Kong SAR, China*

Robyn Anne Stoddard • *National Center for Zoonotic, Vector-Borne, and Enteric Diseases, Centers for Disease Control and Prevention, Atlanta, GA, USA*

Kathleen M. Tatti • *Division of Bacterial Diseases, Centers for Disease Control and Prevention, National Center for Immunization and Respiratory Diseases, Meningitis and Vaccine Preventable Diseases Branch, Atlanta, GA, USA*

KATE TEMPLETON • *Edinburgh Specialist Virology Centre, Directorate of Laboratory Medicine, Royal Infirmary of Edinburgh, Edinburgh, Scotland, UK*
MARIA LUCIA TONDELLA • *Division of Bacterial Diseases, Centers for Disease Control and Prevention, National Center for Immunization and Respiratory Diseases, Meningitis and Vaccine Preventable Diseases Branch, Atlanta, GA, USA*
RENATE JOHANNA VAN DEN BERG • *Department of Medical Microbiology, Centre of Infectious Diseases, Leiden University Medical Centre, Leiden, The Netherlands*
TANJA VOLLMER • *Institut für Laboratoriums- und Transfusionsmedizin , Herz- und Diabeteszentrum Nordrhein-Westfalen, Bad Oeynhausen, Germany*
PAUL S. WALLACE • *Quality control for Molecular Diagnostics, Technology Terrace, Todd Campus, Glasgow, Scotland, UK*
JONAS M. WINCHELL • *Respiratory Diseases Branch, Centers for Disease Control and Prevention, Atlanta, GA, United States*
PETRA F. G. WOLFFS • *Department of Medical Microbiology, Maastricht University Medical Center, Maastricht, The Netherlands*
WING-CHEONG YAM • *Department of Microbiology, Li Ka Shing Faculty of Medicine, The University of Hong Kong, Hong Kong SAR, China*
WATARU YAMAZAKI • *Department of Veterinary Science, Faculty of Agriculture, University of Miyazaki, Miyazaki, Japan*

Chapter 1

Real-Time Quantitative PCR, Pathogen Detection and MIQE

Gemma Johnson, Tania Nolan, and Stephen A. Bustin

Abstract

Nucleic acids are the ultimate biomarker and real-time PCR (qPCR) is firmly established as the method of choice for nucleic acid detection. Together, they allow the accurate, sensitive and specific identification of pathogens, and the use of qPCR has become routine in diagnostic laboratories. The reliability of qPCR-based assays relies on a combination of optimal sample selection, assay design and validation as well as appropriate data analysis and the "Minimal Information for the Publication of real-time PCR" (MIQE) guidelines aim to improve both the reliability of assay design as well as the transparency of reporting, essential conditions if qPCR is to remain the benchmark technology for molecular diagnosis.

Key words: Real-time PCR, Aspergillosis, Diagnosis, DNA, Biomarker

1. Introduction

The last few years have witnessed a prodigious proliferation in the use of the real-time quantitative polymerase chain reaction (qPCR) for diagnostic applications designed to detect and quantify microbial pathogens (1). Key reasons are the characteristic sensitivity, specificity, and wide linear dynamic range of qPCR assays; this development is being accelerated by qPCR integration with nanotechnology, which has resulted in its emergence in novel areas such as high throughput, nanoliter qPCR (2), and microfluidic digital PCR (3). These innovations have extended the scope of qPCR technology by holding out additional advantages such as short assay times, low reagent usage, and exceptionally rapid heating/cooling rates. Importantly, integration of multiple processing modules allows further size reduction and power consumption, making qPCR technology viable for point-of-care diagnostics. Clearly, the combination of nanotechnology and qPCR has huge potential in clinical diagnostics, assuming the availability of adequate and

Mark Wilks (ed.), *PCR Detection of Microbial Pathogens: Second Edition*, Methods in Molecular Biology, vol. 943,
DOI 10.1007/978-1-60327-353-4_1,

appropriate sample material, the need for the technology to accommodate fairly crude biological samples as analytical targets, optimal sampling timing regarding the course of disease and standardization of pre-assay and assay protocols (4, 5). The contribution qPCR can make to improve the diagnosis of life-threatening diseases is clearly illustrated by its application to the detection of invasive aspergillosis (IA), caused by pathogenic *Aspergillus* fungi, where early diagnosis and treatment are essential for adequate therapeutic management. There is an increasing incidence of systemic fungal infections in patients immunocompromised as a result of HIV infection or organ transplantation, patients from intensive care units, and those with prolonged neutropenia after intensified chemotherapy. These include patients with acute leukemia during induction therapy or after transplantation of allogeneic hematologic stem cells. IA has a high mortality rate and, in surviving patients, leads to considerable complications often limiting further anti-neoplastic therapies. The reliability of current diagnostic tools in early detection of fungal infections in neutropenic patients is limited; hence the definitive diagnosis of IA remains a challenge.

2. qPCR and Pathogen Detection

The committee of the European Organization for Research and Treatment of Cancer/Mycosis (EORTC) and Mycoses Study Group of National Institute of Allergy and Infectious Diseases (NIAID) has thus far not recommended the routine use of PCR in the diagnosis of IA (6), mainly because of the absence of standardized protocols (7). However, the feasibility of using qPCR in these patients as a potentially more accurate alternative to conventional diagnostic procedures has been evaluated extensively (8–10). There are a number of compelling reasons for utilizing qPCR as a diagnostic assay for IA:

- A qPCR assay can be completed in minutes. This immediacy allows virtually instantaneous reporting of results.
- Judicious targeting of amplified genomic regions, e.g., the rRNA gene region, maximizes the reliability of the assay. It substitutes the uncertainty of having to identify a single copy of a gene on an infectious fungal particle or single genome with the consistency and reliability of being able to target tens of copies specified within that single genome.
- Primer and probe design is extremely flexible and assays can be tailored to be specific for most fungal genera or species.
- The quantitative potential of qPCR allows determination of fungal load, which can distinguish between colonization and infection.

- Multiplexing capacity allows qPCR assays to detect multiple targets and/or include appropriate controls in a single reaction.

In principle, the only requirements for obtaining quantitative data are a set of reagents (target-specific primers and optional probes, a source of DNA and an enzyme/buffer/dNTP mix), a qPCR instrument that detects the fluorescence emitted by the PCR reaction and software that calculates DNA copy numbers or relative abundance from the quantification cycles recorded by the instrument. In practice, the development and successful application of an optimal qPCR diagnostic assay depends on understanding and addressing several challenges and shortcomings. A systematic review and meta-analysis on the use of PCR tests for the diagnosis of IA highlights the lack of homogeneity of the PCR methods used (11) and another recent study shows that lack of rigorous experimental controls, together with a prevalence of false-positive and -negative results hinders the interpretation of diagnostic performance, thus impeding widespread acceptance of qPCR-based technology (12). False negatives can occur due to suboptimal DNA extraction, for example low recovery of DNA and the presence of PCR inhibitors. Since all fungi produce small, hydrophobic conidia that are difficult to disrupt, it is not surprising that sample preparation is a source of heterogeneity and a critical parameter for obtaining reliable qPCR results (10). Other causes are the presence of large amounts of human genomic DNA competing with the microbial target for amplification and of course suboptimal analytical sensitivity of the qPCR reaction itself. False positives can occur either due to contamination at any time during the pre-assay stage, i.e. during sample collection, DNA extraction, and qPCR set-up, or be the result of poor assay design, resulting from cross-reactivity of the target qPCR assay with other DNA. Hence any serious qPCR assay must incorporate controls to assess for these factors contributing to false-positive and false-negative results, something sadly lacking from many published studies.

3. Quality Control Issues

It is remarkably difficult to make a reaction fail completely but alarmingly simple to produce poor quality data (13). When properly executed, qPCR assays can be highly reproducible; however, since pathogen abundance is frequently low, assay reproducibility is influenced by parameters such as distribution statistics (14), with stochastic sampling effects affecting the reliability of the qPCR data (15). This emphasizes the importance of repetitive testing in clinical samples and one of the strengths of qPCR assays is the ease with which it is possible to obtain quantitative data for every sample, which encourages the use of biological replicates and permits the application of powerful statistical analyses to the quantification procedure.

Like any clinical diagnostic assay, qPCR assays must be properly validated (16) and meet the criteria expected of any laboratory test applied in clinical medicine: (1) standardization of the test across different laboratories, (2) reproducibility of positive and negative predictive values, and (3) reliable sensitivity and specificity. This involves the establishment of a set of quality standards dealing with experimental protocols, the use of appropriate positive and negative control samples, and suitable analysis and reporting guidelines. Standardization, in particular, is all-important (11, 17, 18) as it provides the foundation for robust, reliable, and comparable results that are of critical importance for the consistent management of patients (19).

Specifically, a successful diagnostic qPCR assay requires careful consideration of these issues.

1. Optimal sample quality is a prerequisite for the generation of valid quantitative data (20). Hence sample collection, preparation, and transport and nucleic acid extraction methods are critical parameters in test performance and must be optimized and, ideally, standardized. In principle, extraction of fungal nucleic acids, especially if present in a cell-free state, from bronchoalveolar lavages, blood, and serum is relatively straightforward; however, it is easy to co-purify inhibitors of the PCR that will generate inconsistent and unreliable results. This must be assessed for every sample prior to carrying out any qPCR assay aimed at pathogen detection. Amongst the numerous methods for *Aspergillus* DNA extraction the bead-beating methods appear to be the most successful (21), although no single extraction method is optimal for all fungal species (22). However, extraction of DNA from conidia, combined with the presence of a complex, sturdy cell wall, can require extensive and harsh extraction methods that can damage DNA and RNA to the extent that amplification becomes unreliable.

2. Primer selection is critical since it affects the sensitivity of the qPCR assay. The structure of the nucleic acid target at the primer-binding site must be taken into account, as this affects the accessibility of the target to the primers and extensive secondary structure at a primer-binding site will result in a less efficient and sensitive assay.

3. Regular calibration of the real-time instrument is crucial for obtaining consistent and accurate results. The quantification cycle (C_q) is neither absolute nor invariant, but varies between assays carried out on different days with different reagents or on different instruments. This is because the C_q depends on the instrument's threshold setting, which in turn depends on background fluorescence, which varies with different probes,

chemistries, instruments, and assay protocols. Therefore, samples should not be compared by C_q values (23), but they should be converted to target copy numbers and reported as such.

4. Analytical sensitivity and specificity are critical parameters of any diagnostic qPCR assay. Analytical sensitivity refers to the smallest number of nucleic acid molecules that can be detected and distinguished from a zero result and is best calculated using a standard curve, which defines the range of the assay. It is inappropriate to report results that lie significantly outside the upper and lower concentration of target defined by the standard curve. Analytical specificity is determined by identifying the percentage of samples without the target sequence that generate a positive result. If a well-designed assay is used, this will be zero.

5. There must be agreement between technical replicates (to within ±0.5 C_qs), as this provides important information about the reliability of the assay and its operator. Repeatability is measured as the amount of agreement between replicates tested in different runs on the same instrument in the same laboratory. Reproducibility is determined in several laboratories using the identical assay (protocol, reagents, and controls). It is important to maintain the internal quality control by monitoring the assay for both parameters. If the assay is to be applied in another geographical region and/or population, it might be necessary to revalidate it under the new conditions as mutations within the primer sites, especially at their 3′ ends, or the prevalence of other subtypes, will affect the performance of the assay and render the established validation no longer valid.

6. False-positive results may arise from product carryover from positive samples or, more commonly, from cross-contamination by PCR products from earlier experiments (24). A recent report suggests that nutritional supplements can harbor fungal DNA that can pass into the serum from the intestinal tract and cause false-positive results (25). It is critically important to include negative controls, i.e., samples that are as similar to the test samples as possible but exclude the target. Since false-negative results in an optimized assay are mostly due to inhibitory effects and/or pipetting errors, it is important to always include a positive control with any qPCR assay (26), ideally in the form of a dilution curve. In addition, all samples should be tested for inhibition using a simple "alien" assay (27) and any nucleic acid preparations showing inhibition must be repurified. Dilution curves are useful, as the highest dilutions provide information about the variability of the assay at very low target copy numbers and qPCR results are questionable if they are not supported by data demonstrating the overall sensitivity of

the assay applied (16). Furthermore, running a dilution curve with each assay immediately reveals any problem with that particular run and increases confidence when reporting negative results. The same argument applies to running samples in duplicate or, preferably, in triplicate and in this qPCR is no different from any clinical diagnostic assay.

Robust and precise qPCR usually correlates with high PCR efficiency; consequently the primary aim of assay optimization is to achieve the most efficient qPCR assay optimized at the sensitivity and specificity appropriate for pathogen detection. Detailed optimization requires the consideration of a range of parameters that include the concentration of reaction components, such as salts and oligonucleotides, as well as reaction conditions such as primer annealing temperatures, incubation periods, and even ramp rates. In addition, qPCR results are affected by the fluorescent signal, an important concern when designing multiplex assays. Since variations in experimental protocols lead to highly variable data, it is essential that all relevant experimental conditions and assay characteristics are reported when publishing qPCR results (28).

4. Optimization Parameters

4.1. Dilution Curves

A dilution curve, generated by performing qPCR with a serial dilution of template, is a convenient tool to test assay efficiency. A well-optimized assay should be linear over a range of at least nine logs of template concentration, with efficiency close to 100% and high reproducibility between technical replicates. Amplification efficiency can be determined from the slope of the linear regression of a plot of C_q (*y*-axis) vs. log [quantity]. The template may be any suitable material such as cDNA, genomic DNA, PCR product, or synthetic DNA oligonucleotides that match the sequence of the target amplicons. The advantages of using synthetic oligonucleotides are that (1) one synthesis is sufficient for several million reactions, thus providing a consistent positive control and (2) they can be accurately quantified, thus making it possible to obtain a fairly accurate pathogen copy number value, making comparison of different assays more reliable. However, artificial templates must be handled with extreme caution since they constitute highly concentrated targets that could potentially contaminate all oligonucleotides and reagents if handled carelessly.

4.2. Amplification Efficiency

Amplification efficiency, *E*, is calculated from the slope of the standard curve using the formula: $E = 10^{(-1/\text{slope})}$, which is usually converted into a percentage efficiency (% Efficiency = $(E-1) \times 100\%$). A combination of a good assay and accurate pipetting will generate dilution curves that demonstrate near-perfect doubling (or consistent amplification) with each amplification cycle. Thus, the spacing of the fluorescence curves will be determined by the equation 2^n = dilution factor, where n is the number of cycles between curves. For example, with a tenfold serial dilution of DNA, $2^n = 10$. Therefore, $n = 3.32$ and the C_q values should be separated by approximately 3.32 cycles. If doubling occurs at each cycle, $E = 10^{-(1/-3.32)} = 2$ and the % Efficiency $= (2 - 1) \times 100\% = 100\%$. In practice, amplification efficiency will be around 90–105% and result in gradients of between −3.2 and −3.5. Note that the presence of inhibitors can result in an apparent increase in efficiency. This is because samples with the highest concentration of template also have the highest level of inhibitors, and therefore display a greater lag between C_q values than samples with lower template concentrations and lower levels of inhibitors. As a result, the absolute value of the slope decreases and the calculated efficiency appears to increase. Similarly inaccurate pipetting can lead to data suggestive of higher or lower efficiency and so it is important to assess the correlation coefficient of the line to determine the linearity and reproducibility of the assay. It provides a measure of how well the data fit on a straight line and is influenced by pipetting accuracy and by the range of the assay. During assay evaluation and optimization, three technical replicates of each template dilution should be processed in parallel in order to establish that the assay is reproducible. A stable assay will demonstrate an $R^2 > 0.98$ over at least six logs and with three replicates.

Finally, DNA templates themselves can affect the efficiency and quality of a qPCR assay. A qPCR template may be present at any concentration from a single copy to approximately 10^{11} copies. High concentration of template will inhibit the reaction resulting in reduced yield and inaccurate C_q differences between amplification plots, so inaccurate quantification. Low initial concentration can result in lack of detection of amplified product if the final yield is extremely low.

4.3. Assay Specificity

Reliable pathogen detection requires absolute assay specificity. In a qPCR experiment all detectable products, be they specific or nonspecific, contribute to the final amplification plot and hence any qualitative or quantitative result. This can be a problem when using a generic detection system such as SYBR Green I dye, but can also affect the quality of a probe-based assay designed to detect only a single target sequence. Although nonspecific amplification may not affect the shape of the amplification plot and is not detected by a probe-based assay, it nonetheless affects amplification efficiency and assay sensitivity. Performing a post-reaction melt analysis using SYBR Green I dye during the assay optimization stage can validate the specificity of amplification. An assay with high specificity will produce a single peak at a high temperature with nothing in the no-template controls. If the melting curve has more than one major peak, agarose gel electrophoresis and DNA sequencing should determine the identities of the products. Lowering the primer concentrations and increasing the annealing temperatures will often reduce the amount of nonspecific products.

4.4. Buffer and Sample Considerations

The composition of the reaction buffer has a significant influence on assay specificity, as it influences binding of primer to template. Reaction buffers contain variations on a basic composition consisting of ammonium sulfate, Tris, EDTA, BSA, β-mercaptoethanol, dNTPs, $MgCl_2$, KCl, NaCl, and DNA polymerase. Optimum buffer composition is dependent upon the DNA polymerase used: different enzymes can influence PCR efficiency and therefore product yield. It is generally accepted that *Taq* DNA polymerase performs optimally in a basic buffer of 50 mM KCl and 10 mM Tris–HCl, pH 8.3 (measured at room temperature). Some enzymes have a requirement for added protein (BSA is usually added, when required). Although dNTPs are the standard substrate for DNA polymerases, dUTP may be incorporated into qPCR reactions to provide target for subsequence contamination control steps using Uracil–N-Glycosylase to remove UTP containing templates from reaction mixes prior to amplification. In a standard reaction, the concentration of dNTPs is included in equimolar ratios, usually 200 μM (or up to 500 μM) of each dNTP. Many commercially available buffers may also contain PCR enhancers such as single-stranded binding protein (SSBP), betaine, formamide, or DMSO. The presence of detergent improves the activity of some enzymes, presumably by reducing aggregation.

The salt concentration within the buffer affects the T_m of the primer–template duplex and is required for primer annealing. Concentrations of KCl or NaCl above 50 mM can be inhibitory, while $MgCl_2$ is required as a cofactor for DNA polymerase. The most influential factor effecting free magnesium ions is the concentration of dNTPs in the reaction and so the magnesium ion concentration must exceed the dNTP concentration. Typical reactions contain 1.5 mM $MgCl_2$ in the presence of 0.8 mM dNTPs resulting in approximately 0.7 mM free magnesium. Optimal $MgCl_2$ concentrations can differ for every primer/template combination and this leads to problems if the reactions are to be multiplexed, as there will be significant differences in the efficiency of the individual reaction and, hence, in the yield of the different PCR products. Hence primer optimization is an essential step in multiplex assay development.

4.5. Primer Optimization

Primer optimization serves to drive the kinetics of binding of the primers to the specific template sequence. Annealing is a kinetic result of the annealing temperature of the reaction and also the concentration of the primers. Optimization is a two stage process and should proceed as follows.

1. Optimal annealing temperature T_a is approximately 5°C lower than the T_m (melting temperature) of the primer with the lowest T_m. The T_m for a short oligonucleotide can easily be calculated by the approximation: T_m = 4(number G + number C residues) + 2(number A + number T residues). Taking this temperature as a starting point, different temperatures (±5°C) should be tested in steps of 0.5°C; alternatively, a simpler option is to use a temperature gradient PCR block (as for example is standard on the BioRad CFX). Whilst unexciting, applying the optimal T_a will result in higher specificity and yield, since using too low a T_a results in nonspecific priming, whereas too high a T_a results in inefficient priming and elongation.
2. Primer concentration should be optimized using a primer concentration matrix in which all primer concentrations from 50 to 300 nM are tested against each other and the conditions producing the highest concentration of specific template are selected. When optimizing a multiplex reaction it is essential that each single reaction is of high quality (as defined above) before attempting to combine them. The single most important factor for multiplex success is the primer design and it is recommended that a program such as Beacon Designer is used since this will compare the compatibility of assays in silico. A serial dilution containing each of the targets is then interrogated with the assays in pairs. Any deviation from the original single assay indicates that the oligonucleotides are interfering. Optimization may then include reassessment of oligonucleotide concentration, change in $MgCl_2$ concentration and potentially in the amount of DNA polymerase in the reaction.

5. The MIQE Guidelines

The Minimum Information for the Publication of Quantitative PCR Experiments (MIQE; (http://www.rdml.org/miqe) guidelines have two aims: first, to provide a standardized template for good assay design and second, to encourage transparency of protocols, data analysis, and conclusions. MIQE is part of a drive that promotes minimum guidelines for the reporting of biological experiments (29) and is modeled on similar guidelines such as MIAME (Minimal Information about a Microarray Experiment), developed several years ago (30) and MIAPE (Minimal Information about a Proteomics Experiment) (31). All of these are initiatives developed under the umbrella of the MIBBI (Minimum Information for Biological and Biomedical Investigations) standardization body, which has the goal of unifying all of the standardization guidelines for biological and biomedical research. In addition, the Real-Time PCR Data Markup Language (RDML; http://www.rdml.org) has been developed by a consortium, upon the request of MIBBI, to enable the straightforward exchange of qPCR data and related information between qPCR instruments and third party data analysis software, between colleagues, and with journals or public data repositories (32).

The four key areas of standardization that define any qPCR experiment are (1) study design, (2) technical detail, (3) analysis methods, and (4) statistics. MIQE defines the minimum information required for evaluation of qPCR results and addresses these under a set of nine captions that describe a large number of individual elements. There is a clear hierarchy with some parameters, labeled E (essential) in the published guidelines, indispensable for attaining the ambition of the main aims, whereas other components, labeled D (desirable) more peripheral, yet constituting an effective foundation for the realization of best practice protocols. It is essential to report as much information about sample acquisition, storage, and handling as possible; it is also important to provide details of sample processing procedures, since any sample has to pass through a number of preparative steps prior to the qPCR assay, every one of which can introduce additional variability (33, 34). For pathogen detection, it is critical that the sequences of primers and probes are reported, since an experiment cannot be reproduced if information on one of the principal reagents is lacking. The most commonly used models for the analysis of qPCR data use either the $\Delta\Delta C_q$ (35) or the more generalized efficiency calibrated model (36) and updates or variations continue to be introduced (37, 38). However, confidence interval and statistical significance considerations are still not accorded a high enough priority (39) and there is a tremendous reluctance to use dilution curves to test the amplification efficiencies of individual assays, even though, as discussed above, this method remains by far the easiest, most transparent and informative method for determining amplification efficiency. Furthermore, dilution curves also provide convenient positive controls, can act as inhibition controls, and help define the dynamic range and the limits of detection all at the same time. Ideally such a dilution curve should be run with each sample, as all these parameters could (and probably do) vary between samples.

5.1. Brief Troubleshooting Guide

Below are some basic recommendations for troubleshooting qPCR experiments.

5.1.1. No/Poor/Late Amplification

No amplification is characterized by no significant increase in fluorescence above background. Poor amplification is typified by a very low ΔRn value (<0.05) or by a slope that varies significantly between samples amplifying the same target. Late amplification is anything that generates a $C_q > 35$.

First check qPCR reaction conditions.

- Check amplification products on a gel. If no product is present, repeat the assay checking all reagents are added and that the thermal cycling conditions are correct. If a product is present, the instrument detection settings may be incorrect; for example the wrong filter may have been used to detect the light emitted from the reporter fluorophore.
- Make sure all required reagents, i.e. enzyme(s), reaction buffer, primers, probe, and template were added.
- Check annealing temperature (<60°C), elongation times and cycle number.
- Check that primers or other reactants have been diluted correctly.
- Keep pipettes well calibrated.
- Is the thermal cycler programmed to detect fluorescence during the wrong PCR step?
- Is there too little starting material?
- Is there too much starting material?
- Is the amplicon too long?
- If using homemade buffers ensure that the salt and buffer concentrations are correct.
- Incomplete thawing of frozen buffers will change the salt concentration in the remaining buffer. Increases in $MgCl_2$ concentrations will affect the efficiency of primer binding and may cause the appearance of primer dimers and reduce the efficiency of the PCR reaction.
- If there have been changes in the reaction volumes or number of reactions for which master mix has been prepared, errors may have been made during pipetting. Changing the volumes of the reaction mixtures can cause different amounts of error in the volumes being dispensed. This is partially due to differences in the tolerances between large and small volume pipettes. This error may not be propagated linearly during scale up. Highly sensitive methods such as PCR can significantly magnify these problems. It is best to scale up in stages.
- Increase the number of amplification cycles, especially if the PCR efficiency is low.

Then check the reagents/instrument

- If the assay is probe-based and has previously worked well, the probe may have gone off. For example, it may have been photobleached if it has been left in the light. Always store fluorophore-labeled oligonucleotides in aliquots in the dark at -20°C.
- If the background level of the probe is very high, it is possible that it may have become hydrolyzed, e.g., if subjected to repeated freeze/thaw cycles.

- If the assay is a new assay, it may well be that the probe manufacturer is to blame. Test the quality of the probe using a DNAse I digestion assay; following digestion the fluorescence should greatly increased due to reporter and quencher becoming separated.
- Has the SYBR Green "gone off"? Once SYBR Green has been diluted, it goes off very quickly and can only be kept at 4°C for approximately 2 weeks. It must also be kept in the dark.
- If using a new batch of polymerase, note that different lots of polymerase, even from the same supplier, can have different amounts of specific and exonuclease activity, but still be within the manufacturer's specifications.
- Different thermal cyclers, particularly if they using heating blocks, and thermal cyclers from different manufacturers have different ramping kinetics and heating efficiencies, hence can affect the efficiency of the PCR reaction. Variability in heating efficiency may also occur in different wells within the same block, again particularly so in 96 or 384 well instruments.

5.1.2. Poorly Designed Primers and/or Probe

This results in poor PCR efficiency

- Optimize primer concentrations; this is essential for optimal assay efficiency.
- Confirm annealing temperature is appropriate for primer.
- Are primers binding in a region with secondary structure? If this has not already been checked using Beacon Designer or a similar program. Use alternative primers.
- Is the probe too long?
- Is there probe secondary structure?

5.1.3. Reaction Conditions Not Optimized

This results in poor PCR efficiency

- Is the annealing temperature too high?
- Are the annealing/extension times too short?
- $MgCl_2$ concentration too high/low.

5.1.4. Positive NTC

The master mix may be contaminated with DNA template or PCR product.

- Use fresh aliquots of all reagents including sterile water.
- Only use pipettes, tips, solutions (especially water), and racks dedicated to setting up reactions. Do not use pipettes and accessories that have been exposed to amplicon.
- Decontaminate surfaces, pipets, and racks.
- Change gloves and tips frequently.
- Change location of PCR set-up.

- Check melt curves. If the NTC melt curve is different from the amplicon melt curve, you may still be able to use data.
- An excess of probe can generate an artifactual positive result. This can be determined by leaving out the *Taq* polymerase from the NTC.

5.1.5. Primer Dimers

Primer dimers are more likely to occur in no target controls, or when there is very little target nucleic acid. Even small amounts of target can suppress primer dimer.

- If dimers occur in the presence of normal amounts of target, primers need to be redesigned. Check for absence of complementary sequences at the 3′ ends.

5.1.6. Multiple Peaks in Melt Curve

Improve the stringency by raising the annealing temperature or lowering magnesium concentration.

- Run reaction on gel to check for specificity and consistent presence of additional bands.
- Shoulders in the melt curves do not necessarily mean that the assay is nonspecific. The amplicon may contain AT-rich subdomains. Run reaction on gel to check for specificity.

5.1.7. Standard Curve Is Unreliable ($R^2 < 0.99$)

The low (or high) concentration point(s) of the dilution series can sometimes be removed to improve the R^2 value. However, if the unknowns fall in the low range, the experiment will need to be repeated.

5.1.8. Inhibitor Present

Reagents such as SDS, EDTA, glycerol, sodium pyrophosphate, spermidine, formamide, guanidinium salts, and DMSO can inhibit *Taq* DNA polymerase.

- Dilute DNA sample by 1/10 and 1/100 and repeat the assay, as this may dilute out the inhibitor.
- If this fails, remove inhibitor by ethanol precipitation of the DNA. Include a 70% (v/v) ethanol wash of the DNA pellet. Glycogen (0.25 μg to 0.4 μg/μl) can be included to aid in DNA recovery for small samples.

5.1.9. Erratic Amplification Plots/High Well-to-Well Variation

Use extra care when pipetting solutions containing low amounts of target.

- Is the baseline set using wrong cycle range?
- Is there sample evaporation due to loose lid or poor sealing?
- Was there incomplete mixing of reagents?
- Are there air bubbles at the bottom of the reaction tubes?

- Were the frozen stocks of target, primers or buffer not completely thawed or mixed when used?
- Spikes in the signal can be caused by problems with the lamp, misaligned optics or other mechanical and/or electronic issues.

6. Conclusions

qPCR is a powerful enabling technology that has started to play a central role in clinical diagnostics. However, the combination of ease of use and lack of rigorous standards of practice has resulted in widespread publication of poor data, resulting in inappropriate conclusions. In the context of reliable detection of pathogens in general, and *Aspergillus* in particular, the major limitation of qPCR is the lack of standardization at every stage of the molecular process from sample type and processing to interpretation of results. Nonetheless, when optimized, it is extremely sensitive and highly specific (40).

Any solution to the challenge of how to make PCR-based assays more reliable requires both an appreciation and an understanding of numerous attributes that include statistics, mathematical modeling, technical know-how and a willingness to share this intelligence. MIQE constitutes a reference framework for communication within the research community, instrument, and reagent manufacturers and publishers that promises to deliver guidelines that promote transparency of experiments and confidence in results and conclusions that advance, rather than impede our knowledge.

References

1. Bustin SA, Benes V, Nolan T, Pfaffl MW (2005) Quantitative real-time RT-PCR—a perspective. J Mol Endocrinol 34:597–601
2. Morrison T, Hurley J, Garcia J, Yoder K, Katz A, Roberts D, Cho J, Kanigan T, Ilyin SE, Horowitz D, Dixon JM, Brenan CJ (2006) Nanoliter high throughput quantitative PCR. Nucleic Acids Res 34:e123
3. Zhang C, Xing D (2007) Miniaturized PCR chips for nucleic acid amplification and analysis: latest advances and future trends. Nucleic Acids Res 35:4223–4237
4. Cho YK, Kim J, Lee Y, Kim YA, Namkoong K, Lim H, Oh KW, Kim S, Han J, Park C, Pak YE, Ki CS, Choi JR, Myeong HK, Ko C (2006) Clinical evaluation of micro-scale chip-based PCR system for rapid detection of hepatitis B virus. Biosens Bioelectron 21: 2161–2169
5. Kaigala GV, Huskins RJ, Preiksaitis J, Pang XL, Pilarski LM, Backhouse CJ (2006) Automated screening using microfluidic chip-based PCR and product detection to assess risk of BK virus-associated nephropathy in renal transplant recipients. Electrophoresis 27:3753–3763
6. Ascioglu S, Rex JH, de Pauw B, Bennett JE, Bille J, Crokaert F, Denning DW, Donnelly JP, Edwards JE, Erjavec Z, Fiere D, Lortholary O, Maertens J, Meis JF, Patterson TF, Ritter J, Selleslag D, Shah PM, Stevens DA, Walsh TJ (2002) Defining opportunistic invasive fungal infections in immunocompromised patients with cancer and hematopoietic stem cell transplants: an international consensus. Clin Infect Dis 34:7–14
7. Klingspor L, Loeffler J (2009) Aspergillus PCR formidable challenges and progress. Med Mycol 47(Suppl 1):S241–S247

8. Duval SM, Donnelly JP, Barnes R, Löffler J (2008) PCR-based methods with aspergillosis as a model. J Invasive Fungal Infect 2:46–51
9. Perlin DS, Zhao Y (2009) Molecular diagnostic platforms for detecting Aspergillus. Med Mycol 47(Suppl 1):S223–S232
10. White PL, Bretagne S, Klingspor L, Melchers WJ, McCulloch E, Schulz B, Finnstrom N, Mengoli C, Barnes RA, Donnelly JP, Loeffler J (2010) Aspergillus PCR: one step closer to standardization. J Clin Microbiol 48:1231–1240
11. Mengoli C, Cruciani M, Barnes RA, Loeffler J, Donnelly JP (2009) Use of PCR for diagnosis of invasive aspergillosis: systematic review and meta-analysis. Lancet Infect Dis 9:89–96
12. Khot PD, Fredricks DN (2009) PCR-based diagnosis of human fungal infections. Expert Rev Anti Infect Ther 7:1201–1221
13. Garson JA, Huggett JF, Bustin SA, Pfaffl MW, Benes V, Vandesompele J, Shipley GL (2009) Unreliable real-time PCR analysis of human endogenous retrovirus-W (HERV-W) RNA expression and DNA copy number in multiple sclerosis. AIDS Res Hum Retroviruses 25:377–378
14. de Vries TJ, Fourkour A, Punt CJ, van de Locht LT, Wobbes T, van den BS, de Rooji MJ, Mensink EJ, Ruiter DJ, van Muijen GN (1999) Reproducibility of detection of tyrosinase and MART-1 transcripts in the peripheral blood of melanoma patients: a quality control study using real time quantitative RT PCR. Br J Cancer 80:883–891
15. Peccoud J, Jacob C (1996) Theoretical uncertainty of measurements using quantitative polymerase chain reaction. Biophys J 71:101–108
16. Afzal MA, Osterhaus AD, Cosby SL, Jin L, Beeler J, Takeuchi K, Kawashima H (2003) Comparative evaluation of measles virus-specific RT-PCR methods through an international collaborative study. J Med Virol 70:171–176
17. Niesters HG (2001) Quantitation of viral load using real-time amplification techniques. Methods 25:419–429
18. Niesters HG (2004) Molecular and diagnostic clinical virology in real time. Clin Microbiol Infect 10:5–11
19. Gabert J, Beillard E, van der Velden VH, Bi W, Grimwade D, Pallisgaard N, Barbany G, Cazzaniga G, Cayuela JM, Cave H, Pane F, Aerts JL, De Micheli D, Thirion X, Pradel V, Gonzalez M, Viehmann S, Malec M, Saglio G, van Dongen JJ (2003) Standardization and quality control studies of 'real-time' quantitative reverse transcriptase polymerase chain reaction of fusion gene transcripts for residual disease detection in leukemia - a Europe Against Cancer program. Leukemia 17:2318–2357
20. Muller MC, Hordt T, Paschka P, Merx K, La Rosee P, Hehlmann R, Hochhaus A (2004) Standardization of preanalytical factors for minimal residual disease analysis in chronic myelogenous leukemia. Acta Haematol 112:30–33
21. Griffiths LJ, Anyim M, Doffman SR, Wilks M, Millar MR, Agrawal SG (2006) Comparison of DNA extraction methods for Aspergillus fumigatus using real-time PCR. J Med Microbiol 55:1187–1191
22. Fredricks DN, Smith C, Meier A (2005) Comparison of six DNA extraction methods for recovery of fungal DNA as assessed by quantitative PCR. J Clin Microbiol 43:5122–5128
23. Bustin SA, Nolan T (2004) Pitfalls of quantitative real-time reverse-transcription polymerase chain reaction. J Biomol Tech 15:155–166
24. Kwok S, Higuchi R (1989) Avoiding false positives with PCR. Nature 339:237–238
25. Millon L, Grenouillet F, Crouzet J, Larosa F, Loewert S, Bellanger AP, Deconinck E, Legrand F (2010) False-positive Aspergillus real-time PCR assay due to a nutritional supplement in a bone marrow transplant recipient with GVH disease. Med Mycol 48:661–664
26. Monpoeho S, Coste-Burel M, Costa-Mattioli M, Besse B, Chomel JJ, Billaudel S, Ferre V (2002) Application of a real-time polymerase chain reaction with internal positive control for detection and quantification of enterovirus in cerebrospinal fluid. Eur J Clin Microbiol Infect Dis 21:532–536
27. Nolan T, Hands RE, Ogunkolade BW, Bustin SA (2006) SPUD: a qPCR assay for the detection of inhibitors in nucleic acid preparations. Anal Biochem 351:308–310
28. Bustin SA, Benes V, Garson JA, Hellemans J, Huggett J, Kubista M, Mueller R, Nolan T, Pfaffl MW, Shipley GL, Vandesompele J, Wittwer CT (2009) The MIQE guidelines: minimum information for publication of quantitative real-time PCR experiments. Clin Chem 55:611–622
29. Taylor CF, Field D, Sansone SA, Aerts J, Apweiler R, Ashburner M, Ball CA, Binz PA, Bogue M, Booth T, Brazma A, Brinkman RR, Michael Clark A, Deutsch EW, Fiehn O, Fostel J, Ghazal P, Gibson F, Gray T, Grimes G, Hancock JM, Hardy NW, Hermjakob H, Julian RK Jr, Kane M, Kettner C, Kinsinger C, Kolker E, Kuiper M, Novere NL, Leebens-Mack J, Lewis SE, Lord P, Mallon AM, Marthandan N, Masuya H, McNally R, Mehrle A, Morrison N, Orchard S, Quackenbush J,

Reecy JM, Robertson DG, Rocca-Serra P, Rodriguez H, Rosenfelder H, Santoyo-Lopez J, Scheuermann RH, Schober D, Smith B, Snape J, Stoeckert CJ Jr, Tipton K, Sterk P, Untergasser A, Vandesompele J, Wiemann S (2008) Promoting coherent minimum reporting guidelines for biological and biomedical investigations: the MIBBI project. Nat Biotechnol 26:889–896

30. Brazma A, Hingamp P, Quackenbush J, Sherlock G, Spellman P, Stoeckert C, Aach J, Ansorge W, Ball CA, Causton HC, Gaasterland T, Glenisson P, Holstege FC, Kim IF, Markowitz V, Matese JC, Parkinson H, Robinson A, Sarkans U, Schulze-Kremer S, Stewart J, Taylor R, Vilo J, Vingron M (2001) Minimum information about a microarray experiment (MIAME)-toward standards for microarray data. Nat Genet 29:365–371

31. Taylor CF, Paton NW, Lilley KS, Binz PA, Julian RK Jr, Jones AR, Zhu W, Apweiler R, Aebersold R, Deutsch EW, Dunn MJ, Heck AJ, Leitner A, Macht M, Mann M, Martens L, Neubert TA, Patterson SD, Ping P, Seymour SL, Souda P, Tsugita A, Vandekerckhove J, Vondriska TM, Whitelegge JP, Wilkins MR, Xenarios I, Yates JR III, Hermjakob H (2007) The minimum information about a proteomics experiment (MIAPE). Nat Biotechnol 25:887–893

32. Lefever S, Hellemans J, Pattyn F, Przybylski DR, Taylor C, Geurts R, Untergasser A, Vandesompele J (2009) RDML: structured language and reporting guidelines for real-time quantitative PCR data. Nucleic Acids Res 37:2066–2069

33. Hammerle-Fickinger A, Riedmaier I, Becker C, Meyer HH, Pfaffl MW, Ulbrich SE (2010) Validation of extraction methods for total RNA and miRNA from bovine blood prior to quantitative gene expression analyses. Biotechnol Lett 32(1):35–44

34. Tichopad A, Kitchen R, Riedmaier I, Becker C, Stahlberg A, Kubista M (2009) Design and optimization of reverse-transcription quantitative PCR experiments. Clin Chem 55:1816–1823

35. Livak KJ, Schmittgen TD (2001) Analysis of relative gene expression data using real-time quantitative PCR and the 2(-Delta Delta C(T)) Method. Methods 25:402–408

36. Pfaffl MW (2001) A new mathematical model for relative quantification in real-time RT-PCR. Nucleic Acids Res 29:E45

37. Schmittgen TD, Livak KJ (2008) Analyzing real-time PCR data by the comparative C(T) method. Nat Protoc 3:1101–1108

38. Yuan JS, Wang D, Stewart CN Jr (2008) Statistical methods for efficiency adjusted real-time PCR quantification. Biotechnol J 3:112–123

39. Yuan JS, Reed A, Chen F, Stewart CN Jr (2006) Statistical analysis of real-time PCR data. BMC Bioinformatics 7:85

40. Ramirez M, Castro C, Palomares JC, Torres MJ, Aller AI, Ruiz M, Aznar J, Martin-Mazuelos E (2009) Molecular detection and identification of Aspergillus spp. from clinical samples using real-time PCR. Mycoses 52:129–134

Chapter 2

Overcoming Inhibition in Real-Time Diagnostic PCR

Johannes Hedman and Peter Rådström

Abstract

PCR is an important and powerful tool in several fields, including clinical diagnostics, food analysis, and forensic analysis. In theory, PCR enables the detection of one single cell or DNA molecule. However, the presence of PCR inhibitors in the sample affects the amplification efficiency of PCR, thus lowering the detection limit, as well as the precision of sequence-specific nucleic acid quantification in real-time PCR. In order to overcome the problems caused by PCR inhibitors, all the steps leading up to DNA amplification must be optimized for the sample type in question. Sampling and sample treatment are key steps, but most of the methods currently in use were developed for conventional diagnostic methods and not for PCR. Therefore, there is a need for fast, simple, and robust sample preparation methods that take advantage of the accuracy of PCR. In addition, the thermostable DNA polymerases and buffer systems used in PCR are affected differently by inhibitors. During recent years, real-time PCR has developed considerably and is now widely used as a diagnostic tool. This technique has greatly improved the degree of automation and reduced the analysis time, but has also introduced a new set of PCR inhibitors, namely those affecting the fluorescence signal. The purpose of this chapter is to view the complexity of PCR inhibition from different angles, presenting both molecular explanations and practical ways of dealing with the problem. Although diagnostic PCR brings together scientists from different diagnostic fields, end-users have not fully exploited the potential of learning from each other. Here, we have collected knowledge from archeological analysis, clinical diagnostics, environmental analysis, food analysis, and forensic analysis. The concept of integrating sampling, sample treatment, and the chemistry of PCR, i.e., pre PCR processing, will be addressed as a general approach to overcoming real-time PCR inhibition and producing samples optimal for PCR analysis.

Key words: Amplification facilitators, Clinical diagnostics, Diagnostic PCR, DNA extraction, Forensic analysis, Internal amplification control, Microbial analysis, PCR inhibitors, Pre-PCR processing, Sample treatment, Real-time PCR, Swab techniques, Thermostable DNA polymerase

1. Introduction

The polymerase chain reaction (PCR) (1) is a powerful analytical tool in molecular diagnostics. Only one or a few nucleic acid molecules are required for analysis and denatured and/or partly

Mark Wilks (ed.), *PCR Detection of Microbial Pathogens: Second Edition*, Methods in Molecular Biology, vol. 943, DOI 10.1007/978-1-60327-353-4_2, © Springer Science+Business Media, LLC 2013

degraded DNA can be analyzed, as well as high-molecular-weight DNA. PCR provides fast analysis in a matter of hours, compared with the days required in culture-based microbial methods. In real-time PCR the growth of the amplification product is monitored continuously, through the detection of fluorescently labeled probes or dyes intercalating with DNA. This removes the need for gel electrophoresis, further shortening the analysis time, and allowing PCR to be used for nucleic acid quantification. Table 1 provides examples of applications of real-time PCR.

Thermostable DNA polymerases are a vital component of PCR. The polymerase from *Thermus aquaticus* (*Taq*) and its commercial derivatives, i.e. mutants or chemically enhanced variants, are most widely used because of their tolerance to high temperatures and good processivity. Polymerases from a range of other organisms are also readily available, such as *Thermus thermophilus* (*Tth*), *Thermus flavus* (*Tfl*), and *Pyrococcus furiosus* (*Pfu*). Some manufacturers have made use of the different properties of polymerases by providing mixtures of enzymes from different organisms.

PCR is extremely sensitive and reproducible when pure DNA samples are analyzed. However, the amplification efficiency (AE) and detection limit are sometimes changed by molecules interfering with the DNA polymerase or affecting the nucleotides and nucleic acids (2). Several substances have been identified as PCR inhibitors and a few have been partially characterized regarding their molecular PCR-inhibitory mechanism(s) (Table 2). PCR

Table 1
Different applications of diagnostic real-time PCR

Discipline	Target cell type[a]	Concentration/ number of target molecules[b]	Sample homogeneity	Degree of PCR inhibition	Reference
Archeological analysis	Human cells, animal cells	Low	Medium	High	(12, 30)
Clinical diagnostics	Human cells, microorganisms	High/medium	Medium	Medium	(31–33)
Environmental analysis	Micro- and macro-organisms	Low/medium	Low	High	(34–37)
Food analysis	Micro- and macro-organisms	Low	Low	High	(38–41)
Forensic science	Human cells, animal cells	Low	Low	High	(42–46)

Assays for the detection and/or quantification of nucleic acids
[a]The target cells in clinical and food analysis are often pathogenic organisms, and in archeological, environmental, and forensic analysis partly degraded nucleic acids from macroorganisms
[b]Concentration of target nucleic acids in pathogens or relevant cells in the sample

Table 2
Overview of ions and molecules inhibiting PCR

Type of inhibitor	Molecule or ion	Source	Mechanism/s[a]	Reference
Polymerase inhibitors	Al^{3+}	Sampling using aluminum-shafted swabs	Alters ion composition	(27)
	Alginate	Sampling with calcium alginate swabs	Adsorption of Mg^{2+} or entrapment of polymerase	(27)
	Bile salts (cholic and deoxycholic acid)	Feces	Direct effect on polymerase	(15, 61)
	Calcium ions	Milk	Competition with the polymerase cofactor Mg^{2+}	(24)
	Collagen	Bone	Alteration of ion composition by binding cations	(62)
	EDTA	Anticoagulant	Chelation of Mg^{2+}	(63)
	$FeCl_3$		Release of iron ions	(51)
	Free radicals	UV treatment of PCR tubes	Reaction with polymerase	(64–66)
	Fulvic acid	Soil	Binding to polymerase	(51)
	Heme	Blood	Release of iron ions and competition with template	(18, 48)
	$KAc/K_2Cr_2O_7$	DNA extraction/ preservative	Alteration of ion composition	(63, 67)
	Lactoferrin	Blood	Release of iron ions	(18)
	LiCl	Growth medium	Alteration of ion composition	(63)
	Melanin	Skin, hair	Binding to polymerase	(11)
	$MgCl_2$	Growth medium	Alteration of level of Mg^{2+} in PCR	(63)
	Myoglobin	Muscle tissue	Release of iron ions	(22)
	NaCl		Alteration of ion composition	(51, 63)
	NaOH	DNA extraction	Degradation of DNA, pH-mediated denaturation of polymerase	(63)
	NH_4Ac	DNA extraction	Alteration of ion composition	(63)
	Phenol	Soil, DNA purification	Denaturation of polymerase or binding to the polymerase via hydrogen bonds	(21)
	Phytic acid	Feces	Chelation of Mg^{2+} or change in ion content if present as salt	(68)

(continued)

Table 2 (continued)

Type of inhibitor	Molecule or ion	Source	Mechanism/s[a]	Reference
	Polysaccharides	Feces	Binding to polymerase	(69)
	Proteinases (plasmin)	Milk	Degradation of polymerase	(23)
	Reverse transcriptase	RT-PCR	Competition with DNA polymerase and/or formation of complex with ssDNA	(49, 70)
	Tannic acid	Soil	Binding to polymerase	(51)
	Urea	Urine	Prevents non-covalent bonding, acting directly on polymerase or hindering primer annealing	(54, 71)
Nucleic acid inhibitors	Bilirubin	Feces	Competition with template	(18, 51)
	Cellulose, nitrocellulose	Sampling filters	Binding to DNA	(72)
	Ethanol	DNA extraction	Precipitation of DNA	(63)
	Ethidium bromide	DNA extraction	Binding to DNA	(63)
	Formaldehyde	Preservative	Interference with DNA and DNA polymerase	(67)
	Heparin	Anticoagulant	Binding to DNA, competition with template and direct interaction with polymerase	(18, 73)
	Immunoglobulin G	Blood	Binding to ssDNA	(16)
	Isopropanol	DNA extraction	Precipitation of DNA	(63)
	PEG	DNA extraction	Precipitation of DNA	(63)
	SYBR Green I	Detection dye	Binding to dsDNA with high affinity	(57)
	SYTOX Orange	Detection dye	Binding to dsDNA with high affinity	(57)
	TO-PRO-3	Detection dye	Binding to dsDNA with high affinity	(57)
Fluorescence inhibitors	Humic compounds	Soil	Fluorescence quenching, direct interaction with polymerase and primer annealing	(19, 20, 51, 52, 59)
	Polymeric surfaces	Miniaturized real-time PCR instruments	Binding of detection dye, e.g., SYBR Green	(60)

[a]Confirmed or probable mechanism. Inhibitors affecting the ion composition of the reaction could act as polymerase inhibitors, nucleotide/nucleic acid inhibitors, and/or fluorescence inhibitors

inhibitors are especially prominent in food and forensic analysis, where the amount of target cells/nucleic acids is often small and sample matrices are diverse and complicated. Therefore, the issue of PCR inhibitors has been discussed mainly in these areas (3–7). However, inhibition must be considered in all kinds of PCR-based diagnostic analysis, in particular when quantifying nucleic acids. The effect of inhibitors is complex. Different DNA polymerases and their buffer systems have different abilities to maintain functionality in the presence of PCR-inhibitory molecules (8, 9); samples with low levels of target nucleic acids are more severely affected than those with more DNA (10), and amplification of longer amplicons is more easily inhibited than that of shorter ones (11). The dNTP sequences of the target amplicon and primers affect tolerance of PCR to inhibition (12), and there is some evidence that amplicons with a lower GC content and primers with lower Tm are more affected by inhibitors than amplicons with a higher GC content and primers with higher Tm (13). Inhibitors may be thermolabile, or present at low amounts, both retarding the reaction (12). This results in a delay in amplification, but the PCR efficiency is not affected, making it difficult to detect the presence of inhibitors. Understanding PCR inhibition is made even more difficult by the fact that other factors can cause similar problems. DNA degradation can also lead to lower amplification yields than expected, and DNA lesions, such as *cis–syn* thymidine dimers, have been shown to lower the sensitivity of real-time PCR (14) or completely block amplification.

Real-time PCR inhibitors can be categorized into three groups, depending on their PCR-inhibitory mechanism(s), namely (1) DNA polymerase inhibitors, (2) nucleotide/nucleic acid inhibitors, and (3) fluorescence inhibitors. DNA polymerase inhibitors can either affect the enzyme directly, e.g., by degrading or denaturing it, or indirectly, e.g., by chelating the essential Mg^{2+} ions. Nucleotide or nucleic acid inhibitors can block amplification by binding to single- or double-stranded DNA or by destroying it. Fluorescence inhibitors affect the detection of amplicons in real-time PCR. They may form precipitates blocking fluorescence or interact with the fluorophore. Cell lysis inhibitors have been proposed as a fourth group of PCR inhibitors (5). These supposedly come into play when whole cells are used in the PCR reaction, i.e. no cell lysis or DNA extraction is performed prior to PCR. However, to our knowledge no one has been able to distinguish between inhibitors affecting cell lysis and those affecting the DNA polymerase and/or the nucleic acids.

PCR-inhibiting molecules originate from one or more of the following steps in sample processing: the sample matrix itself, the methods and materials used for sampling, or the sample treatment process. Inhibitors present in the sample matrices themselves include bile salts and polysaccharides in feces (15), heme, immunoglobulin G and lactoferrin in blood (16–18), humic acids in soil (19–21), melanin in skin and hair (11), myoglobin in muscle tissue (22), plasmin and Ca^{2+} ions in milk (23, 24), particulate matter in indoor air (25), and indigo dyes in denim fabric (26). Inhibitors arising from the sampling process may originate from materials on which the sample is found or from the equipment used for sampling. The alginate of calcium-alginate swabs used, for example, for the collection of nasopharyngeal samples, is a potent PCR inhibitor (27). Possible inhibitors arising from sample treatment are phenol from organic DNA purification (21), proteases used for DNA extraction (28), and growth media used for pre-enrichment of microorganisms (29).

For good performance of PCR all the steps in sample processing must be optimized bearing in mind the type of sample and the DNA target in question. The concept of pre-PCR processing implies combining sampling, sample treatment and PCR chemistry optimally in each particular case (7). Different applications place different demands on pre-PCR processing. In clinical analysis, PCR samples are often fairly homogeneous, for example, blood, cerebrospinal fluid, and urine. Sampling is performed directly on the patient and the sampling medium best suited for the analysis can be used, minimizing PCR-inhibiting substances. In contrast, in forensic DNA analysis, crime scene sampling involves a range of different cell types and sample matrices that may inhibit PCR. Not only are more PCR inhibitors present, but also sample treatment may be more complex, costly, and time consuming.

2. The Nature of PCR Inhibitors

2.1. DNA Polymerase Inhibitors

Interactions between proteins and DNA polymerases are a major cause of inhibition. For example, human blood contains an excess of proteins compared with the amount of DNA. In one study, 1 μl of blood was found to contain about 35 ng of DNA and about 150 μg of proteins (47). Many of these proteins interfere with the polymerization process in PCR.

Heme, part of hemoglobin, has been identified as a key PCR inhibitor in human blood (17, 18). It is thought to affect PCR by releasing iron ions, affecting the ion balance, and thereby disturbing polymerase activity as well as primer and probe annealing. Heme can therefore be regarded as a universal PCR inhibitor. This compound is also found in proteins such as myoglobin, cytochrome *b*, and catalase and could therefore cause inhibition in tissues other than blood. Inhibition due to heme has been found to be most notable in dried blood stains, while fresh blood shows comparably low degrees of inhibition. This suggests that the inhibition resulting from heme arises from a degenerated heme complex. Hemoglobin that had been digested with proteinase K was shown to inhibit PCR, whereas non-digested hemoglobin did not (48), further strengthening the notion that the form of heme is important regarding its PCR-inhibiting activity. The iron-releasing molecule lactoferrin is another key inhibitor in blood, affecting PCR in a manner similar to heme (18).

Different polymerases show considerable differences in their resistance to inhibitors in blood. In the first study systematically comparing different polymerases with respect to inhibitor tolerance, *Taq* and Ampli*Taq* Gold were found to be inhibited by 0.004% (v/v) blood, whereas r*Tth* and *Tfl* functioned in 20% (v/v) blood (8). r*Tth* is also considerably less susceptible to inhibition by the heme-containing myoglobin in muscle samples than *Taq* (22). Even different batches of the same brand of *Taq* polymerase have been shown to give different detection limits with the same amounts of blood (18).

The protein melanin, present in skin and hair, is known to completely inhibit *Taq* (11). Melanin binds reversibly to the *Taq* polymerase, thereby hindering its activity. When performing reverse transcription PCR (RT-PCR), the reverse transcriptase enzyme may inhibit PCR, possibly by competing with the DNA polymerase (49). Since *Tth* has reverse transcription activity, it can be used for both steps of RT-PCR to circumvent PCR inhibition. A nucleotide analogue, acyclovir triphosphate, is used to treat patients who have immunodeficiency symptoms or prior to transplantation. This molecule inhibits viral DNA polymerase by premature chain termination and has also been found to inhibit PCR (50).

Humic, fulvic, and tannic acids, together with other polyphenolic compounds, are potent PCR inhibitors present in soil (19, 20, 51). Humics probably affect PCR both by binding to the polymerase via hydrogen bonds and by changing the melting temperature of dsDNA, preventing primer annealing (52). A method of dealing with inhibition by humic substances in soil by simply increasing the concentration of *Taq* polymerase has been proposed (53). However, increasing the amount of polymerase is costly and not advisable as a standard method. Replacing the polymerase may be a better approach, since *Tth* has been found to withstand phenol considerably better than *Taq* (21).

Polyvalent as well as monovalent metal ions are important in PCR. Mg^{2+} is a cofactor for the DNA polymerase, and the overall ion content should create a favorable ion environment for denaturation and primer and probe annealing, as well as for the polymerase. Ca^{2+} ions in milk compete with Mg^{2+} for the binding sites on the polymerase, thereby inhibiting PCR (24). K^{+} ions are often used in the PCR buffer to ensure a suitable ion content. Changing the amount of K^{+} ions or introducing other ions, e.g. Na^{+}, can cause inhibition (8). Different polymerases can cope with different amounts of ions. r*Tth* retains its activity at almost twice the K^{+} ion content which causes inactivation of *Taq* (8). Salt crystals altering the ion content of the PCR reaction are probable inhibitors in urine (54).

2.2. Nucleotide/Nucleic Acid Inhibitors

Nucleic acid inhibitors are molecules that affect the melting temperature or conformation of DNA in such a way that primer annealing or extension is inhibited, either completely or partially. Cations are potent nucleic acid inhibitors (see above). Nontarget DNA may inhibit amplification by sterically preventing the primers from annealing or by providing nonspecific binding sites for the primers (55). Immunoglobulin G in blood plasma is believed to form a complex with ssDNA, thereby inhibiting PCR (16). The effect was found to be greater when immunoglobulin G was heated to 95°C together with the DNA template. Similar complexes can be formed with ssDNA and other proteins during heating and denaturation, which could explain the inhibition seen with reverse transcriptase (16). This should be kept in mind when using boiling as a step in DNA extraction, as in Chelex® extraction (56). The enzyme r*Tth* can amplify DNA in the presence of undiluted immunoglobulin G, which is not possible with *Taq*, Ampli*Taq* Gold, *Pwo* or *Tfl* (16). The DNA-intercalating dye SYBR Green I, used for real-time PCR target detection, has been shown to partly inhibit PCR when used at the recommended concentration (57). This is explained by the high affinity of SYBR Green to dsDNA. Dyes with lower affinity, such as SYTO-13 and SYTO-82, showed no PCR-inhibiting properties in the same study.

2.3. Fluorescence Inhibitors

The advent of real-time PCR allowed the introduction of a new set of PCR-inhibiting molecules, i.e., those that directly affect the detection of fluorescence. The choice of detection chemistry and probe technology influences fluorescence inhibition. Hydrolysis probes (*Taq*Man) emit fluorescence when hydrolyzed by the DNA polymerase. Changes in the ion content of the reaction could affect probe hybridization, thus causing inhibition. Molecules inhibiting the polymerization of DNA polymerases could also lower their 5′–3′ exonuclease activity, thereby inhibiting the hydrolysis of *Taq*Man probes (58), leading to a double inhibition effect. Different fluorochromes have different biophysical properties, which probably affects their susceptibility to quenching inhibitors in the sample matrix. However, this has not yet been extensively studied. Humic acids quench the fluorescence of SYBR Green, probably by binding to the fluorophore or by collisional quenching (59). Surface-bound SYBR Green fluoresces most, and humics may prevent surface binding. Humic acids sequester ethidium bromide, thereby reducing the amount available for DNA interaction. Humics may thus affect the analysis using classic PCR as well as real-time PCR. SYBR Green has been shown to adsorb onto polymeric tubes, which could be used to create miniature PCR systems (60), resulting in a lower signal. The effect increases with greater tube length and lower volume and would lead to fluorescence inhibition in a miniaturized real-time PCR assay.

3. Quantifying the Level of PCR Inhibition

In diagnostic PCR it is vital to know if a sample is affected by PCR inhibitors. The risk of false-negative results is one of the major drawbacks of diagnostic PCR (74). A low level of inhibition could affect the efficiency of amplification and the detection limit, leading to underestimation of DNA concentrations. In forensic analysis it is important to distinguish between samples with little DNA and those with reasonable amounts of DNA and also containing PCR inhibitors, in order to perform the appropriate analysis. Samples with too little DNA should be concentrated, while those with inhibitors must be purified. Methods of measuring the quantity and quality of DNA based on optical density (OD) are not ideal for estimating the level of PCR inhibitors, since there is no direct correlation between OD and successful PCR amplification (10, 75). OD ratios at different wavelengths such as OD_{260}/OD_{280}, mainly predict the presence of proteins, while other substances not detected may interfere significantly with PCR. For example, RNA samples that were found to be pure when analyzed with UV spectrophotometry and different types of RNA chips, showed inhibition using real-time PCR (76). This indicates that the classical techniques are less sensitive to inhibitors and should be complemented with PCR assays. Also, one-third of the scientists using RT-PCR do not check the quality of the RNA before cDNA synthesis (77), further stressing the importance of real-time PCR quality control to avoid false-negative results.

The end-point fluorescence of real-time PCR is usually lowered by fairly small amounts of inhibitors (78). However, lowering of the end-point fluorescence does not necessarily affect the detection limit of the assay (79), and it should therefore not be used to estimate inhibition. Quantification cycle (C_q) shifts are better means of measuring the effects of inhibitors. C_q is derived from the amplification curve, generally using the threshold method or the second derivative maximum method. There are three methods of calculating PCR inhibition (1) using an assay with an internal amplification control (IAC), (2) calculating the amplification efficiency, or (3) modeling the amplification curve.

3.1. Internal Amplification Control

The fastest and most straightforward way of monitoring PCR inhibition in routine analysis is by using an IAC, which involves adding a known amount of a DNA fragment, which is amplified simultaneously with the target (80–82). The presence of inhibitors results in either complete failure to amplify the IAC, or slower amplification than expected, evidenced by a higher C_q value than that of a pure sample. The use of IACs is strongly recommended in order to avoid false-negative results (83) and is a legal requirement in the detection of pathogens in food in Europe (EN ISO 22174). The notation used differs, and IC (internal control) (84), IPC (internal PCR control) (82), and internal standard DNA (25) are also used. Note that the abbreviation IPC is also used for internal process control (see below).

IACs can either have primer sites equivalent to the target, so that only one primer pair is required to amplify the two fragments (85) or different primer sites, creating a need for separate primer pairs (86). In the first case there is competition for the primers. Incorporating a second primer pair makes the assay more complex. Several systems employing specific IACs have been developed during the past few years for a range of purposes (81, 82). A universal IAC for hydrolysis probe assays, with specific primers, has been proposed (86). It uses a fabricated DNA fragment and has been applied in assays to detect agents posing a biological threat, such as *Bacillus anthracis* and *Clostridium botulinum*. An IAC with primer sites for five different human virus detection primers has been developed (85). It can be used in multiplex assays amplifying any of the five targets and the IAC with the same primers.

If the concentration of the IAC is too high, the amplification of the target may be inhibited by IAC DNA. The appropriate number of IAC molecules depends on the assay and must be titrated during assay optimization. Low numbers of IAC molecules, such as 20 (84), 58 (85), or 100 molecules (41) per reaction, have been successfully applied.

In order to detect all levels of inhibition, the amplification of the IAC must be at least as sensitive to PCR inhibitors as the target. The size of the IAC is important in this respect, since shorter fragments are usually more readily amplified in the presence of inhibitors than longer ones (11). An IAC amplicon longer than the target is therefore recommended (85).

Since the amplification of target DNA competes with IAC amplification when using one set of primers, the value of C_q of the IAC is elevated if the amount of target DNA is high. Therefore, the direct use of C_q shifts is not an ideal measure of inhibition. Hudlow et al. (46) proposed an equation incorporating the effect of target DNA concentration on IAC C_q. The equation can be used to predict the possibility of successful short tandem repeat (STR) analysis, e.g., for a forensic DNA sample. The difference between the C_q of the IAC in the sample and the C_q of the IAC in a pure reaction is calculated. This value, called "delta C_q," is divided by the amount of DNA in ng, giving a normalized inhibition factor (NIF). NIF values have been shown to be well correlated to the possibility of obtaining complete STR profiles; values over 1 indicate unsuccessful STR typing (46).

If a real-time quantitative PCR assay is used to optimize a subsequent PCR reaction, such as a multiplex human identity PCR for forensic or parental investigations, the relative amounts of sample in both assays must be considered. The quantification kit Quantifiler™ Human (Applied Biosystems, Foster City, CA) is more sensitive to hematin inhibition than the Identifiler® STR analysis kit (kits inhibited by 16 and 20 μM hematin, respectively) (82), suggesting that all the inhibitors affecting the Identifiler analysis should also be seen in the results using the Quantifiler kit. However, in the Quantifiler kit the sample is only 2 of the 25 μl reaction volume, compared with 10 of 25 μl in the Identifiler kit. The level of PCR inhibitors in a real sample is therefore up to five times larger in an analysis using the Identifiler kit than in one using Quantifiler. A sample that appears to be uninhibited according to the IAC using Quantifiler can give a completely inhibited blank profile when using Identifiler or comparable systems, such as SGM Plus® (Applied Biosystems) (unpublished data, Swedish National Laboratory of Forensic Science). The sample-to-reaction-volume ratio should preferably be the same for all PCR-based analyses that are performed on the same sample.

When using an assay without an IAC, negative samples can be confirmed by spiking the reaction with an alien plasmid (26) or target DNA molecules, e.g., 50 copies (40). Roussel et al. (10) spiked DNA extracts from mouse stomachs with *Helicobacter pylori* DNA. The resulting C_q values were compared with those from the same amount of DNA molecules in water, giving an inhibition ratio.

Alien DNA, or cells, can be introduced before the sample is collected or before DNA extraction. An internal process control monitors not only the PCR but also the analysis steps prior to PCR, such as sampling and sample treatment. Murphy et al. (87) used a genetically modified *Escherichia coli* strain as an internal process control, which was added to pathogen-containing food samples prior to sampling. Lenticule discs were used to encapsulate *E. coli* to enable the addition of equal amounts to all samples. *E. coli* is detected using specific primers and can therefore be used universally in process control. Juen and Traugott (88) developed a multiplex with primers for both prey and predator for the detection of *Amphimallon solstitiale* in the gut content of soil-living invertebrates. A negative result from prey DNA amplification is a sign of inhibition, or failure in the DNA extraction.

3.2. Amplification Efficiency (AE)

The AE may be calculated from the slope of a standard curve obtained from the analysis of a dilution series of DNA, using the formula $AE = 10^{(-1/\text{slope})} - 1$. A slope of −3.32 indicates the ideal efficiency of 1.0 (exponential amplification). The deviation of the calculated value of AE from 1.0 provides a measure of the level of PCR inhibition in the sample. The most systematic way of measuring inhibition using the AE is to spike the sample with different amounts of pure DNA from a source other than the target and generate a standard curve based on this alien DNA (34). In this way, the level of PCR inhibitors is kept constant for all DNA dilutions and an inhibited sample would give an AE of less than 1.0. This method can be used to compare different DNA extraction methods with regard to their ability to produce inhibitor-free extracts (38). The use of alien DNA removes the effect of the DNA yield, which would make the results biased. Zebra fish DNA (78) and potato DNA (76) have been as used, and these are suitable for testing extracts from all organisms and plants, except for zebra fish and potato, respectively.

Another way of measuring inhibition through the AE is to dilute the sample directly and create a standard curve based on the target DNA. However, the inhibitors are also diluted, which leads to a larger inhibition effect for high concentration dilutions than for lower ones. This gives a steeper standard curve slope, and the AE for a sample containing inhibitors could deviate from the expected low value, and instead exceed 1.0. If the objective is to thoroughly investigate the inhibition effect that a certain sample matrix has on PCR, it is better to use alien DNA than a dilution series of the sample itself.

Pure DNA is often used to obtain standard curves for absolute quantification. For the quantification to be correct, the AE of the sample must be the same as for the standard DNA. This is not the case for partially inhibited samples. A new data analysis method, taking the slope of the amplification curve, which is affected by inhibitors, into account has been proposed as a way of dealing with this problem (89). There, C_q values are defined as the intersection of the *x*-axis and the tangent to the inflection point of the amplification curve. This alternative method was found to give better consistency between inhibited and pure samples than both the threshold method and the second derivative maximum method.

3.3. Modeling

The PCR inhibition of a sample can be investigated by mathematical interpretation of the amplification curve using computer modeling (29, 90). The shape of a curve that is affected by inhibitors differs from that given by a pure sample (43) due to differences in the kinetics of the reaction. An inhibited sample generally gives a flatter, "less exponential" curve. The end-point fluorescence value can also be lowered by inhibitors. Tissue-specific inhibition in bovine RNA analysis has been confirmed using mathematical modeling of the amplification curve (91).

In absolute quantification the DNA concentration will be underestimated if the AE of the sample is lower than that of the DNA standards. Using a mathematical model, it has been shown that differences in AE greater than 0.2 caused more than 30% underestimation of the DNA in samples from genetically modified organisms (38). Accurate, absolute quantification of DNA using real-time PCR therefore requires that the AE of the external standard and samples are the same or at least that the differences in efficiency are known. This can be achieved through computer modeling. A model can be fitted to experimental data to analyze a part of the amplification curve, and a sample-specific AE can be calculated, without the laborious work of spiking the sample with alien DNA and obtaining a standard curve as described above. The resulting AE is compared with the AE of the pure standard DNA. Samples with efficiencies outside a predetermined interval are considered outliers and cannot be confidently quantified using the standard curve (92).

4. Amplification Facilitators and PCR Buffer Systems

The difference in AE between different polymerases can to some extent be explained by differences in the buffers, the presence of PCR facilitators, the pH, and salt concentration (29). The resistance to inhibitors of a given polymerase can therefore be altered by using different buffers (9, 93). Both the AE and the detection window can be improved by changing the buffer (9). Several compounds have been shown to facilitate PCR (Table 3). Facilitators were first used to increase the specificity and fidelity of PCR, but some have also shown the capacity to relieve PCR inhibition (94). Recently, manufacturers have started adding PCR facilitators to buffers for commercial DNA polymerases. The *Tth* buffer (Roche Diagnostics, Mannheim, Germany) and the Klen*taq* buffer (DNA Polymerase Technologies, St. Louis, MO) both contain the detergent Tween 20, the *Tth* buffer also contains BSA. CertAmp buffer (Biotools, Madrid, Spain) contains glycerol. The effect of PCR facilitators is dependent on the facilitator concentration, and overloading will result in the inhibition of amplification, as has been shown for BSA, Tween 20, Triton X-100 (63), formamide, and glycerol (95). Synergistic effects between different types of facilitators should not be expected (94), on the contrary combining facilitators can cause inhibition (95). Categorization of PCR facilitators into five groups has been proposed (7): (1) proteins, (2) nonionic detergents, (3) organic solvents, (4) biologically compatible solutes, and (5) polymers.

4.1. Proteins

Bovine serum albumen (BSA) is the most commonly used PCR facilitator. BSA is a transport protein that binds fatty acids (lipids) and organic molecules. Its excellent binding capacity makes it suitable for reducing various types of inhibition in *in vitro* amplification. BSA binds the inhibitors heme and melanin (11), in the latter case preventing it from binding to the polymerase. BSA may also act as a competitive target for proteases, thereby sheltering the polymerase from these. When BSA is present, phenols preferentially bind to these molecules instead of to the polymerase. A range of different BSA concentrations has been used to deal with inhibition

Table 3
PCR facilitators used to alleviate inhibition

Type of facilitator	Facilitator	Concentration	Observed effect	Reference
Protein	BSA	0.1–1.28 g/l	Relieves inhibition due to bile salts, feces, $FeCl_3$, fulvic acid, hemoglobin, humic acid, immunoglobulin G, lactoferrin, meat, melanin, tannic acid, waste water sludge, 50-year-old saliva stains	(11, 16–18, 45, 51, 61, 88, 94, 96)
Protein	Gp32	0.1–0.15 g/l	Relieves inhibition due to feces, $FeCl_3$, fulvic acid, hemoglobin, humic acid, lactoferrin, meat, tannic acids	(16, 51, 94)
Protein	A-macroglobulin	0.04 g/l	Relieves inhibition due to proteases in milk	(23)
Protein	Casein	0.01% (w/v)	Relieves inhibition caused by bile salts	(61)
Protein	Lima bean trypsin inhibitor	0.02 g/l	Relieves inhibition caused by blood	(8)
Protein	Phytase	50 U/ml	Relieves inhibition due to phytic acids in feces	(68)
Protein	Proteinase inhibitor	1×	Relieves inhibition in samples of feces	(94)
Protein	Soybean trypsin inhibitor	0.04 g/l	Relieves inhibition due to proteases in milk	(23)
Nonionic detergent	Tween 20	0.5–2.5%	Relieves inhibition in samples of feces, phenolic compounds, and plant polysaccharides	(94, 99, 100)
Organic solvent	DMSO	2–10%	Rescue of failed amplification	(103, 104)
Organic solvent	Formamide	0.01% (w/v)	Relieves inhibition due to bile salts	(61)
Biologically compatible solute	Betaine	11.7% (w/v)	Relieves inhibition due to hemoglobin	(16, 94)
Polymer	PEG 400	5–15%	Relieves inhibition due to blood and plant polysaccharides	(94, 99, 103)

in different assays (Table 3). In the *Taq*-mediated detection of hydrogenase A associated with *Clostridia* in sludge waste water and manure, 100 ng BSA/µl gave more efficient and specific amplification, and a better detection limit, than both lower and higher concentrations (96). When comparing different PCR facilitators for blood, feces, and meat samples, with the enzymes *Taq* and r*Tth*, BSA was found to have the best performance and reduced inhibition in all these types of sample (94). Tween 20, glycerol, polyethylene glycol (PEG), dextran, formamide, and dimethyl sulphoxide (DMSO) had no effect on inhibition in this study. The protein Gp32 shows similar inhibition-alleviation properties to BSA (51, 94). However, the use of Gp32 instead of BSA is not recommended since it is far more costly. Specific enzymes have been successfully used to reduce inhibition from some compounds, such as phytase for phytic acid (68) and heparinase for the anticoagulant heparin (73). Skim-milk alleviates inhibition of *Taq* amplification of plant samples containing polyphenolic compounds (97) and humic material containing environmental samples (98) and also increases the specificity of amplification (98). The active substance is probably the protein casein, and it is believed to function in the same manner as BSA, preferentially binding compounds that would otherwise bind to the polymerase lowering its efficiency.

4.2. Nonionic Detergents

The detergents Tween 20 and Triton X-100 are frequently used as amplification facilitators, to reduce inhibition in samples from feces (94), plant polysaccharides (99), and phenolic compounds (100), although the mechanism is unclear. However, Triton X-100 at a concentration of 2% v/v was found to decrease the specificity of a *Taq* polymerase assay, indicating that care should be taken not to overload the reaction. Tween 20 did not show this effect (101).

4.3. Organic Solvents

DMSO and formamide are commonly used for PCR facilitation. Formamide denatures DNA and can facilitate amplification for assays with insufficient thermal denaturation (102), or if a high GC content elevates the melting temperature, making denaturation more difficult. Formamide also has some inhibitor-alleviation capabilities, as has been shown for bile salts (61). The effect of organic solvent facilitators is based on their ability to destabilize DNA.

4.4. Biologically Compatible Solutes

Betaine and glycerol are frequently used facilitators. Betaine increases the thermostability of proteins and has been reported to reduce inhibition in blood and meat samples (94).

4.5. Polymers

PEG and dextran are the most common polymer PCR facilitators. PEG stabilizes the DNA polymerase and has been used to reduce inhibition in samples of blood, feces (94), and plant polysaccharides (99).

5. Sampling

Sampling is a key step in diagnostic PCR analysis. Good sampling should provide a sample that is representative of the material to be tested, maximize sample uptake and minimize PCR inhibitor uptake. In clinical diagnostics, the size of the sample is usually not a limiting factor, and sampling is straightforward and reproducible, e.g., the sampling of venous blood. In microbial detection for environmental or food analysis, as well as forensic DNA analysis, sampling is more complicated. The amount of target material can be very low, and the background may be any type of surface or liquid, such as animal carcasses or waste water. In practice, one cannot be certain that amplifiable DNA is actually present in the sample until it has been subjected to PCR. Great care is needed during sampling in order to minimize the risk of false-negative results.

5.1. Standardization

Standardization of sampling is vital to obtain reliable and reproducible PCR results. However, standardized methods are lacking in several scientific fields. In microbial food and feedstuff testing, there are specific international standards governing the sampling of different substrates, e.g., carcasses (105) and horizontal surfaces (106). The standards define where samples should be taken, the size of the sampling surface, approved sampling methods and how each method is to be applied, including how to hold the swab, how many times the area should be covered, and which buffer to use to moisten the sampling swab or material. These standards also govern the transportation and handling of samples before analysis. Prior to the terrorist actions involving *B. anthracis* spores in the USA in 2001, there were no standardized sampling methods for biocontaminants on different surfaces (107). The events clearly illustrated the need for standardization, and significant work has since been carried out to develop and evaluate sampling methods for bacterial spores on different surfaces, both for PCR (108, 109) and classical microbial analysis (107–111).

There are no sampling standards in forensic science; partly because of the wide variety of materials and tissues investigated and also for historical reasons. Efforts have been focused on developing sensitive and highly discriminatory PCR-based identification systems, while little has been done to develop standardized sampling methods.

Several sampling techniques used for diagnostic PCR, e.g. swabbing and excision, are based on protocols for classical, non-PCR analysis methods such as cultivation-based microbial detection (112) and immunoassays (113). This can be problematic since the biochemical and physical process of PCR is very different from that of classical methods. The material of the sampling swab itself may inhibit PCR, as is the case with calcium-alginate swabs and Al^{3+}-releasing aluminum swab shafts (27). Cotton has been shown to inhibit PCR (101), but the pure cotton used in dedicated PCR swabs supplied by several manufacturers should not cause inhibition.

5.2. Direct Sampling

In direct sampling, the material carrying the target cells is placed directly into a tube for sample treatment, e.g., excision of meat and soil sampling. This gives a relatively high amount of target cells, but a high level of PCR inhibitors may also be released from the background material. Excision of beef carcasses for pathogen detection gives higher microbial yields than surface swabbing using cotton swabs for cultivation methods (114), but maceration releases a great deal of tissue debris into the extract, inhibiting PCR. Excision can only be performed on a rather small area, it is time consuming, requires special skills, and destroys the sample. Thus, excision is not a suitable sampling method for online diagnostic PCR analysis. Direct sampling of soil for microbial diagnostics also shows substantial levels of inhibition (19, 20), and extensive sample treatment is needed. In forensics, clothing such as suede and denim introduce PCR inhibitors when used directly for analysis (115).

In clinical analysis, e.g., the screening of blood from newborns, filter paper provides a stable sampling matrix without the need for cooling. Filter paper has been used for many years in various enzyme and immunoassays, as well as being a good medium for PCR-based analysis (113). FTA® cards (GE Healthcare, Little Chalfont, UK) are chemically treated filter paper, on which the cells are lysed and the proteins degraded directly on contact, and DNA is immobilized within the paper structure. By shielding DNA from oxidation and nucleases FTA cards allow for long-term storage at room temperature (116).

5.3. Swabbing Techniques

Surface swabbing can be used instead of direct sampling in order to lower the amount of PCR inhibitors released from the substrate. Polyurethane sponge swabbing has been shown to give yields comparable to those from excision for beef, pork, and lamb carcasses (117). Cotton swabbing of stains on suede has been shown to reduce the amount of PCR inhibitors in the sample, compared with direct sampling (unpublished data, Swedish National Laboratory of Forensic Science).

Swabbing using a moistened cotton swab has been used for microbial cultivation methods for many years (112). However, cotton is not an ideal material for sensitive analysis because of its high cell absorption; furthermore there are difficulties in standardizing sampling methods and the reproducibility is low (112). Despite these drawbacks, it is still probably the single most popular method of diagnostic PCR sampling. Various liquids have been used to moisten swabs before sampling. Physiological saline was found to improve cell recovery for several kinds of crime scene stains compared with water (unpublished data, Swedish National

Laboratory of Forensic Science), and ethanol is sometimes used for sampling of contact traces (118). Moistening swabs with extraction buffer was shown to increase DNA yield in the sampling of saliva from skin (119). Surfactants improve the collection of spores from nonporous surfaces (120), further showing that both the collection material and the buffer must be appropriate for the sample in question.

The double-swab technique (114), or wet and dry swab method (105), is a slight modification of cotton swab sampling. The surface is first swabbed with a moistened cotton swab, and then with a dry one to soak up excess fluid. As for the single-swab method, the double-swab technique was first used for microbial cultivation methods (114) and then employed in PCR analysis. The method is commonly used in forensics, e.g., for collecting saliva from bite marks (121) and contact stains from a range of different surfaces (122).

Nylon flocked swabs are a newer sampling medium, developed especially for PCR analysis. Flocked swab sampling give RNA yields comparable to nasal secretions for clinical virus detection and also allow for a more standardized sampling procedure (123). Thanks to their design, with thousands of short nylon fibers extending from the swab head enabling sampling of only the outermost cells, less PCR inhibitors are probably introduced than in the case of nasal secretions and other swab materials. Also, the target cells remain on the surface of the fibers and are easily released.

For the detection of biocontaminants on surfaces, large areas must be sampled to avoid false-negative results. Neither ordinary cotton swabs nor swabs made of other materials are suitable for collecting samples from large surfaces. A commercially available biological sampling kit (BiSKit) containing a thick foam material enables both wet and dry sampling of a surface area of 1 m^2 for subsequent PCR analysis (109). BiSKit foam has been shown to perform better than cotton and foam swabs, which only enable swabbing of around 100 cm^2 of both metal and wood laminate surfaces. Gauze can also be used for swabbing of large surfaces, e.g., for carcass sampling (105).

Swabbing involves the use of water or another liquid to release cells, but soluble inhibitors are released at the same time. Dry sampling, e.g., tape lifting, is a way of avoiding the collection of inhibitors with the sample. A piece of tape, e.g., hydrophilic adhesive tape (HAT) (124), is pressed against the substrate a number of times, usually until it no longer adheres. Cells are transferred to the tape, which is placed in a tube for DNA extraction. Tape lifting is ideal for contact DNA on nonporous surfaces. In the investigation of a bank robbery in Sweden, one of the perpetrators had put his hand on the shoulder of one of the bank employees. A complete DNA profile of the suspect was obtained using tape lifting and he was apprehended (125). Tape lifting has been shown to give a higher DNA yield and visibly purer extracts than the double-swab technique and excision in forensic DNA analysis of shoe insoles (126).

6. Sample Treatment

The objectives of sample treatment are to prepare the sample for PCR amplification by (1) concentrating target nucleic acids, (2) removing/neutralizing PCR inhibitors, and (3) making the sample more homogeneous to ensure better repeatability in the amplification. Sample treatment may involve cell separation, cell lysis, and DNA purification. Direct cell lysis followed by DNA purification is the most common approach; several DNA extraction methods combine these two steps (Table 4). To enable sensitive microbial detection using PCR, a pre-enrichment step may also be necessary before nucleic acid purification. Generally, PCR is regarded as extremely sensitive, but in the context of finding one *Salmonella* bacterium in 25 g of minced meat, which is the legislative requirement for food safety in Europe, PCR is in fact rather insensitive compared with culture-based methods (74).

Table 4
Sample treatment methods

Analysis step	Method	Reference
Cell separation	Aqueous two-phase system (solubility separation)	(15, 154)
Cell separation	Flotation (buoyant density separation)	(129–131)
Cell separation	Flow cytometry	(67)
Cell separation	Immuno-magnetic affinity separation	(67, 128)
Cell separation	Laser capture microdissection	(132–135)
DNA extraction	Agarose gel diffusion of inhibitors (size separation)	(155)
DNA extraction	Chelex DNA extraction	(56, 156)
DNA extraction	Detergent treatment (non-lysis DNA release)	(101)
DNA extraction	Phenol chloroform purification (solubility separation)	(136, 137)
DNA extraction	Protease treatment (cell lysis, protein degradation)	(28, 152)
DNA extraction	Silica beads (DNA binding)	(145, 146, 148, 149)
DNA extraction	Solid-phase extraction (protein adsorption)	(26, 47, 98, 142, 143)
DNA purification	Coated activated charcoal (protein adsorption)	(140)
DNA purification	Dilution	(19, 20, 54)
DNA purification	Filter purification	(139)
DNA purification	NaOH treatment	(115)

6.1. Cell Separation

The separation of target cells from the surrounding matrix is often complicated and time consuming. However, since nucleic acid inhibitors are removed before they can interfere with the released DNA molecules, it may be beneficial for certain types of samples and complex sample matrices. Whole cells also provide a more stable environment than free DNA. Affinity bead separation is a classical way of separating a certain type of cell from other cells and the sample matrix. The beads are either coated with generic proteins or specific antibodies, depending on the level of specificity required. Dynabeads® coated with lectin have been shown to provide successful purification of gram-positive bacteria from meat samples (127). Antibody separation is very specific and has been successfully applied to cell separation in oocyst samples (67), as well as in the isolation of *Listeria* in milk (128). However, the specificity can be a drawback, as PCR simply confirms the antibody separation and the capacity of PCR to distinguish the target from a DNA background is not used.

Flow cytometry is another very specific cell separation method that has been used with success for PCR analysis (67). Both antibody separation and flow cytometry are quite expensive and perform poorly in the presence of complex sample matrices. Flotation is an efficient separation system that exploits the difference in the buoyant densities of cells and particles to separate whole cells from PCR-inhibitory substances and free nucleic acids (129). It is robust and specific, and also laborious in its present form. Flotation isolates target cells in a thin band using centrifugation in a medium such as Percoll® (130). The method has been shown to effectively remove PCR inhibitors from various food samples (131). Laser capture microdissection (LCM) utilizes an infrared laser beam to lift cells from a surface, preferably a microscope slide (132). LCM has been used in forensic DNA analysis to separate sperm cells from epithelial cells (133) and to separate leukocytes from buccal cells, in order to avoid mixed DNA profiles (134). The technique has also enabled DNA typing of crime scene samples mixed with soil, by separating cells from the matrix (135). LCM requires expensive equipment and is time consuming, but can be valuable in solving serious crimes.

6.2. DNA Extraction and Purification

There are many rather expensive commercial DNA extraction kits whose compositions have not been disclosed. Most less costly methods are based on either Chelex 100 resin or phenol chloroform. Chelex is a chelating styrene divinylbenzene copolymer that binds polyvalent metal ions that may otherwise catalyze DNA degradation at high temperatures and at low ion contents (56). Water or an aqueous buffer is added to the sample, and the cells are removed from the substrate by vortexing. The cells are then pelleted using centrifugation, and water-soluble inhibitors can be discarded with the supernatant. However, disrupted cells and free

DNA are lost in the process, lowering the yield. The sample is heated to 56°C to achieve lysis and boiled to degrade the proteins, both in one tube without sample transfer. The Chelex method is described as "quick and dirty," i.e., it is a rapid, cheap method that can be applied to most types of sample, but it is not a powerful inhibitor remover. Care should be taken not to include Chelex beads in the PCR as they would chelate the necessary Mg^{2+} ions. Since centrifugation is a vital step in the Chelex method, it is difficult to automate the procedure.

Phenol chloroform purification, or organic purification as it is also called, is a powerful method for obtaining inhibitor-free extracts (136, 137). It has long been the preferred method of extracting DNA from "dirty" samples, such as bone and soil. One drawback is that phenol is toxic, and working with it thus constitutes a health hazard. Phenol chloroform effectively removes inhibitors that are soluble in the alcohol phase, e.g. many types of proteins. For inhibitors with solubilities similar to DNA, such as humic substances, polysaccharides, hemoglobin, and urea, another purification strategy is needed. For example, alkaline and acid hematin have been successfully removed from DNA extracts from blood using phenol chloroform, but complexes of ferric heme and serum proteins were simultaneously purified with DNA in the water phase and inhibited PCR (48). Washing with water, exclusion by size or binding of proteins to a solid phase such as silica beads, are possible ways of removing water-soluble inhibitors. Phenol inhibits PCR (21) and must be completely removed from the DNA template after purification. Protease treatment, usually employing proteinase K, is a common step in phenol chloroform purification and is applied in many commercial DNA extraction kits. The proteases mediate cell lysis and degrade cell proteins and proteins bound to DNA, which could otherwise interfere with the PCR (28). Collagenase has been suggested as an alternative to proteinase K in phenol purification of human bones, as human collagen was identified as a key inhibitor of such samples (62). Ethanol and isopropanol are commonly used for the precipitation of DNA after treatment with phenol chloroform, e.g., when purifying environmental soil samples (138). A combination of PEG and NaCl could provide an alternative to the alcohols, since this has been shown to produce extracts from soil with less humic substances and more DNA than when using isopropanol (98).

If the initial DNA extraction using phenol chloroform or Chelex fails to remove inhibitors, the template can be further purified. Filtration and dilution are two quick and simple DNA purification methods. In filtration, the sample, together with an aqueous buffer, is washed through a filter with pores that allow the smaller inhibitor molecules to pass through, while retaining the larger DNA molecules. Microcon filter tubes (Millipore, Billerica, MA) have been used to purify forensic DNA extracts from cigarette butts (139). Dilution, i.e., simply adding water or a buffer to the extracts, has been successfully used to circumvent inhibition from humic substances (19, 20, 54), urine (54) and reverse transcriptase used in cDNA formation (49). Other DNA purification methods of varying complexity are also available, depending on the sample in question. Sodium hydroxide (NaOH) treatment alleviates inhibition in blood samples from various substrates, including soil and wood (115). NaOH denatures the DNA strands, releasing PCR inhibitors that bind to double-stranded DNA, and alkaline conditions are believed to inactivate protein inhibitors. Because of the risk of DNA degradation, NaOH treatment is not recommended for samples with minute amounts of DNA. Activated charcoal coated with *Pseudomonas fluorescens* has been found to adsorb water-soluble PCR inhibitors (140). Using uncoated charcoal resulted in complete adsorption of all DNA. Chemical flocculation, developed as a method of removing suspended organic solids from waste water, has been successfully applied as a DNA purification step for soil samples (141).

An alternative to phenol chloroform purification is solid-phase extraction, a method in which particulate matter, such as silica beads, is used to adsorb aqueous proteins (142). Small size and large surface area make adsorption a rapid process. The negatively charged bead surface binds positively charged proteins with high affinity, whereas the negatively charged DNA is repelled. Solid-phase extraction can be performed in two ways: (1) by adding the beads to the template tube, which is then agitated allowing the proteins to become adsorbed, and then removing the proteins by filtration or centrifugation or (2) by passing the sample through a column packed with beads upon which the proteins are adsorbed. Sepharose 6B affinity beads have been used to purify extracts from indigo dyes in denim (26), bone samples and saliva on envelopes (143), by adding the resin to the samples to allow inhibitor adsorption, and then removing them by centrifugation. A great deal of DNA is lost in this process, and caution should therefore be exercised when applying the method (143). The column approach is better suited for small volumes and samples with high protein concentrations. Sepharose 4B beads are used in spin column purification, having both size-exclusion and adsorption capabilities. Polyvinyl polypyrrolidone (PVPP) is another frequently used spin column medium, which forms a complex together with PCR-inhibiting phenolic compounds via hydrogen bonds (144). The complex is then precipitated and removed. Arbeli and Fuentes (98) showed that a combined Sepharose-PVPP spin column gave more efficient soil sample purification than separate Sepharose and PVPP columns. The Sepharose first separates humic compounds from the DNA by size exclusion, and the PVPP then absorbs most of the remaining humics. Others have found PVPP methods to be unreliable in removing PCR-inhibiting compounds (141). PVPP treatment has also been shown to lower the DNA yield (138).

Solid-phase extraction is quick and safe, in contrast to phenol chloroform, and the risk of phenol impurities in the extract is removed. Today, several column-based solid-phase extraction kits are commercially available. In a recent survey, 71% of real-time PCR users stated that they used spin-column-based methods for RNA extraction, while only 8% used home-made reagents and protocols such as phenol chloroform (77). The future for solid-phase extraction could lie in automated miniaturized systems. These could make DNA purification quicker, by reducing the running time, and cheaper, since smaller amounts of sample and chemicals are needed. Analysis could also be performed in the field. Silica and silica-coated beads are promising media for use in miniaturized systems. Wen et al. (47) developed a dual-phase microchip. As the lysed blood sample flows through the chip, proteins, such as hemoglobin and lactoferrin are bound to the silica beads. The DNA is bound to a monolithic column and subsequently released.

In the presence of a chaotropic salt, such as guanidinium thiocyanate, DNA binds to silica particles. The chaotropic salt also lyses cells and inactivates proteins. These properties can be used for efficient DNA purification. When DNA binds and forms a complex with the silica, unbound material such as cell debris and proteins can be washed away in a series of washing steps using washing buffers and/or ethanol (145). Incubating the silica–DNA complex in a low-ionic-strength buffer releases the DNA into solution. The method has been improved by introducing magnetic silica-coated beads. The magnetic force can be used either to move the beads from tube to tube during washing or to immobilize the beads in one tube while changing the washing liquid, simplifying sample handling. The technique is suitable for automation, since no centrifugation is needed. Several commercially available kits can be used either manually or with pipetting robots. Promising miniaturized systems with DNA-binding silica beads have also been developed (146).

6.3. Automation

Automation of DNA extraction is important for increasing sample throughput and ensuring the quality of laboratory analysis. Automating the extraction of DNA from heterogeneous, single-source samples in simple sample matrices, such as fresh venous blood in clinical analysis, is quite straightforward. A cheap, fast, and simple two-step method of automating the analysis of blood samples on filter paper punches in 96-well plates has been developed by Lin et al. (113). Leukocytes are released from the filter paper using methanol and lysed by heating in Tris buffer. When lysed in water a large amount of heme was released, but using Tris buffer pure extracts were produced. DNA binding FTA filter paper punches can be used directly in PCR analysis, after a simple washing procedure using, as in an automated system for forensic DNA reference samples (147).

Automation of DNA extraction from complicated and heterogeneous samples and sample matrices, such as forensic crime scene samples, is more complicated. The methods available today are mainly commercial kits relying on DNA-binding magnetic beads, which require extensive pretreatment, e.g. lysis and removal of the substrate in order to prevent the pipettes from becoming clogged. Biorobot M-48 (Qiagen, Hilden, Germany) utilizes a closed system of magnetic silica-coated beads and a DNA purification chemical in pre-filled cartridges. In a study by Nagy et al. (148) the Biorobot M-48 gave higher DNA yields than phenol chloroform extraction, and produced inhibitor-free extracts from a range of sample and substrate types. However, Kishore et al. (149) found it necessary to modify the system using carrier RNA to give results comparable to those obtained using the manual phenol chloroform method. Biorobot EZ1 (Qiagen) and Maxwell® 16 (Promega, Madison, WI) are automated, small-scale, desktop plug-and-play systems for low- to medium-throughput analysis. The EZ1 has been extensively evaluated and has been shown to produce results comparable to those obtained with phenol chloroform and Chelex extraction (118, 149, 150). Open robotic systems, such as the Hamilton Star (Hamilton, Reno, NV), Freedom EVO® (Tecan, Männedorf, Switzerland), and versions of BioMek (Beckman Coulter, Fullerton, CA) have been used together with automatable magnetic bead systems such as DNA IQ™ (Promega) and ChargeSwitch® (Invitrogen, Carlsbad, CA). In a forensic DNA laboratory, the Freedom EVO and DNA IQ were found to generally perform slightly worse than phenol chloroform extraction for crime scene samples, but produced acceptable DNA yields and low amounts of PCR inhibitors in most samples (151).

Apart from the standard magnetic bead procedures, there are a number of promising new methods that can improve DNA purification in samples. A simple, automatable closed-tube method using a novel protease (EA1) has been developed by Moss et al. (152). A buffer containing protease is added to the sample, which is heated to 75°C for 15 min to achieve cell lysis and protein breakdown, and then to 94°C for 15 min to break down protease remains. The method gave better results than the Chelex system for crime scene blood stains on different materials. Although the extracts were visibly red, the inhibitor heme had been broken down and inactivated. In samples that had been subjected to different kinds of stress the method also produced higher DNA yields than Chelex, where free DNA is washed away. The EA1 method understandably failed in the analysis of samples from denim and cigarette butts, since these substrates have an abundance of non-protein PCR inhibitors.

Synchronous coefficient of drag alteration (SCODA) is a novel and possibly automatable technique for DNA purification in complex samples, e.g., in forensics (153). A rotating electric field separates the DNA molecules from the cell lysate and other particulate matter on an agarose gel, using the long-charged property of DNA as the basis for separation. Direct release of DNA mediated by a detergent is a possible alternative to maceration for analysis of plants. An automatable 96-well DNA extraction system suitable for different types of plant samples has been developed, using Tween 20 as detergent (101).

7. Pre-PCR Processing

Knowledge concerning PCR-inhibitory mechanisms at the molecular level is vital for the development of accurate and efficient real-time PCR-based systems for rapid diagnostics. To achieve the full potential of diagnostic PCR, the issue of inhibitors must be thoroughly addressed. Since inhibiting molecules can be introduced at any stage of the process, coming from the sample itself, the sampling equipment or the procedure used for sample preparation, all steps must be verified and controlled for each type of sample. Pre-PCR processing includes all the steps prior to the detection/quantification of the PCR product, i.e., sampling, sample treatment, and PCR. Choosing the appropriate thermostable DNA polymerase and using PCR facilitators are central components of this integrated concept (7). PCR inhibitors are defined by the PCR chemistry, in other words, the performance of PCR in the presence of inhibiting molecules depends on the choice of thermostable DNA polymerase, buffer, and facilitators (8, 9, 94). Therefore, the first step of pre-PCR processing must be the identification of the PCR chemistry best suited for the sample and assay to be used. Minimizing the effect of inhibitors in this fashion does not affect the sample itself, in contrast to sample treatment. Extensive DNA extraction and purification should be avoided if possible, since this may lower the amount of target DNA. If the inhibitory effects are not eliminated, sampling and sample treatment must be improved. In the scientific community, sampling is often performed in "the usual way." The development of sampling methods has not followed the development of PCR instruments and PCR-based analysis methods. Most of the techniques in use today, e.g., cotton swabbing and direct sampling, were developed for classical analysis methods (112). As the result of the analysis depends greatly on the sample and the way in which it is collected, more effort should be devoted to the development of sampling strategies.

The idea of pre-PCR processing is to generate PCR-compatible samples allowing high analytical precision. Simon et al. (100) were able to amplify inhibited *Listeria monocytogenes* samples in two separate ways: by employing silica column separation and by adding the PCR facilitator Tween 20. Juen and Traugott (88) reported successful amplification of 98% of their severely inhibited soil samples when adding BSA as a facilitator, and applying DNA purification using silica beads only gave a success rate of 56%, while, at the same time, over 90% of the DNA was lost. In some cases, the combination of improved sample treatment and PCR chemistry is required to produce acceptable results. Abu Al-Soud et al. (61) alleviated the inhibition caused by bile salts in *Helicobacter* detection by combining dilution and heating with the addition of the PCR facilitators casein and formamide. The combined effect was greater than when using only dilution and heating or facilitator addition. The concept of pre-PCR processing can be the basis for fast and efficient analysis systems, such as a newly developed 12-h method for *Salmonella* detection, combining pre-enrichment, sample treatment, and an optimized quantitative real-time PCR assay (39). Reducing the amount of manual handling in the pre-PCR processing and integrating this into an automated system are future challenges.

Acknowledgments

This work was in part financed by the support of the Swedish Research Council for the Environment, Agricultural Sciences and Spatial Planning, FORMAS.

References

1. Mullis KB, Faloona FA (1987) Specific synthesis of DNA in vitro via a polymerase-catalyzed chain reaction. Methods Enzymol 155:335–350
2. Rådström P, Löfström C, Lövenklev M, Knutsson R, Wolffs P (2003) Strategies for overcoming PCR inhibition. PCR primer. Cold Spring Harbor Laboratory Press, New York
3. Lantz PG, Hahn-Hägerdal B, Rådström P (1994) Sample preparation methods in PCR-based detection of food pathogen. Trends Food Sci Technol 5:384–389
4. Bessetti J (2007) An introduction to PCR inhibitors. Promega Profiles in DNA 10:9–10
5. Wilson IG (1997) Inhibition and facilitation of nucleic acid amplification. Appl Environ Microbiol 63:3741–3751
6. Butler JM (2005) Forensic DNA typing; biology, technology, and genetics of STR markers. Elsevier Academic, New York
7. Rådström P, Knutsson R, Wolffs P, Lövenklev M, Löfström C (2004) Pre-PCR processing: strategies to generate PCR-compatible samples. Mol Biotechnol 26:133–146
8. Abu Al-Soud W, Rådström P (1998) Capacity of nine thermostable DNA polymerases to mediate DNA amplification in the presence of PCR-inhibiting samples. Appl Environ Microbiol 64:3748–3753
9. Wolffs P, Grage H, Hagberg O, Rådström P (2004) Impact of DNA polymerases and their buffer systems on quantitative real-time PCR. J Clin Microbiol 42:408–411
10. Roussel Y, Wilks M, Harris A, Mein C, Tabaqchali S (2005) Evaluation of DNA extraction methods from mouse stomachs for the quantification of *H. pylori* by real-time PCR. J Microbiol Methods 62:71–81
11. Eckhart L, Bach J, Ban J, Tschachler E (2000) Melanin binds reversibly to thermostable DNA polymerase and inhibits its activity. Biochem Biophys Res Commun 271:726–730
12. Pruvost M, Geigl E-M (2004) Real-time quantitative PCR to assess the authenticity of

ancient DNA amplification. J Archaeol Sci 31:1191–1197
13. Huggett JF, Novak T, Garson JA et al (2008) Differential susceptibility of PCR reactions to inhibitors: an important and unrecognised phenomenon. BMC Res Notes 1:70
14. Sikorsky JA, Primerano DA, Fenger TW, Denvir J (2004) Effect of DNA damage on PCR amplification efficiency with the relative threshold cycle method. Biochem Biophys Res Commun 323:823–830
15. Lantz PG, Matsson M, Wadström T, Rådström P (1997) Removal of PCR inhibitors from human faecal samples through the use of an aqueous two-phase system for sample preparation prior to PCR. J Microbiol Methods 28:159–167
16. Abu Al-Soud W, Jönsson LJ, Rådström P (2000) Identification and characterization of immunoglobulin G in blood as a major inhibitor of diagnostic PCR. J Clin Microbiol 38:345–350
17. Akane A, Shiono H, Matsubara K, Nakamura H, Hasegawa M, Kagawa M (1993) Purification of forensic specimens for the polymerase chain reaction (PCR) analysis. J Forensic Sci 38:691–701
18. AbuAl-Soud W, Rådström P (2001) Purification and characterization of PCR-inhibitory components in blood cells. J Clin Microbiol 39:485–493
19. Tsai YL, Olson BH (1992) Rapid method for separation of bacterial DNA from humic substances in sediments for polymerase chain reaction. Appl Environ Microbiol 58:2292–2295
20. Tsai YL, Olson BH (1992) Detection of low numbers of bacterial cells in soils and sediments by polymerase chain reaction. Appl Environ Microbiol 58:754–757
21. Katcher HL, Schwartz I (1994) A distinctive property of *Tth* DNA polymerase: enzymatic amplification in the presence of phenol. Biotechniques 16:84–92
22. Belec L, Authier J, Eliezer-Vanerot MC, Piedouillet C, Mohamed AS, Gherardi RK (1998) Myoglobin as a polymerase chain reaction (PCR) inhibitor: a limitation for PCR from skeletal muscle tissue avoided by the use of *Thermus thermophilus* polymerase. Muscle Nerve 21:1064–1067
23. Powell HA, Gooding CM, Garrett SD, Lund BM, McKee RA (1994) Proteinase inhibition of the detection of *Listeria monocytogenes* in milk using the polymerase chain reaction. Lett Appl Microbiol 18:59–61
24. Bickley J, Short JK, McDowell DG, Parkes HC (1996) Polymerase chain reaction (PCR) detection of *Listeria monocytogenes* in diluted milk and reversal of PCR inhibition caused by calcium ions. Lett Appl Microbiol 22:153–158
25. McDevitt JJ, Lees PSJ, Merz WG, Schwab KJ (2007) Inhibition of quantitative PCR analysis of fungal conidia associated with indoor air particulate matter. Aerobiologia 23:35–45
26. Shutler GG, Gagnon P, Verret G et al (1999) Removal of a PCR inhibitor and resolution of DNA STR types in mixed human-canine stains from a five year old case. J Forensic Sci 44:623–626
27. Wadowsky RM, Laus S, Libert T, States SJ, Ehrlich GD (1994) Inhibition of PCR-based assay for *Bordetella pertussis* by using calcium alginate fiber and aluminum shaft components of a nasopharyngeal swab. J Clin Microbiol 32:1054–1057
28. McHale RH, Stapleton PM, Bergquist PL (1991) Rapid preparation of blood and tissue samples for polymerase chain reaction. Biotechniques 10(20):22–23
29. Knutsson R, Löfström C, Grage H, Hoorfar J, Rådström P (2002) Modeling of 5 nuclease real-time responses for optimization of a high-throughput enrichment PCR procedure for *Salmonella enterica*. J Clin Microbiol 40:52–60
30. Malmström H, Storå J, Dalén L, Holmlund G, Götherström A (2005) Extensive human DNA contamination in extracts from ancient dog bones and teeth. Mol Biol Evol 22:2040–2047
31. Noordhoek GT, Weel JF, Poelstra E, Hooghiernstra M, Brandenburg AH (2008) Clinical validation of a new real-time PCR assay for detection of enteroviruses and parechoviruses, and implications for diagnostic procedures. J Clin Virol 41:75–80
32. Ten Hove RJ, van Lieshout L, Brienen EA, Perez MA, Verweij JJ (2008) Real-time polymerase chain reaction for detection of Isospora belli in stool samples. Diagn Microbiol Infect Dis 61(3):280–283
33. Verweij JJ, Mulder B, Poell B, van Middelkoop D, Brienen EAT, van Lieshout L (2007) Real-time PCR for the detection of *Dientamoeba fragilis* in fecal samples. Mol Cell Probes 21:400–404
34. Volkmann H, Schwartz T, Kirchen S, Stofer C, Obst U (2007) Evaluation of inhibition and cross-reaction effects on real-time PCR applied to the total DNA of wastewater samples for the quantification of bacterial antibiotic resistance genes and taxon-specific targets. Mol Cell Probes 21:125–133
35. Pan M, McBeath AJA, Hay SJ, Pierce GJ, Cunningham CO (2008) Real-time PCR assay for detection and relative quantification

of *Liocarcinus depurator* larvae from plankton samples. Mar Biol 153:859–870

36. Ramirez NE, Sreevatsan S (2006) Development of a sensitive detection system for *Cryptosporidium* in environmental samples. Vet Parasitol 136:201–213
37. Demontis MA, Cacciola SO, Orru M et al (2008) Development of real-time PCR systems based on SYBR Green I and *Taq*Man technologies for specific quantitative detection of *Phoma tracheiphila* in infected citrus. Eur J Plant Pathol 120:339–351
38. Cankar K, Stebih D, Dreo T, Zel J, Gruden K (2006) Critical points of DNA quantification by real-time PCR - effects of DNA extraction method and sample matrix on quantification of genetically modified organisms. BMC Biotechnol 6:37
39. Josefsen MH, Krause M, Hansen F, Hoorfar J (2007) Optimization of a 12-hour *Taq*Man PCR-based method for detection of *Salmonella* bacteria in meat. Appl Environ Microbiol 73:3040–3048
40. Hein I, Flekna G, Krassnig M, Wagner M (2006) Real-time PCR for the detection of Salmonella spp. in food: An alternative approach to a conventional PCR system suggested by the FOOD-PCR project. J Microbiol Methods 66:538–547
41. Rodriguez-Lazaro D, Pla M, Scortti M, Monzo HJ, Vazquez-Boland JA (2005) A novel real-time PCR for *Listeria monocytogenes* that monitors analytical performance via an internal amplification control. Appl Environ Microbiol 71:9008–9012
42. Swango KL, Timken MD, Chong MD, Buoncristiani MR (2006) A quantitative PCR assay for the assessment of DNA degradation in forensic samples. Forensic Sci Int 158:12–26
43. Andreasson H, Gyllensten U, Allen M (2002) Real-time DNA quantification of nuclear and mitochondrial DNA in forensic analysis. Biotechniques 33:402–404, 407–411
44. Evans JJ, Wictum EJ, Penedo MC, Kanthaswamy S (2007) Real-time polymerase chain reaction quantification of canine DNA. J Forensic Sci 52:93–96
45. Andreasson H, Allen M (2003) Rapid quantification and sex determination of forensic evidence materials. J Forensic Sci 48:1280–1287
46. Hudlow W, Chong M, Swango K, Timken M, Buoncristiani M (2008) A quadruplex real-time qPCR assay for the simultaneous assessment of total human DNA, human male DNA, DNA degradation and the presence of PCR inhibitors in forensic samples: a diagnostic tool for STR typing. Forensic Sci Int Genet 2:108–125
47. Wen J, Guillo C, Ferrance JP, Landers JP (2007) Microfluidic-based DNA purification in a two-stage, dual-phase microchip containing a reversed-phase and a photopolymerized monolith. Anal Chem 79:6135–6142
48. Akane A, Matsubara K, Nakamura H, Takahashi S, Kimura K (1994) Identification of the heme compound copurified with deoxyribonucleic acid (DNA) from bloodstains, a major inhibitor of polymerase chain reaction (PCR) amplification. J Forensic Sci 39:362–372
49. Suslov O, Steindler DA (2005) PCR inhibition by reverse transcriptase leads to an overestimation of amplification efficiency. Nucleic Acids Res 33:e181
50. Yedidag EN, Koffron AJ, Mueller KH et al (1996) Acyclovir triphosphate inhibits the diagnostic polymerase chain reaction for cytomegalovirus. Transplantation 62:238–242
51. Kreader CA (1996) Relief of amplification inhibition in PCR with bovine serum albumin or T4 gene 32 protein. Appl Environ Microbiol 62:1102–1106
52. Sutlovic D, Gamulin S, Definis-Gojanovic M, Gugic D, Andjelinovic S (2008) Interaction of humic acids with human DNA: proposed mechanisms and kinetics. Electrophoresis 29:1467–1472
53. Sutlovic D, Definis Gojanovic M, Andelinovic S, Gugic D, Primorac D (2005) Taq polymerase reverses inhibition of quantitative real time polymerase chain reaction by humic acid. Croat Med J 46:556–562
54. Mahony J, Chong S, Jang D et al (1998) Urine specimens from pregnant and nonpregnant women inhibitory to amplification of *Chlamydia trachomatis* nucleic acid by PCR, ligase chain reaction, and transcription-mediated amplification: identification of urinary substances associated with inhibition and removal of inhibitory activity. J Clin Microbiol 36:3122–3126
55. Tebbe CC, Vahjen W (1993) Interference of humic acids and DNA extracted directly from soil in detection and transformation of recombinant DNA from bacteria and a yeast. Appl Environ Microbiol 59:2657–2665
56. Walsh PS, Metzger DA, Higuchi R (1991) Chelex 100 as a medium for simple extraction of DNA for PCR-based typing from forensic material. Biotechniques 10:506–513
57. Gudnason H, Dufva M, Bang DD, Wolff A (2007) Comparison of multiple DNA dyes

for real-time PCR: effects of dye concentration and sequence composition on DNA amplification and melting temperature. Nucleic Acids Res 35:e127

58. Burkhart CA, Norris MD, Haber M (2002) A simple method for the isolation of genomic DNA from mouse tail free of real-time PCR inhibitors. J Biochem Biophys Methods 52:145–149
59. Zipper H, Buta C, Lammle K, Brunner H, Bernhagen J, Vitzthum F (2003) Mechanisms underlying the impact of humic acids on DNA quantification by SYBR Green I and consequences for the analysis of soils and aquatic sediments. Nucleic Acids Res 31:e39
60. Gonzalez A, Grimes R, Walsh EJ, Dalton T, Davies M (2007) Interaction of quantitative PCR components with polymeric surfaces. Biomed Microdevices 9:261–266
61. Abu Al-Soud W, Ouis IS, Li DQ, Ljungh A, Wadström T (2005) Characterization of the PCR inhibitory effect of bile to optimize real-time PCR detection of *Helicobacter* species. FEMS Immunol Med Microbiol 44:177–182
62. Scholz M, Giddings I, Pusch CM (1998) A polymerase chain reaction inhibitor of ancient hard and soft tissue DNA extracts is determined as human collagen type I. Anal Biochem 259:283–286
63. Rossen L, Norskov P, Holmstrom K, Rasmussen OF (1992) Inhibition of PCR by components of food samples, microbial diagnostic assays and DNA-extraction solutions. Int J Food Microbiol 17:37–45
64. Burgess LC, Hall JO (1999) UV light irradiation of plastic reaction tubes inhibits PCR. Biotechniques 27:252, 254–256
65. Tamariz J, Voynarovska K, Prinz M, Caragine T (2006) The application of ultraviolet irradiation to exogenous sources of DNA in plasticware and water for the amplification of low copy number DNA. J Forensic Sci 51:790–794
66. Fox DH, Huang CK, Du J, Chang TY, Pan Q (2007) Profound inhibition of the PCR step of CF V3 multiplex PCR/OLA assay by the use of UV-irradiated plastic reaction tubes. Diagn Mol Pathol 16:121–123
67. Johnson DW, Pieniazek NJ, Griffin DW, Misener L, Rose JB (1995) Development of a PCR protocol for sensitive detection of *Cryptosporidium oocysts* in water samples. Appl Environ Microbiol 61:3849–3855
68. Thornton CG, Passen S (2004) Inhibition of PCR amplification by phytic acid, and treatment of bovine fecal specimens with phytase to reduce inhibition. J Microbiol Methods 59:43–52
69. Monteiro L, Bonnemaison D, Vekris A et al (1997) Complex polysaccharides as PCR inhibitors in feces: *Helicobacter pylori* model. J Clin Microbiol 35:995–998
70. Fehlmann C, Krapf R, Solioz M (1993) Reverse transcriptase can block polymerase chain reaction. Clin Chem 39:368–369
71. Khan G, Kangro HO, Coates PJ, Heath RB (1991) Inhibitory effects of urine on the polymerase chain reaction for cytomegalovirus DNA. J Clin Pathol 44:360–365
72. Bej AK, Mahbubani MH, Dicesare JL, Atlas RM (1991) Polymerase chain reaction-gene probe detection of microorganisms by using filter-concentrated samples. Appl Environ Microbiol 57:3529–3534
73. Satsangi J, Jewell DP, Welsh K, Bunce M, Bell JI (1994) Effect of heparin on polymerase chain reaction. Lancet 343:1509–1510
74. Vaneechoutte M, Van Eldere J (1997) The possibilities and limitations of nucleic acid amplification technology in diagnostic microbiology. J Med Microbiol 46:188–194
75. Gryson N, Dewettinck K, Messens K (2007) Influence of cocoa components on the PCR detection of soy lecithin DNA. Eur Food Res Technol 226:247–254
76. Nolan T, Hands RE, Ogunkolade W, Bustin SA (2006) SPUD: a quantitative PCR assay for the detection of inhibitors in nucleic acid preparations. Anal Biochem 351:308–310
77. Bustin SA (2005) Real-time, fluorescence-based quantitative PCR: a snapshot of current procedures and preferences. Expert Rev Mol Diagn 5:493–498
78. Flekna G, Schneeweiss W, Smulders FJM, Wagner M, Hein I (2007) Real-time PCR method with statistical analysis to compare the potential of DNA isolation methods to remove PCR inhibitors from samples for diagnostic PCR. Mol Cell Probes 21:282–287
79. Schneeweiss W, Stanek C, Wagner M, Hein I (2007) Inhibitor-free DNA for real-time PCR analysis of synovial fluid from horses, cattle and pigs. Vet Microbiol 121:189–193
80. Reiss RA, Rutz B (1999) Quality control PCR: a method for detecting inhibitors of *Taq* DNA polymerase. Biotechniques 27:920–922, 924–926
81. Niederstätter H, Köchl S, Grubwieser P, Palvic M, Steinlechner M, Parson W (2007) A modular real-time PCR concept for determining the quantity and quality of human nuclear

and mitochondrial DNA. Forensic Sci Int Genet 1:29–34
82. Green RL, Roinestad IC, Boland C, Hennessy LK (2005) Developmental validation of the Quantifiler real-time PCR kits for the quantification of human nuclear DNA samples. J Forensic Sci 50:809–825
83. Hoorfar J, Cook N, Malorny B et al (2003) Making internal amplification control mandatory for diagnostic PCR. J Clin Microbiol 41:5835
84. Rosenstraus M, Wang Z, Chang SY, DeBonville D, Spadoro JP (1998) An internal control for routine diagnostic PCR: design, properties, and effect on clinical performance. J Clin Microbiol 36:191–197
85. Maaroufi Y, de Bruyne JM, Duchateau V, Scheen R, Crokaert F (2006) Development of a multiple internal control for clinical diagnostic real-time amplification assays. FEMS Immunol Med Microbiol 48:183–191
86. Hartman LJ, Coyne SR, Norwood DA (2005) Development of a novel internal positive control for *Taq*Man based assays. Mol Cell Probes 19:51–59
87. Murphy NM, McLauchlin J, Ohai C, Grant KA (2007) Construction and evaluation of a microbiological positive process internal control for PCR-based examination of food samples for *Listeria monocytogenes* and *Salmonella enterica*. Int J Food Microbiol 120:110–119
88. Juen A, Traugott M (2006) Amplification facilitators and multiplex PCR: Tools to overcome PCR-inhibition in DNA-gut-content analysis of soil-living invertebrates. Soil Biol Biochem 38:1872–1879
89. Guescini M, Sisti D, Rocchi MB, Stocchi L, Stocchi V (2008) A new real-time PCR method to overcome significant quantitative inaccuracy due to slight amplification inhibition. BMC Bioinformatics 9:326
90. Kontanis EJ, Reed FA (2006) Evaluation of real-time PCR amplification efficiencies to detect PCR inhibitors. J Forensic Sci 51:795–804
91. Tichopad A, Didier A, Pfaffl MW (2004) Inhibition of real-time RT-PCR quantification due to tissue-specific contaminants. Mol Cell Probes 18:45–50
92. Chervoneva I, Hyslop T, Iglewicz B et al (2006) Statistical algorithm for assuring similar efficiency in standards and samples for absolute quantification by real-time reverse transcription polymerase chain reaction. Anal Biochem 348:198–208
93. Yang YG, Kim JY, Song YH, Kim DS (2007) A novel buffer system, AnyDirect, can improve polymerase chain reaction from whole blood without DNA isolation. Clin Chim Acta 380:112–117
94. Abu Al-Soud W, Rådström P (2000) Effects of amplification facilitators on diagnostic PCR in the presence of blood, feces, and meat. J Clin Microbiol 38:4463–4470
95. Ahokas H, Erkkila MJ (1993) Interference of PCR amplification by the polyamines, spermine and spermidine. PCR Methods Appl 3:65–68
96. Wang MY, Olson BH, Chang JS (2007) Improving PCR and qPCR detection of hydrogenase A (hydA) associated with *clostridia* in pure cultures and environmental sludges using bovine serum albumin. Appl Microbiol Biotechnol 77:645–656
97. De Boer SH, Ward LJ, Li X, Chittaranjan S (1995) Attenuation of PCR inhibition in the presence of plant compounds by addition of BLOTTO. Nucleic Acids Res 23:2567–2568
98. Arbeli Z, Fuentes CL (2007) Improved purification and PCR amplification of DNA from environmental samples. FEMS Microbiol Lett 272:269–275
99. Demeke T, Adams RP (1992) The effects of plant polysaccharides and buffer additives on PCR. Biotechniques 12:332–334
100. Simon MC, Gray DI, Cook N (1996) DNA Extraction and PCR methods for the detection of *Listeria monocytogenes* in cold-smoked salmon. Appl Environ Microbiol 62:822–824
101. Xin ZG, Velten JP, Oliver MJ, Burke JJ (2003) High-throughput DNA extraction method suitable for PCR. Biotechniques 34:820–826
102. Comey CT, Jung JM, Budowle B (1991) Use of formamide to improve amplification of HLA DQ alpha sequences. Biotechniques 10:60–61
103. Pomp D, Medrano JF (1991) Organic solvents as facilitators of polymerase chain reaction. Biotechniques 10:58–59
104. Sidhu MK, Liao MJ, Rashidbaigi A (1996) Dimethyl sulfoxide improves RNA amplification. Biotechniques 21:44–47
105. ISO/FDIS17604. Microbiology of food and animal feeding stuffs-Carcass sampling for microbiological analysis. INTERNATIONAL STANDARD ISO/FDIS 17604 2003
106. ISO18593. Microbiology of food and animal feeding stuffs-horizontal methods for sampling techniques from surfaces using contact plates and swabs. INTERNATIONAL STANDARD ISO 18593 2004
107. Hodges LR, Rose LJ, Peterson A, Noble-Wang J, Arduino MJ (2006) Evaluation of a macrofoam swab protocol for the recovery of *Bacillus anthracis* spores from a steel surface. Appl Environ Microbiol 72:4429–4430

108. Buttner MP, Cruz P, Stetzenbach LD, Cronin T (2007) Evaluation of two surface sampling methods for detection of *Erwinia herbicola* on a variety of materials by culture and quantitative PCR. Appl Environ Microbiol 73:3505–3510
109. Buttner MP, Cruz P, Stetzenbach LD, Klima-Comba AK, Stevens VL, Emanuel PA (2004) Evaluation of the biological sampling kit (BiSKit) for large-area surface sampling. Appl Environ Microbiol 70:7040–7045
110. Brown GS, Betty RG, Brockmann JE et al (2007) Evaluation of a wipe surface sample method for collection of *Bacillus* spores from nonporous surfaces. Appl Environ Microbiol 73:706–710
111. Brown GS, Betty RG, Brockmann JE et al (2007) Evaluation of rayon swab surface sample collection method for *Bacillus* spores from nonporous surfaces. J Appl Microbiol 103:1074–1080
112. Patterson JT (1971) Microbiological assessment of surfaces. J Food Technol 6:63–72
113. Lin Z, Suzow JG, Fontaine JM, Naylor EW (2005) A simple automated DNA extraction method for dried blood specimens collected on filter paper. JALA 10:310–314
114. Snijders JMA, Janssen MHW, Gerats GE, Corstiaensen GP (1984) A comparative study of sampling techniques for monitoring carcass contamination. Int J Food Microbiol 1:229–236
115. Bourke MT, Scherczinger CA, Ladd C, Lee HC (1999) NaOH treatment to neutralize inhibitors of *Taq* polymerase. J Forensic Sci 44:1046–1050
116. Hsiao KM, Lin HM, Pan H et al (1999) Application of FTA sample collection and DNA purification system on the determination of CTG trinucleotide repeat size by PCR-based Southern blotting. J Clin Lab Anal 13:188–193
117. Pearce RA, Bolton DJ (2005) Excision vs. sponge swabbing - a comparison of methods for the microbiological sampling of beef, pork and lamb carcasses. J Appl Microbiol 98:896–900
118. Anslinger K, Bayer B, Rolf B, Keil W, Eisenmenger W (2005) Application of the BioRobot EZ1 in a forensic laboratory. Leg Med (Tokyo) 7:164–168
119. von Wurmb-Schwark N, Malyusz V, Fremdt H, Koch C, Simeoni E, Schwark T (2006) Fast and simple DNA extraction from saliva and sperm cells obtained from the skin or isolated from swabs. Leg Med (Tokyo) 8:177–181
120. Rose L, Jensen B, Peterson A, Banerjee SN, Arduino MJ (2004) Swab materials and *Bacillus anthracis* spore recovery from nonporous surfaces. Emerg Infect Dis 10:1023–1029
121. Sweet D, Lorente M, Lorente JA, Valenzuela A, Villanueva E (1997) An improved method to recover saliva from human skin: the double swab technique. J Forensic Sci 42:320–322
122. Pang BC, Cheung BK (2007) Double swab technique for collecting touched evidence. Leg Med (Tokyo) 9:181–184
123. Walsh P, Overmyer CL, Pham K et al (2008) Comparison of respiratory virus detection rates in infants and toddlers using flocked swabs, saline aspirates and saline aspirates mixed in UTM-RT. J Clin Microbiol 46(7):2374–2376
124. Li RC, Harris HA (2003) Using hydrophilic adhesive tape for collection of evidence for forensic DNA analysis. J Forensic Sci 48:1318–1321
125. Schöld M, Dufva C, Forsslund S, Jangblad A. The hand on the shoulder: a case report where Low Copy Number (LCN) DNA analysis was vital for solving a robbery. Proceedings of the fourth European Academy of Forensic Science Conference, Helsinki, Finland, 2006
126. Bright JA, Petricevic SF (2004) Recovery of trace DNA and its application to DNA profiling of shoe insoles. Forensic Sci Int 145:7–12
127. Grant KA, Dickinson JH, Payne MJ, Campbell S, Collins MD, Kroll RG (1993) Use of the polymerase chain reaction and 16 S rRNA sequences for the rapid detection of *Brochothrix* spp. in foods. J Appl Bacteriol 74:260–267
128. Nogva HK, Rudi K, Naterstad K, Holck A, Lillehaug D (2000) Application of 5 -nuclease PCR for quantitative detection of *Listeria monocytogenes* in pure cultures, water, skim milk, and unpasteurized whole milk. Appl Environ Microbiol 66:4266–4271
129. Wolffs P, Knutsson R, Norling B, Rådström P (2004) Rapid quantification of *Yersinia enterocolitica* in pork samples by a novel sample preparation method, flotation, prior to real-time PCR. J Clin Microbiol 42:1042–1047
130. Lindqvist R (1997) Preparation of PCR samples from food by a rapid and simple centrifugation technique evaluated by detection of *Escherichia coli* O157:H7. Int J Food Microbiol 37:73–82
131. Wolffs P, Norling B, Hoorfar J, Griffiths M, Rådström P (2005) Quantification of *Campylobacter* spp. in chicken rinse samples

by using flotation prior to real-time PCR. Appl Environ Microbiol 71:5759–5764
132. Curran S, McKay JA, McLeod HL, Murray GI (2000) Laser capture microscopy. Mol Pathol 53:64–68
133. Elliott K, Hill DS, Lambert C, Burroughes TR, Gill P (2003) Use of laser microdissection greatly improves the recovery of DNA from sperm on microscope slides. Forensic Sci Int 137:28–36
134. Anoruo B, van Oorschot R, Mitchell J, Howells D (2007) Isolating cells from non-sperm cellular mixtures using the PALM microlaser micro dissection system. Forensic Sci Int 173:93–96
135. Lambie-Anoruo BL, Prince DV, Koukoulas I, Howells DW, Mitchell RJ, van Oorschot RAH (2006) Laser microdissection and pressure catapulting with PALM to assist typing of target DNA in dirt samples. Int Cong Ser 1288:559–561
136. Hochmeister MN, Budowle B, Borer UV, Eggmann U, Comey CT, Dirnhofer R (1991) Typing of deoxyribonucleic acid (DNA) extracted from compact bone from human remains. J Forensic Sci 36:1649–1661
137. Sambrook J, Fritsch EF, Maniatis T (1989) Molecular cloning: a laboratory manual. Cold Spring Harbor Laboratory Press, New York
138. Zhou J, Bruns MA, Tiedje JM (1996) DNA recovery from soils of diverse composition. Appl Environ Microbiol 62:316–322
139. Watanabe Y, Takayama T, Hirata K et al (2003) DNA typing from cigarette butts. Leg Med (Tokyo) 5(Suppl 1):S177–S179
140. Abolmaaty A, Gu W, Witkowsky R, Levin RE (2007) The use of activated charcoal for the removal of PCR inhibitors from oyster samples. J Microbiol Methods 68:349–352
141. Braid MD, Daniels LM, Kitts CL (2003) Removal of PCR inhibitors from soil DNA by chemical flocculation. J Microbiol Methods 52:389–393
142. McCormick RM (1989) A solid-phase extraction procedure for DNA purification. Anal Biochem 181:66–74
143. Soerensen E, Hansen SH, Eriksen B, Morling N (2003) Applications of thiopropyl Sepharose 6B for removal of PCR inhibitors from DNA extracts from different sources. Int Cong Ser 1239:821–823
144. Gustavson KH (1954) Note on the fixation of vegetable tannins by polyvinylpyrrolidone. Sven Kem Tidskr 66:359–362
145. Boom R, Sol CJ, Salimans MM, Jansen CL, Wertheim-van Dillen PM, van der Noordaa J (1990) Rapid and simple method for purification of nucleic acids. J Clin Microbiol 28:495–503
146. Tian H, Huhmer AF, Landers JP (2000) Evaluation of silica resins for direct and efficient extraction of DNA from complex biological matrices in a miniaturized format. Anal Biochem 283:175–191
147. Hedman J, Albinsson L, Ansell C et al (2008) A fast analysis system for forensic DNA reference samples. Forensic Sci Int Genet 2:184–189
148. Nagy M, Otremba P, Kruger C et al (2005) Optimization and validation of a fully automated silica-coated magnetic beads purification technology in forensics. Forensic Sci Int 152:13–22
149. Kishore R, Reef Hardy W, Anderson VJ, Sanchez NA, Buoncristiani MR (2006) Optimization of DNA extraction from low-yield and degraded samples using the BioRobot EZ1 and BioRobot M48. J Forensic Sci 51:1055–1061
150. Montpetit SA, Fitch IT, O'Donnell PT (2005) A simple automated instrument for DNA extraction in forensic casework. J Forensic Sci 50:555–563
151. Andradas Heranz J, García Poveda E, Alvarez Fernández F (2008) DNA Extraction using the Tecan Freedom EVO® 200 and DNA IQ™ system. Promega Profiles in DNA 11:7–8
152. Moss D, Harbison SA, Saul DJ (2003) An easily automated, closed-tube forensic DNA extraction procedure using a thermostable proteinase. Int J Legal Med 117:340–349
153. Broemling D, Pel J, Gunn D et al (2007) An instrument for automated purification of nucleic acids from contaminated forensic samples. JALA 13:40–48
154. Lantz PG, Tjerneld F, Borch E, Hahn-Hägerdal B, Rådström P (1994) Enhanced sensitivity in PCR detection of *Listeria monocytogenes* in soft cheese through use of an aqueous two-phase system as a sample preparation method. Appl Environ Microbiol 60:3416–3418
155. Moreira D (1998) Efficient removal of PCR inhibitors using agarose-embedded DNA preparations. Nucleic Acids Res 26:3309–3310
156. Sweet D, Lorente M, Valenzuela A, Lorente JA, Alvarez JC (1996) Increasing DNA extraction yield from saliva stains with a modified Chelex method. Forensic Sci Int 83:167–177

Chapter 3

Quality in the Molecular Microbiology Laboratory

Paul S. Wallace and William G. MacKay

Abstract

In the clinical microbiology laboratory advances in nucleic acid detection, quantification, and sequence analysis have led to considerable improvements in the diagnosis, management, and monitoring of infectious diseases. Molecular diagnostic methods are routinely used to make clinical decisions based on when and how to treat a patient as well as monitor the effectiveness of a therapeutic regime and identify any potential drug resistant strains that may impact on the long term patient treatment program. Therefore, confidence in the reliability of the result provided by the laboratory service to the clinician is essential for patient treatment. Hence, suitable quality assurance and quality control measures are important to ensure that the laboratory methods and service meet the necessary regulatory requirements both at the national and international level. In essence, the modern clinical microbiology laboratory ensures the appropriateness of its services through a quality management system that monitors all aspects of the laboratory service pre- and post-analytical—from patient sample receipt to reporting of results, from checking and upholding staff competency within the laboratory to identifying areas for quality improvements within the service offered. For most European based clinical microbiology laboratories this means following the common International Standard Organization (ISO9001) framework and ISO15189 which sets out the quality management requirements for the medical laboratory (BS EN ISO 15189 (2003) Medical laboratories—particular requirements for quality and competence. British Standards Institute, Bristol, UK). In the United States clinical laboratories performing human diagnostic tests are regulated by the Centers for Medicare and Medicaid Services (CMS) following the requirements within the Clinical Laboratory Improvement Amendments document 1988 (CLIA-88). This chapter focuses on the key quality assurance and quality control requirements within the modern microbiology laboratory providing molecular diagnostics.

Key words: Quality assurance, Quality control, External quality assessment, Proficiency testing, Validation and verification

1. Introduction

Twenty-five years ago, the clinical microbiological laboratory's test portfolio was primarily based on tests for the identification of infectious organisms using culture techniques and immunoassay. Physicians generally viewed diagnostic virology in particular as simply

Mark Wilks (ed.), *PCR Detection of Microbial Pathogens: Second Edition*, Methods in Molecular Biology, vol. 943, DOI 10.1007/978-1-60327-353-4_3, © Springer Science+Business Media, LLC 2013

a confirmatory tool; a method of resolving discrepancy within a physician's initial clinical findings; or just a means of supporting a course of action taken. The introduction of molecular techniques such as viral load determination placed the emphasis on rapid diagnosis and created an environment where molecular diagnostic testing for infectious diseases became fundamental to the decision-making process. This provided the opportunity for more rational use of diagnostic tests within the clinical microbiology laboratory which in turn complemented novel antimicrobial development and advances in disease management strategies. The rate of innovation within molecular diagnostics has continued with the development of technologies such as real-time PCR. This has meant that laboratories are potentially able to provide assays "on demand" and in response to the most clinically relevant pathogens at a given time and in terms of infectivity, viral load, patient group susceptibility (for example transplant patients), and geographical pandemic (such as influenza virus). The rapid introduction of molecular techniques into the clinical microbiology laboratory often meant that both quality assurance and quality control practices were neglected. Particular concerns regarding the specificity, sensitivity, and potential contamination risk of molecular assays were recognized early after their introduction into the routine clinical microbiology laboratory through early molecular proficiency testing studies. In recent years, as the role of the modern clinical microbiology laboratory has become more dynamic and the general awareness of infectious disease healthcare issues has increased, clinical microbiology laboratories are under increasing pressure to become fully accountable for the quality of the services they provide. Consequently, although the principals of quality management and quality assurance are well defined, they have only really been applied within the clinical microbiology laboratory in the context of molecular diagnostics over the last 10 years. As a result, there has been a steady improvement in the performance of molecular assays within the clinical microbiological laboratory. This has also been due to the increased availability of manufactured reagents and a growing expertise in molecular diagnostics within the clinical laboratory. Nevertheless, with limited molecular International standards and/or reference materials available quality assurance and quality control practice remain underdeveloped and are often inappropriately addressed within the clinical microbiology laboratory (1).

2. Quality Assurance

All medical laboratories should adhere to defined Quality Assurance practices and in many countries these requirements are mandatory and they support the clinical microbiology laboratory's annual accreditation and/or certification process for the service it provides

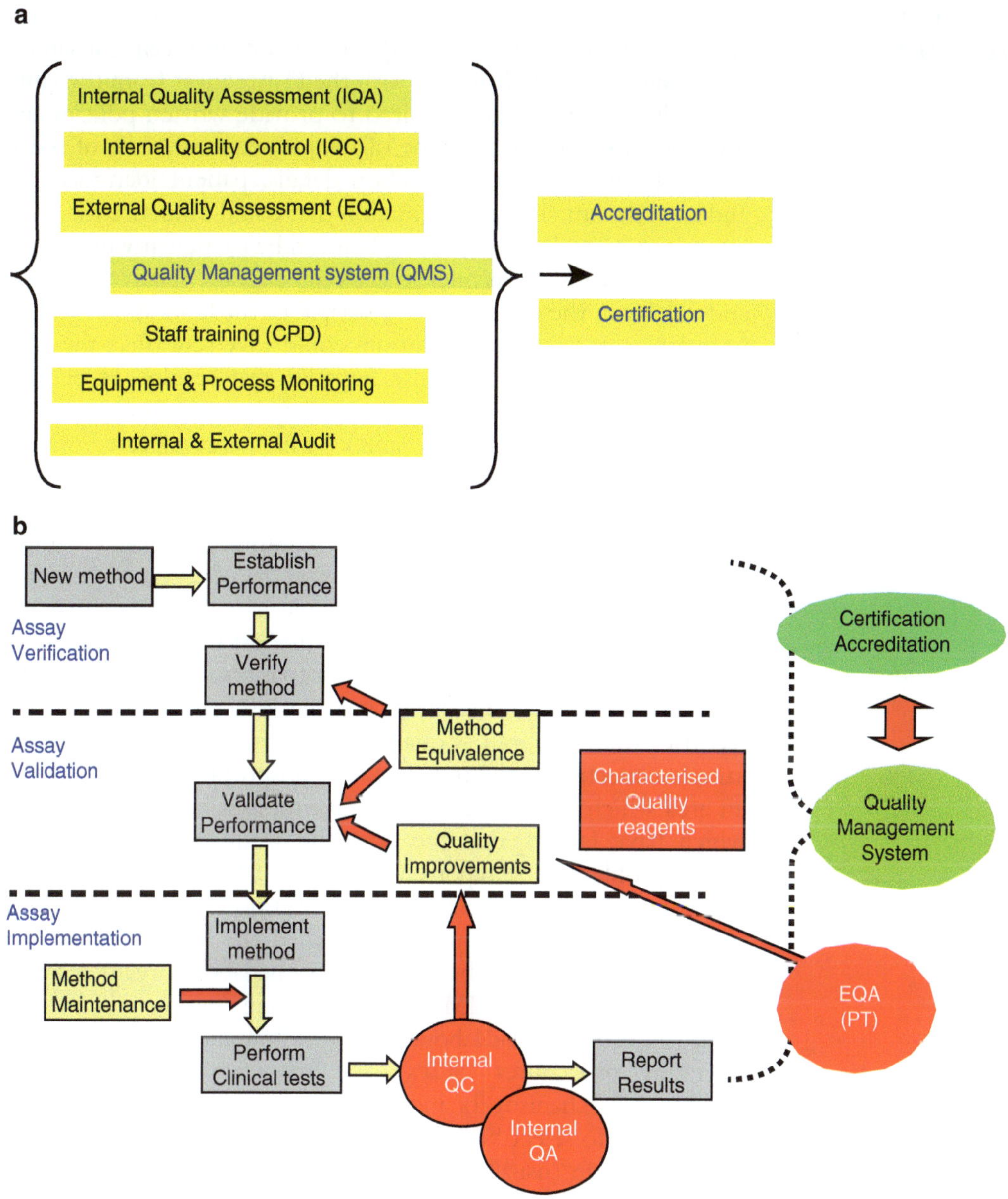

Fig. 1. (**a**) "Quality Assurance" with the clinical microbiology laboratory. (**b**) "Quality Assurance" in the clinical microbiology laboratory.

as a diagnostic laboratory. Quality Assurance ensures both competence and confidence in the service provided and reliability in the patient test results reported. It is an ongoing process that requires input from all laboratory staff on a daily basis. The key components of quality assurance are generally defined within the laboratories quality manual (Fig. 1). They are an integral part of a laboratory's Quality Management System (QMS) and in the medical profession also assure the safety and effectiveness of the service.

2.1. Internal Quality Assessment

Internal Quality Assessment (IQA) is the mechanism for monitoring all the laboratory's activities and processes from receipt of sample to issuing of report. It begins with the test request from the clinician. All laboratories are required to provide written policies and procedures regarding the type of sample required, date of specimen collection and transportation details, patient identification, and any relevant clinical information that could influence the laboratory testing to be undertaken. This can be particularly important for molecular testing as nucleic acids can degrade relatively quickly and a delay in the delivery of the sample to the laboratory or inappropriate transportation conditions could adversely affect the sample integrity and lead to false negative test results. Hence, all samples should be inspected upon receipt and before testing to ensure that they are suitable. Samples which are suspected of contamination or show signs of lipemia or hemolysis should be excluded from testing as these may interfere with the molecular assay test performance. Under these circumstances a new sample should be requested. If it is not possible to obtain a new sample and the laboratory must test the sample it is important that the condition of the sample is noted with the test result and report so that any anomaly can be reviewed prior to releasing or rejecting the final test result. This also helps provide essential traceability and feedback to the laboratory quality team on any potential pretesting sample collection, handling, and/or transportation issues. This is an area which needs effective monitoring as a number of reports which have investigated the type and frequency of laboratory testing problems have reported that up to 75% of errors occur prior to testing in the sample collection and transportation stages (2–5).

Once the patient sample has reached the laboratory the test request information will be logged on to the laboratory information management system (LIMS) and the patient sample will be processed according to the laboratories workflow and relevant standard operating procedures (SOPs). An important Internal Quality Assurance mechanism for monitoring error within the laboratory workflow is to select "IQA samples" randomly from a representative proportion (usually between 0.5 and 1.0%) of the laboratory clinical sample workload. This approach is often called split sampling. A patient sample is randomly selected, split and a new request form created with a laboratory IQA code. This generated IQA sample is then processed alongside the patient test sample. The results are reported for both samples and the IQA sample results are compared with the patient sample. Any inconsistencies within the result and report for both IQA and patient are investigated and corrective action taken through the laboratories quality system.

For some microbiological targets the split sample approach may be impractical and inappropriate as it the frequency of testing for that particular pathogen may be too low. It may also be difficult to obtain suitable clinical material due to sample viability or the infrequency or atypical nature of the target pathogen. For example strains of *Chlamydia trachomatis* containing specific cryptic plasmids or a particular HIV-1 drug resistant strain. In these situations a laboratory may choose to use previously tested, characterized, and stored clinical samples or "spiked" cultured specimens and also previous EQA panel materials to complement its daily IQA procedures. Often these types of materials will be used to investigate day-to-day differences in laboratory processes, operator performance, and staff training requirements.

Maintaining and monitoring the integrity of the molecular facilities within the clinical microbiological laboratory are also a vital component of IQA. Most modern molecular facilities employ a unidirectional workflow with separate and defined working areas for molecular reagent preparation, sample handling and nucleic acid extraction, reaction template addition, and molecular amplification and detection. However, laboratory space constraints are regularly a problem and some laboratories are only able to establish minimum separation of activities which can put an additional burden on the molecular facilities in terms of risk of contamination. Hence, rigorous and disciplined laboratory practices are essential to ensure contamination control and environmental sampling such as swabbing or wipe testing of molecular equipment including bench tops, microfuges, water baths, pipettes, and molecular platforms followed by testing for infectious disease nucleic acids on a regular basis is essential.

Other factors that are monitored through the laboratory's IQA processes include time to report results, assay validation and verification, staff training, and continuous professional development (CPD).

2.2. Internal Quality Control

There are many factors that can affect the quality and outcome of a test result. One as mentioned above is the primary condition of the test specimens. Other factors include a failure of the test system, the controls used, potential staff training issues, the interpretation of results, and also mistranscription of results prior to reporting. The aim of quality control is to ensure that the test results generated on a day to day and run to run basis are acceptable. A clinical microbiology laboratory's quality control practice refers to the measures that must be included during each assay run in order to verify the test result. The clinical microbiology laboratory must therefore establish QC criteria for each of the assays that it routinely runs as part of its diagnostics service. In general a commercially available kit will be provided with a set of internal and external controls that are to be included in each test run.

The intended use of these controls is to monitor the performance of the assay within the limits defined by the manufacturer. The external controls supplied by the manufacturer may simply be positive and negative samples or quantifiable kit controls which are optimized and intended for use only with the manufacturers test kit. Internal controls are usually incorporated into the molecular test either by the assay manufacturer and form an integral part of the test system and protocol. Or in the case of laboratory developed tests (LDTs) or "in-house" developed assays the laboratory normally determines the internal control system and defines the control limits itself.

Ideally, the internal control is added to the sample at the beginning of the isolation process this enables monitoring of the entire molecular process from sample pretreatment through the extraction stage to amplification and detection. The nature of the internal control used differs and opinions vary as to what control material and approach is best. A homologous internal control is one where the primer binding sites are identical to the target sequence and the internal region of the control is same size and nucleic acid composition as the target but contains a unique probe binding region in order to distinguish it from the target sequence during the detection phase. Hence, the amplification of both the internal control and target pathogen is in theory equivalent, although in practice the concentrations of the internal control are optimized to ensure that it does not compete out the target template within the reaction. In general homologous internal controls consist of plasmid constructs, single stranded oligonucleotides, and in some commercially available kits contain modified lambda phage particles or bacteriophages in which cloned microbial nucleic acids are packaged in a bacteriophage such as MS2 (6, 7).

In contrast, within a heterologous internal control system the primer binding sites, internal target regions, and probe binding region are different from the target sequence, and they require different primers and probes. Heterologous internal control amplification is, as a result, independent of the target pathogen although the control is normally optimized in order to have extraction and amplification efficiencies similar to that of the target pathogen. Heterologous internal controls are also often modified bacteriophages, and there are a number of commercial products available (7, 8). In addition, within the virology field whole nonhuman-related viruses are used as internal controls because the whole virus is believed to mimic the viral target pathogens more closely than a synthetic nucleic acid construct or a recombinant bacteriophage control. Examples include the murine picornavirus and encephalomyocarditis viruses, as well as the phocine viruses such as DNA virus Phocine HerpesVirus type 1 (PLHV-1) and RNA virus Phocine distemper virus (PDV) (8). The phocine virus is added to each patient sample at the beginning of the isolation procedure at concentrations of approximately 2,500–5,000 copies/ml or cycle threshold (CT) values around 34. By quantifying the phocine virus it can then be tracked throughout the extraction and amplification process and allows the laboratory to control for both inhibition (false negative results) and extraction efficiency. In addition the phocine distemper virus can also be used to monitor the initial reverse transcriptase step within RNA viral target assays.

It is important that the clinical microbiology laboratory establishes its own QC criteria for each of the assays that it runs regardless of whether it is using a commercial or “in-house” developed assay. This is usually done by the inclusion of some additional QC samples which are often referred to as external controls or External Run Controls (ERC). External controls are independent of the molecular assay kit used and are aimed to resemble a clinical sample in terms of format and composition. They are often used to monitor the whole test process and are frequently included by the laboratory on a run to run basis. Evaluating the results obtained from an independent external control and the results obtained from the manufacturer supplied kit control allows the laboratory to check the consistency of performance between test kit batches, reagent lots, and potential assay drift.

There are a number of commercial sources which provide independent external controls for molecular quality control purposes. However, the range of materials available to the clinical microbiology laboratory is still relatively limited and the materials often lack molecular characterization and standardization. Hence, many laboratories will tend to use residual clinical samples which can have a limited life span carry an additional risk of contamination, and can be inappropriate for long term QC monitoring.

Whatever the source of material the laboratory chooses for its quality control purposes, it is important that the material is tested in conjunction with known clinical samples in order to establish the relationship between the clinical test result and the results obtained for a quality control. This performance should be documented within the QC procedures and reviewed regularly in line with the laboratories quality management system and any regulatory requirements.

Many clinical microbiology laboratories performing molecular diagnostics also choose to monitor the run to run performance of their assays. Both the results obtained from the internal controls and external control samples can be used within an internal quality control (IQC) program or run control program. This approach has routinely been performed within other clinical laboratory services such as clinical chemistry but is still relatively new to many clinical microbiology laboratories performance molecular assays. Each time the quality controls are tested the results are collated and plotted within a control chart. The control chart is usually a Shewhart type chart or more often a Levey–Jennings chart, where the *x*-axis is date and time and/or the number of control run observations and the *y*-axis is the concentration in log copies/ml, International Units/ml, or cycle thresholds, or even a laboratory specific unitage such as PCR units/ml. Example control charts for

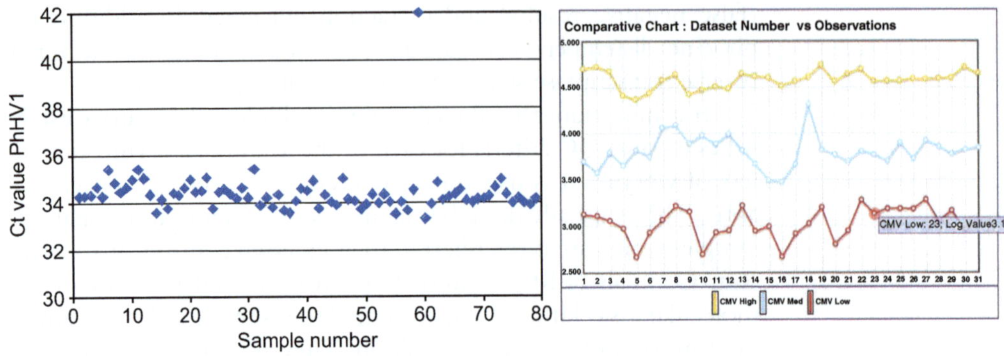

Fig. 2. Examples of control chart for internal and external molecular controls. (**a**) *PhHV1*: reported $_{t}$ values for the seal virus internal control (reproduced courtesy of Prof H.G.M. Niesters, UMCG, The Netherlands). (**b**) *CMV*: copies/ml values for 3 whole virus CMV external controls high, medium, and low (QCMD IQC reference database).

internal and external molecular controls are shown in Fig. 2. The control limits can be determined based on an initial set up phase where the quality control samples are run repeatedly (usually 20 times) in order to establish the mean and standard deviation (SD). Alternatively, the mean and standard deviation can be determined by calculating a rolling mean based on the last set of observations reported. The data is plotted to provide a visual indication of laboratory control performance with time.

Control rules such as those defined by Westgard (further information can be obtained from www.westgard.com) are often used in order to support the validation of patient results. The Westgard rules outline performance limits based on the control observations and are aimed at detecting both random and systematic errors. Examples of the most common Westgard rules used within the clinical laboratory setting are provided in Table 1. If a "warning rule" is broken then the laboratory should investigate the likely reason and taken any necessary corrective action. If a "mandatory rule" is broken the general practice is for the test run results to be rejected and the clinical samples repeat tested after the source of the control error has been identified and the appropriate corrective action taken.

There are various software packages of differing degrees of complexity available for reporting and monitoring quality control data, ranging from simple spreadsheets (such as "How to create a control chart": www.wikhow.com, accessed 2012) to fully functional commercial software packages. Some examples can be accessed via the internet and the clinical microbiology laboratory needs to define which is most appropriate to its requirements. If used appropriately chart controls can be a useful tool and help support a laboratory's QC regulatory requirements. Using an IQC control chart approach assists with the required QC review process and can help identify important trends or shifts in expected results over time which may impact on the assay performance and hence the patient test results.

Table 1
Examples of the most common Westgard rules used within the clinical medical laboratory

Type of rule	Description of rule	Action to be taken
Warning	If a single value of the control sample exceeds the mean by 2 SD ($p<0.005$)	Review and monitor
Warning	If two consecutive values for the control sample exceed the target value on the same side of the mean by 2 SD	An early indication of systematic error
Warning	If four consecutive values for the control sample exceed the same limit described above	Further indication of a systematic error
Mandatory	The control value exceeds the target value by 3 SD ($p<0.01$:99.7% of values should lie within 3 SD)	An indication that the instrument needs recalibration or maintenance
Mandatory	Where the control sample is tested in duplicate, the difference in SD between the duplicates should not exceed 4 SD	This indicates that the assay run is out of control limits
Mandatory	Ten consecutive values are on the same side of the mean or target value	Indicative of a systematic error. This can also arise when a new test batch is introduced or there have been changes in the calibration of test system

2.3. External Quality Assessment or Proficiency Testing

External Quality Assessment (EQA) or Proficiency testing (PT) was first introduced in the 1940s when Sunderman and colleagues sent around the first test panel of plasma samples to laboratories in Philadelphia (9, 10). Their focus was clinical chemistry and their pioneering work set the scene for the development of EQA programs covering a wide range of medical disciplines including nucleic acid tests (NATs). Their reasons for undertaking this work was the apparent variation in results observed by physicians, who when sending the same plasma sample to two different laboratories would often receive divergent results. The results of Sunderman's first study were so surprising, in their demonstration of the variation in results between laboratories, that the Committee on Laboratories of the Pennsylvania Medical Society asked Sunderman and his colleagues to expand the studies to cover the most common chemical measurements. The results of this study were published in 1947 (11). The terminology used to describe such studies varies by geographical region. In the US they are generaly referred to as Proficiency testing (PT), while in Europe they are termed External Quality Assessment Studied (EQA) or Ring Trials. The term EQA stresses the more educational aspects, which are a common feature of these studies in Europe (9).

The development of NATs in the mid- to late-1980s as research tools used only in specialist research laboratories quickly led to the development of the first medical tests. The first PCR kits for the detection of human pathogens appeared in 1992 with the Roche Amplicor™ kits for *C. trachomatis* and HIV-1 (12). At the same time many laboratories begun to develop and use their own in-house developed PCR assays. Early quality studies demonstrated that there was much variation in the results generated by these new tests. For example Noordhoek undertook a quality study covering the abilities of seven laboratories in the detection of *Mycobacterium bovis* (BCG). The results of this study showed high levels of variation. False positive rates generally ranged from 3 to 20%, with one notable case of 77%. Additionally the sensitivity of the assays varied markedly with 2–90% correct results observed on samples containing 1,000 *M. bovis* cells (12). There was a clear requirement for the establishment of EQA programs to assist laboratories in monitoring the performance of these new technologies. In Europe, an initial Quality Control Concerted Action (EU-QCCA) Program was established in the mid-1990s and distributed molecular EQA panels for a number of the more common human pathogens such as HIV-1, HBV, and *C. trachomatis* (13). The EU-QCCA was the forerunner to the independent International organization Quality Control for Molecular Diagnostics whose core focus remains the provision of molecular EQA for infectious disease diagnostics (http://www.qcmd.org, accessed 2012). In the United States, The College of American Pathologists (CAP) is responsible for the distribution of a number of mandatory molecular EQA programs (http://www.cap.org/apps/cap.portal, accessed 2010). These are provided as CLIA-approved PT programs and under CLIA-88 the EQA programs are seen as the major quality indicator of laboratory test performance (14). In Europe most EQA providers work to national codes of practice that fulfil the regulatory requirements as outlined in ISO17043 and additional aspects within the medical laboratories quality standard ISO15189 or equivalent. This includes ILAC G13-1:2007 (15, 16). There are also many useful guidelines documents such as the Clinical and Laboratory Standards Institutes (CLSI) document which specifically deals with molecular proficiency testing : MM14-A, *Proficiency Testing (External Quality Assessment) for Molecular Methods; Approved Guideline* to MM14-A2.

Many European EQA organizations are listed on the *Network of Quality Assessment and EQA Providers in Europe* (EumetaLAB) Web site (http://www.ec-4.org/eqanetwork/eqaproviderslist.html, accessed 2010). There is also a list of international EQA providing organization available through the CDC Web site (http://wwwn.cdc.gov/mlp/pdf/EQA/eqa_list.pdf accessed 2010).

2.3.1. Why Should Laboratories Participate in EQA?

EQA forms an important part of the laboratory quality assurance systems (17) and provides the laboratory with an independent evaluation of its performance. In many countries participation in EQA is mandatory for each of the assays that the laboratory is providing patient test results. Where there is no formal EQA available, laboratory's accredited to ISO15189 (under section 5.6) are encouraged to participate in inter-laboratory comparisons and increasingly EQA is seen as essential in the demonstration of a laboratories the ability to perform a test reliably (18). In many countries the clinical microbiology laboratory is required to report EQA results to the national laboratory accreditation body as part of its annual laboratory assessment. For example in the UK the laboratory accreditation process is undertaken through Clinical Pathology (UK) Ltd (http://www.cpa-uk.co.uk/, accessed 2010) and UK EQA are reviewed periodically through an independent body the National Quality Assurance Advisory Panel (NQAAP) which is a joint working group affiliated to the National Health Laboratory Service. The NQAAP are responsible for reviewing laboratory EQA poor performance.

While the details of molecular EQA schemes provided to the clinical microbiology laboratory may vary they all generally involve the laboratory receiving materials from the EQA provider in a blinded fashion. These materials are tested and the results returned to the EQA provider for independent evaluation. The laboratory is then provided with a report that details their performance and provides a summary of the performance of all who participated in the EQA program. Guidance may also be provided on the minimum levels of performance required to demonstrate proficiency. The results of EQA programs are also used to highlight problems within the medical laboratory, leading to the initiation of corrective actions; identification of inter-laboratory/technology differences; evidence of competence to laboratory customers; and for educational purposes (15).

2.3.2. The Design and Objectives of Molecular Microbiology EQA Programs

The design of EQA programs varies widely both between EQA providers and based on the pathogen in question. Additionally there are often specialized requirements that are specific to the testing being performed such as DNA sequencing for the determination of HIV-1 drug resistance (19). Some EQA providers follow a single sample strategy that although may cover some of the regulatory aspects leaves little room for laboratory education, which is an important part of the EQA process (9, 13). Other EQA providers such as QCMD provide laboratories with extended EQA panels that cover both the regulatory and educational aspects. Extended panels (typically containing 8–12 samples) allow for the inclusion of different sample types that focus on many different aspects of molecular assay performance (Table 2).

Table 2
Example EQA Panel specifications for the QCMD herpes simplex virus EQA in 2010

Sample	Sample content	Sample matrix	Sample conc. (copies/ml)	Stock dilution	Sample type
HSV DNA 10-07	HSV-1	VTM	9,016	1.00×10^{-5}	Core
HSV DNA 10-02	HSV-1	VTM	968	1.00×10^{-6}	Core
HSV DNA 10-06	HSV-1	VTM	129	1.00×10^{-7}	
HSV DNA 10-10	HSV-1	VTM	116	1.00×10^{-7}	
HSV DNA 10-01	HSV-2	VTM	4,710	1.00×10^{-4}	Core
HSV DNA 10-05	HSV-2	VTM	1,294	1.00×10^{-5}	Core
HSV DNA 10-09	HSV-2	VTM	178	1.00×10^{-6}	
HSV DNA 10-03	HSV-2	VTM	69	1.00×10^{-7}	
HSV DNA 10-04	VZV	VTM			Core
HSV DNA 10-08	Negative	VTM			Core

Details of the QCMD 2010 HSV EQA panel. Sample: QCMD panel sample codes for the samples distributed to participants. Sample content: viral content of the panel samples. Sample matrix: Virus Transport Medium. Sample conc.: consensus values calculated from all of the data returned by participants, once outliers had been removed. Stock dilution: serial dilution of a stock. Sample type: panel samples classified as core proficiency samples, which participants were required to report correctly in order to demonstrate an acceptable level of proficiency. Panel samples HSVDNA10-06 and -10 were duplicate samples. The strains used in the panel were HSV-1 MacIntyre (ATCC no. VR-539) and HSV-2 MS (ATCC no. VR-540). Varicella Zoster virus (VZV) strain 9/84 was present at a concentration of approximately 1×10^5 copies/ml

In the example shown, the EQA program contained two different types of HSV virus (types 1 and 2). The program also consisted of an extended dilution series for both virus types. This allows laboratory performance to be evaluated over a range of virus concentrations. For HSV type 1 all panel samples were present in clinically relevant concentrations (20). For HSV type 2 panel samples HSVDNA10-01 and -05 were present in clinically relevant concentrations (20) and samples HSVDNA10-09 and -03 were included to assess analytical sensitivity.

The inclusion of a duplicate sample (in this case samples HSVDNA10-06 and -10) allows reproducibility to be evaluated within the context of the assays clinical range. The specificity of a laboratory's molecular HSV1 and 2 assay is evaluated through the inclusion of a sample containing Varicella Zoster virus (VZV strain 9/84) present at a concentration of approximately 1×10^5 copies/ml (HSVDNA10-04). A negative sample (HSVDNA10-08) is included in order to assess false positivity potential as a result of assay contamination.

2.3.3. Feedback to Laboratories

According to ISO 17043 the reports provided to participants must be clear, comprehensive, and include an overall summary of the performance as well as individually focused information that provides the laboratory with an idea of their performance within that particular EQA distribution and how it compares to previous EQA distributions. The aims and objectives of the EQA program should also be clearly explained, along with information regarding the origin of the materials and how they were prepared. The use of statistics in the analysis of the EQA results must also be defined so that it provides the laboratory with sufficient feedback with which to review their performance and make appropriate informed decisions. Commentary should be provided, which includes coverage of any general trends in the EQA results (such as differences in performance between different molecular technologies, or the incidence of false positive results reported), and recommendations made based on the outcomes of the EQA program results and analysis.

The majority of molecular EQA programs within the clinical microbiology area tend to focus on qualitative molecular analysis and the presence or absence of the target pathogen (Table 3 shows a summary of the qualitative results for the 2010 QCMD EQA program for HSV1/2). However, with the growing use of molecular techniques such as real-time PCR, quantitative analysis has become a central requirement of the molecular EQA. This is particularly so for microbiological targets where microbial load determination is a standard diagnostic laboratory test parameter. The pathogens of most importance from a quantification perspective are the chronic viral pathogens HIV, Hepatitis B and C viruses, Human Cytomegalovirus, and Epstein–Barr Virus and target pathogens from specific patient groups such as transplant and the

Table 3
Number of correct qualitative results per panel member and technology group

Sample	Sample content	Sample conc. (copies/ml)	Total datasets n=227		PCR								NA SBA	
					Conventional				Real time					
					Commercial n=9		In-house n=13		Commercial n=79		In-house n=121		n=5	
			n	%	n	%	n	%	n	%	n	%	n	%
HSV DNA 10-07	HSV-1	9.016	223	98.2	9	100.0	12	92.3	77	97.5	120	99.2	5	100
HSV DNA 10-02	HSV-1	968	218	96.0	9	100.0	11	84.6	75	94.9	118	97.5	5	100
HSV DNA 10-06	HSV-1	129	170	74.9	5	55.5	7	53.8	61	77.2	94	77.7	3	60
HSV DNA 10-10	HSV-1	116	173	76.2	6	66.7	6	46.2	62	78.5	96	79.3	3	60
HSV DNA 10-01	HSV-2	4710	222	97.8	9	100.0	13	100.0	76	96.2	119	98.3	5	100
HSV DNA 10-05	HSV-2	1294	217	95.6	8	88.9	11	84.6	75	94.9	118	97.5	5	100
HSV DNA 10-09	HSV-2	178	175	77.1	5	55.6	8	61.5	55	69.6	103	85.1	4	80
HSV DNA 10-03	HSV-2	69	58	25.6	1	11.1	4	30.8	18	22.8	35	28.9	0	0
HSV DNA 10-04	VZV		223	98.2	9	100.0	11	84.5	79	100.0	119	98.3	5	100
HSV DNA 10-08	Negative		221	97.4	9	100.0	13	100.0	77	97.5	117	96.7	5	100

Sample: QCMD panel sample codes for the samples distributed to participants. Sample content: viral content of the panel samples. Sample conc.: consensus values calculated from all of the data returned by participants, once outliers had been removed. The values are not technology specific. Total datasets: number and percentage of datasets reporting the correct qualitative result for each panel sample. A breakdown of the results for all datasets is also provided based on technology type immune-suppressed. Important target pathogens in these areas include BK and JV virus as well as bacterial targets such as *Chlamydia pneumoniae* and relatively new molecular targets such as *Aspergillus*. In addition analysis of genotypic and drug resistance variants based on nucleic acid sequencing and mutation detection techniques is of increasing importance (19).

The EQA data are presented for each sample within the distribution, ordered by virus type and concentration. The consensus of the quantitative results returned by participants, once outlying values have been removed is used to determine an EQA score (21). This provides the laboratory with a detailed overview of the results returned by all participants that took part within the EQA. Additional guidance to the participating laboratory is provided in the form of a scoring system for qualitative results which is based on the premise of penalty points, received for each incorrect result reported. This approach is used by a number of EQA providers and the scoring schemes are generally weighted in such a manner that laboratories that miss a sample which was detected by the vast majority of other laboratories receive a greater penalty than those missing a sample where a greater proportion of laboratories reported the same result (Tables 4 and 5).

Quantitative data for molecular EQA programs in the clinical microbiology area can be analyzed based on the distance of the laboratories result from consensus values. Two consensus values are used. The first is based on all of the data returned by all labora-

Table 4
Scoring system for qualitative EQA data

	Participant's result		
Sample status	**Negative**	**Not determined**	**Positive**
Frequently detected	3	3	0
Detected	2	2	0
Infrequently detected	1	1	0
Negative	0	3	3

Sample status is assigned by peer-group consensus, based on the qualitative results returned by all participants. It is not a measure of the "strength" of a positive sample, nor is it technology-dependent, and is used solely for the scoring of the EQA data. The rationale for the sample status is as follows. Frequently detected: More than 95% of datasets recorded the correct positive result. Detected: Between 65 and 95% of datasets recorded the correct positive result. Infrequently detected: Less than 65% of datasets recorded the correct positive result. Negative: A panel sample that does not contain the target and produces an unequivocal negative result. The scores awarded for qualitative EQA data are based on the sample status, where 0 is "highly satisfactory" and 3 is "highly unsatisfactory." Color has been included as an extra visual aid

Table 5
Qualitative performance scores per technology type

		Total				PCR																NA SBA			
		All technologies				Conventional								Real time											
		n=227				Commercial *n*=9				In-house *n*=13				Commercial *n*=79				In-house n=121				*n*=5			
Sample	Sample status	0	1	2	3	0	1	2	3	0	1	2	3	0	1	2	3	0	1	2	3	0	1	2	3
HSV DNA 10-07	Frequently detected	223	0	0	4	9	0	0	0	12	0	0	1	77	0	0	2	120	0	0	1	5	0	0	0
HSV DNA 10-02	Frequently detected	218	0	0	9	9	0	0	0	11	0	0	2	75	0	0	4	118	0	0	3	5	0	0	0
HSV DNA 10-06	Detected	170	0	57	0	5	0	4	0	7	0	6	0	61	0	18	0	94	0	27	0	3	0	2	0
HSV DNA 10-10	Detected	173	0	54	0	6	0	3	0	6	0	7	0	62	0	17	0	96	0	25	0	3	0	2	0
HSV DNA 10-01	Frequently detected	222	0	0	5	9	0	0	0	13	0	0	0	76	0	0	3	119	0	0	2	5	0	0	0
HSV DNA 10-05	Frequently detected	217	0	0	10	8	0	0	1	11	0	0	2	75	0	0	4	118	0	0	3	5	0	0	0
HSV DNA 10-09	Detected	175	0	52	0	5	0	4	0	8	0	5	0	55	0	24	0	103	0	18	0	4	0	1	0
HSV DNA 10-03	Infrequently detected	58	169	0	0	1	8	0	0	4	9	0	0	18	61	0	0	35	86	0	0	0	5	0	0
HSV DNA 10-04	Negative	223	0	0	4	9	0	0	0	11	0	0	2	79	0	0	0	119	0	0	2	5	0	0	0
HSV DNA 10-08	Negative	221	0	0	6	9	0	0	0	13	0	0	0	77	0	0	2	117	0	0	4	5	0	0	0

Sample: QCMD panel sample codes for the samples distributed to participants. Sample status: the sample status assigned to each panel sample. Total. All technologies: number of datasets awarded each score (0–3). A breakdown of the results for all datasets is also provided based on technology type tories, irrespective of technology type used. The second consensus is technology type specific, for example NASBA versus PCR. The mean and standard deviation are calculated accordingly and although outliers are removed for the calculation of the assigned consensus values, they are included in the data analysis and scored accordingly (21). In the QCMD proficiency scheme achieving a low points score within a round of the EQA is an indication of a good performance. Zero points are awarded if the quantitative value returned is within one standard deviation from the mean. One point is awarded if the quantitative value was between 1 and 2 standard deviations, two points if the value is within 2 and 3 standard deviations, and three points for quantitative values more than 3 standard deviations from the mean (Tables 6 and 7).

In many mandatory EQA programs a clinical microbiology laboratories performance is considered in relation to the consensus 80% of all the participating laboratories in the EQA and/or the reference laboratories results within the EQA (14, 22). Failure to obtain an overall score of 80% is generally considered unsatisfactory and laboratories which fail successive proficiency distributions can be subject to strict regulatory sanctions through their respective national laboratory accreditation bodies unless a fitting reason for the failure can be provided. It is also important that the clinical microbiology laboratory maintain their EQA documentary records. Some regulatory authorities recommend a minimum of 2 years from the date of participation in the EQA distribution. The length that the EQA records are held should be consistent within the laboratories quality management and document control records.

2.3.4. How EQA Results Are Used by the Clinical Microbiology Laboratory

EQA results can assist senior laboratory personnel in the identification of quality-related issues. However, laboratories must not overreact if they fail an EQA round but should take time to review their results in line with other quality measures that they undertake. Initial focus should be on determining whether the failure is due to a transcriptional error during the results of results process. Reference should also be made to the known limits of the assay before deciding whether there is an issue that requires further investigation. Laboratories should make maximum use of the EQA report documents and review their results in line with other laboratories using similar assays or technologies.

If no transcriptional errors are found and the EQA results are in disagreement with the known limits of the assay, then the focus should turn to the laboratory's internal quality control (IQC) information. When both IQC and EQA are found to be in agreement (i.e., both showing a problem), then the laboratory should begin a corrective action procedure in line with ISO 15189. The focus of this investigation should include a review of the performance of different aspects of the assay, such as the extraction procedure or internal calibration. When the IQC results are as expected

Table 6
Quantitative scores by technology type in comparison to the consensus concentration

Sample	Consensus log_{10} virus concentration		Mean (±1 SD) (copies/ml)	Total														
				All technologies $n=37$					Real time									
									Commercial $n=27$					In-house $n=10$				
	Mean	SD		0	1	2	3	LOD/NR	0	1	2	3	LOD/NR	0	1	2	3	LOD/NR
HSVDNA 10-07	3.955	0.464	9,016 (3,097–26,242)	23	10	2	0	2	14	9	2	0	2	9	1	0	0	0
HSVDNA 10-02	2.986	0.498	968 (308–3,048)	21	11	1	0	4	14	8	1	0	4	7	3	0	0	0
HSVDNA 10-06	2.109	0.581	129 (34–490)	20	8	1	0	8	11	7	1	0	8	9	1	0	0	0
HSVDNA 10-10	2.066	0.700	116 (23–583)	19	3	2	0	13	12	3	2	0	10	7	0	0	0	3
HSVDNA 10-01	3.673	0.851	4,710 (664–33,420)	21	10	1	1	4	16	6	1	1	3	5	4	0	0	1
HSVDNA 10-05	3.112	0.534	1,294 (378–4,426)	21	4	1	2	9	12	4	1	2	8	9	0	0	0	1
HSVDNA 10-09	2.251	0.627	178 (42–755)	16	3	1	0	17	9	2	1	0	15	7	1	0	0	2
HSVDNA 10-03	1.837	0.479	69 (23–207)	9	2	0	0	26	5	2	0	0	20	4	0	0	0	6

Sample: QCMD panel sample codes for the samples distributed to participants. Consensus log_{10} virus concentration: the mean quantitative value and standard deviation value for each panel sample expressed in log_{10} units and calculated once outlying values had been removed. Mean (±1 SD) copies/ml: The mean and standard deviation values converted from log_{10} to standard notation. Shows the mean and the range based on ±1 standard deviation. Total. All technologies: number of datasets awarded each score (0–3). LOD/NR refers to datasets where an upper or lower limit of detection was reported or no value was reported (these data were excluded from the scoring). SD refers to the Standard Deviation. A breakdown of the results for all datasets is also provided based on technology type

Table 7
Quantitative scores by commercial and in-house PCR technology consensus concentrations

Sample	Tech. consensus log $_{10}$virus concentration		Mean (±1 SD) (copies/ml)		Real time Commercial n=27					Tech. consensus log $_{10}$virus concentration		Mean (±1 SD) (copies/ml)		Real time In-house n=10				
	Mean	SD			0	1	2	3	LOD/ NR	Mean	SD			0	1	2	3	LOD/ NR
HSV DNA 10-07	3.917	0.523	8.260	(2,477–27,542)	17	7	1	0	2	4.051	0.248	11,246	(6,353–19,907)	7	3	0	0	0
HSV DNA 10-02	2.996	0.534	991	(290–3,388)	15	7	1	0	4	2.962	0.395	916	(369–2,275)	6	4	0	0	0
HSV DNA 10-06	2.161	0.636	145	(33–527)	12	6	1	0	8	2.011	0.450	103	(36–289)	6	4	0	0	0
HSV DNA 10-10	2.035	0.800	108	(17–684)	12	4	1	0	10	2.14	0.301	138	(69–276)	5	2	0	0	3
HSV DNA 10-01	3.399	0.966	2,506	(271–23,174)	18	5	1	0	3	4.404	0.350	25,351	(11,324–56,754)	6	2	1	0	1
HSV DNA 10-05	2.996	0.608	991	(244–4,018)	14	3	1	1	8	3.344	0.327	2,208	(1,040–4,688)	8	0	1	0	1
HSV DNA 10-09	2.178	0.710	151	(29–773)	9	2	1	0	15	2.36	0.467	229	(78–671)	5	3	0	0	2
HSV DNA 10-03	1.788	0.479	61	(20–185)	6	1	0	0	20	1.922	0.178	84	(55–126)	–	–	–	–	–

Sample: QCMD panel sample codes for the samples distributed to participants. Consensus $\log_{10}$ virus concentration: the mean quantitative value and standard deviation value for each panel sample expressed in $\log_{10}$ units and calculated once outlying values had been removed. Mean (±1 SD) copies/ml: The mean and standard deviation values converted from $\log_{10}$ to standard notation. Shows the mean and the range based on ±1 standard deviation. The number of datasets awarded each score (0–3) is then presented. LOD/NR refers to datasets where an upper or lower limit of detection was reported or no value was reported (these data were excluded from the scoring). SD refers to the Standard Deviation

but the EQA results indicate a possible issue, then the laboratory should in the first instance contact the EQA provider to seek advice. A corrective action should be initiated as above and the laboratory should work with the EQA provider to resolve the issue.

The EQA provider can generally assist the laboratory in a number of ways. The laboratory may request additional materials to test and then review the results of this analysis with the assistance of the EQA provider. The laboratory may also seek advice from the scientific advisers that most EQA organizations will use to support the EQA programs. From an educational perspective and where appropriate, additional assistance may include being put in contact with another laboratory using the same assay so that methods and common issues can be discussed.

Participation in EQA programs is designed to support and assist the clinical microbiology laboratories quality requirements. Laboratories should ensure regular "quality" meetings with staff in order to review the results of EQA programs and to discuss possible issues and actions needs. This help to ensure that all members of staff are familiar with the laboratories quality practices and also foster a culture that quality is the responsibility of all as well as identifying areas for continuous improvement.

3. Staff Training requirements and Laboratory Quality Documentation

The modern clinical microbiology laboratory engaged in molecular testing operates within a highly technology and regulatory driven environment. The head of the laboratory must ensure that they have sufficient staff levels and the competence with which to provide the molecular testing service to the appropriate regulatory standards. It is therefore essential that all staff are well trained and competent in good molecular laboratory practice. Staff training and competency assessment are also key elements within a laboratories accreditation to both the ISO standard 15189 and CLIA-88 regulatory requirements. The clinical microbiology laboratory must have defined training guidelines and each laboratory employee should be provided with a clear training plan. This should cover the preliminary laboratory training requirements, initial duties and responsibilities, as well as a further personal development which also actively supports the employee's Continued Professional Development (CPD) requirements. This is often initiated through a "new employee induction process" where a training checklist or matrix for the employee is established in conjunction with the staff member assigned supervisory responsibility. Considerations will include background knowledge and qualifications, previous molecular laboratory experience, and an understanding of laboratory quality requirements. All members of laboratory staff occupied in molecular testing are required to demonstrate competence. This can be achieved through practical laboratory assessments where the test results obtained by staff members on known samples or additional QC samples introduced into the routine workload are used to monitor performance. Previous EQA program samples are also a useful tool for monitoring and training staff members. This training may also be complemented with a written assessment covering general molecular laboratory practices as well as standard and equipment operating procedures that staff members will actively be involved in. Staff members will be required to pass these assessments prior to undertaking formal laboratory work. The results of the initial training and any practical and written assessments must then be recorded within the individual staff members training records. This will provide documentation that the staff member is competent in the molecular diagnostic testing methods they are trained in. The training records also provide a record that the staff members have read and understood all the associated laboratory policies and procedures (SOPs and EOPs) and are considered proficient to conduct their agreed role within the laboratory. The competency of all staff must also be reviewed on a regular basis. This is often covered through 6 monthly performance and development reviews. These reviews help identify any potential problems as well as further training and development requirements.

It is also useful to have an employee "handbook" which covers the key components of the laboratory or departments operations and it provides a useful general training tool for "new employees" and a reminder to others. The format and composition will depend on the type of laboratory organization and in many cases may be dictated by the organization the laboratory is affiliated to. However, the handbook should include an outline of the organization including the management structure with key roles and responsibilities. It should also include the basic "do's and dont's" when working in the molecular laboratory, key health and safety aspects as well as the quality policy of the laboratory. Where appropriate the handbook may also include basic human resource items such as IT policies; staff performance review policies; policies on intellectual property within the laboratory/organization; and sickness and holiday policies.

4. Equipment Qualification and Process Monitoring

Fundamental to the maintenance and daily running of the molecular facility is the regular monitoring of the equipment and processes used within it. All equipment, instrumentation, and assay test systems must be shown to be fit for intended purpose. However, despite numerous guidelines on equipment qualification and process monitoring these areas frequently remain overlooked and laboratories are often unclear what is required. As a result the areas are one of the most regularly cited deviations following regulatory inspections.

4.1. What Is Equipment Qualification?

Equipment qualification is the course of action a laboratory takes to ensure that the equipment it is uses is fit for its chosen purpose. The laboratory will generally include equipment qualification within its validation plan. The initial qualification process will consist of any training and preliminary calibration requirements which are at first conducted in conjunction with the equipment manufacturer and supplier. Within industry the 4Q's approach has became the standard (23). The first step in the process is to define the functional and operational specifications of the equipment and this is referred to as Design Qualification (DQ). Within this step the laboratory may also include the rationale for the selection of the equipment which can include aspects such as cost and footprint size. The second step in the process is Installation Qualification (IQ), and this establishes that the equipment meets specifications and is suitable for operation. The third phase in the equipment qualification process is Operational Qualification (OQ), and it is the process of demonstrating the operation of the equipment with the specified laboratory setting. The fourth and final phase is known as Performance Qualification (PQ), and here the aim is to show that the equipment performs to the appropriate specifications and intended use.

Within the clinical microbiology laboratory these four phases are often broadly covered under assay validation and/or verification. For the most common molecular laboratory equipment such as pipettes, microfuges it may be appropriate to simply follow the recommendations provided by the equipment manufacturer. For example, the calibration of pipettes can easily and accurately be performed by the laboratory using gravimetric calibration which is also the accepted standard for pipette calibration.

However, where specialist equipment such as generic molecular platforms (e.g., thermal cyclers) have a range of utilities within the laboratory and can be used in combination with different extraction protocols, different test kits and assays it may be necessary for the clinical microbiology laboratory to establish its own basic requirements for equipment qualification and process monitoring. In these cases, the laboratory will define the equipment qualification criterion in accordance with the manufacturers' equipment specifications and recommendations; the laboratories intend use, and the regulatory requirements. The aim will be at qualifying the laboratory equipment and monitoring performance to ensure the validity of data for both the individual equipment components as well as the whole test system.

Key factors for consideration include the frequency and extent of use of the equipment; maintenance and safety testing; annual servicing, and specific calibration requirements.

4.2. Calibration

Calibration procedures assure that the individual equipment components and the whole test system will produce results in line with the defined specification and where appropriate reportable range for the molecular assay. For many commercial molecular assays calibration materials are provided with the test kit and the manufacturer's instruction manual will determine the recommended frequency for calibration. It is the responsibility of the laboratory to ensure that the manufacturers recommended calibration procedures are followed. Hence, it is important to remember that if a laboratory does not follow the manufacturers' recommendations and takes the decision to change the approach to calibration or any other aspect of the manufacturers' protocol, the laboratory may well be invalidating the manufacturers warrant for use. Therefore, if a laboratory considers that modifying the manufacturers recommended protocol is in the interests of the "intended use" it is important that the laboratory documents the changes appropriately and communicates this with manufacturer.

Where an assay has been developed by the laboratory or no calibration material is provided by the kit manufacturer the laboratory must define its own calibration protocol and material requirements. The laboratory must consider the frequency for calibration and the type of material. If possible calibration material which is traceable to an International standard or reference material should be used in preference to stored clinical samples. However, this type of material is not always readily available within the molecular infectious disease area. Both the FDA and CAP provide guidelines for the calibration of in-house developed assays (23). They also recommend that recalibration at least every 6 months along with other criteria. Calibration is a central factor in the quality management requirements of a laboratory and contributes to both equipment qualification and also assay validation requirements.

5. Assay Verification and Validation

The introduction and implementation of a new technology, kit method or amended protocol needs to be accompanied by appropriate verification and validation studies prior to its implementation into routine laboratory use. In many countries this is also a regulatory requirement, and in the United States this is governed by the Clinical Laboratory Improvement Amendments (CLIA-88) federal regulatory standards. The performance characteristics that must be established include accuracy, precision, reportable range, reference interval, analytical sensitivity, and analytical specificity. The challenge to the clinical microbiology laboratory is to understand what is needed for a molecular test system when faced with these regulatory requirements and to determine the types of studies and course of action necessary to meet these requirements.

The terms verification and validation are often used interchangeably when describing the steps in the assessment, evaluation, and implementation of an assay, test system into the clinical microbiology laboratory. In general verification can be defined as the process of testing and reviewing an assay's performance in relation to its known or reported performance such as against the manufacturers' performance specifications. Whereas validation is typically used to describe the evaluation of a test method in order to determine its fitness for use within the clinical microbiology laboratory. In addition, validation is often viewed as a cycle in which everyday performance characteristics for the assay are established, optimized, reviewed, and improved.

Approaches to verification and validation also vary considerably and to a large extent are often dependent on the type and purpose of the assay, the commercial manufacturer, and the clinical microbiology laboratories available time and resources. They tend to aim at achieving minimal regulatory requirements.

For FDA approved and IVD/CE labelled assays, the manufacturer is responsible for defining a basic performance for the assay or test system. Therefore, when a laboratory is using an FDA or IVD/CE commercial assay, verification is used to confirm that the assay performance obtained by the laboratory is equivalent to that described and determined by the assay manufacturer. Under these circumstances the FDA specify that the clinical laboratory must verify the accuracy, precision, and reportable range of the assay as presented in the claims within the manufacturers instruction manual. The verification provides documented evidence to determine that the commercial or in-house assay and equipment are performing within the stated manufacturers specifications.

Whereas for assays or test systems that have been developed in-house by the clinical microbiology laboratory or developed by another third-party laboratory and transferred to the clinical microbiology laboratory the verification protocol is often extended. It will also include parallel testing of the new test method against the original assay or an assay of known performance characteristics. In both these instances the initial studies confirm the performance characteristics in relation to those defined by the manufacturer or developed by the laboratory in terms of:

Accuracy. The closeness of agreement between values obtained from a series of test results. Where possible within the assay verification phase the test results are compared to "known" and/

or an accepted reference value. For example, those provided within a commercial kit manufacturers instruction manual or those of a known reference material or standard.

Precision. The ability to consistently produce the same result for a given test sample. This is also called repeatability and reproducibility. Although repeatability tends to be used to describe repeat testing under unchanged conditions for example within a single assay run (inter-assay). Whereas, reproducibility is more often used to describe repeat testing of the same test sample in different runs across a time period, possibly using different operators (intra-assay).

The basic accuracy of the assay can be expressed qualitatively as the percentage recovery as (detected value/expected reference value) × 100. The precision of the assay can be characterized by the Standard deviation (SD) and the coefficient of variation (CV) as defined in %CV (coefficient of variation) and (SD/mean) × 100. It is worth noting that for real-time PCR values if the SD is >0.250, the power to discriminate between a twofold dilution is less than 95%.

For many quantitative molecular assays it is difficult to establish the analytical accuracy of the assay as there are no reference standards available. So often for in-house developed assays the clinical laboratory has to rely on its own internal reference as a guide for the analytical performance and look to bench-mark this against other clinical laboratories in the field.

Linearity and reportable range. This is the ability of the assay to obtain test results which are directly proportional to the concentration of the analyte within a given range. The reportable range is the interval between the upper and lower concentrations of an analyte for which the assay has acceptable levels of precision, accuracy, and linearity. Within CLIA'88 (42 CFR 493 section 493.2) reportable range is defined as *the span of test result values over which the laboratory can establish or verify the accuracy of the instrument or test system measurement response.* The CAP guidelines refer to reportable range as Analytical Measurement Range (AMR).

For quantitative molecular tests or test systems it is necessary to establish the analytical measuring range or linear range of the assay. For determining linear range a log or 0.5 log dilution series of a known reference material or working standard is prepared in an appropriate matrix. Each sample is tested repeatedly (usually between three and six times) and the mean result and standard deviation determined. The results are plotted, linearity and range is determined using linear regression analysis.

The limit of detection (*LOD*) or limit of quantification (LOQ) is defined as the lowest quantity of a substance that can be distinguished from the absence of that substance (a blank value)

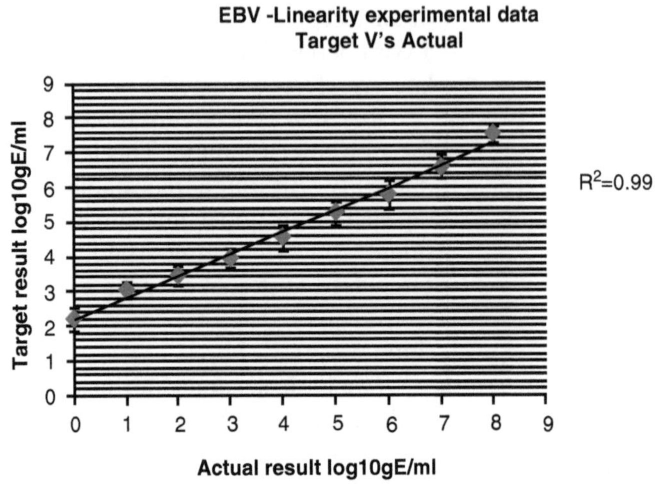

Fig. 3. Linear relationship between target values and actual values from a nine member dilution series from 10E8 to 10E0 gE/ml.

within a stated confidence limit. The limit of detection can also be determined using a serial dilution of prequalified standard or reference material spiked into an appropriate clinical matrix. The dilutions are replicate tested and the results analyzed using PROBIT analysis.

In conjunction with linearity, for laboratories using PCR by plotting C_t values against log-transformed concentrations of serial tenfold log dilutions (usually covering 5–7 logs) of a reference material containing the target nucleic acid a standard curve is obtained (Fig. 3). The standard curve can be used to establish the dynamic range of the assay and to calculate the slope and efficiency of the PCR assay in terms of r and R^2 coefficients. The efficiency of the PCR reaction is calculated using the following equation: $E = 10(-1/\text{slope}) - 1$. In a fully optimized assay where there is doubling of the target sequence (amplicon) at each cycle the efficiency should be between 90 and 110%. Ideally, the slope of a standard curve should be −3.32, $R^2 > 0.99$. The y-intercept value corresponds to the C_t value for a single copy of the target and a value of around 40 (C_t) is generally a good indication of assay sensitivity.

Analytical sensitivity is defined as the lowest amount of target nucleic acid that still returns a true positive result. For Qualitative assays the sensitivity will be the number of true positive samples correctly identified by the assay divided by the total number of true positive samples, expressed as a percentage. In comparison, for Quantitative assays analytical sensitivity is defined from the limit of detection and where appropriate confidence interval (CI) can be calculated in order to support the chance of obtaining a true positive result.

Analytical specificity is defined as the ability to return a true negative result when the target nucleic acid is not present. Therefore, specificity is the number of true negative samples correctly identified by the test assay divided by the total number of true negative samples. It is also expressed as a percentage. For quantitative molecular assays a result lower than the reported range will be defined as "below the limit of detection" and is analytically negative.

During the initial assay verification and validation phase the laboratory also needs to consider any potential interfering or cross-reacting substances that may be present in the specimen or matrix. This can be a particular problem with molecular assays aimed at detecting and differentiating related bacterial or viral strains. For example the intention of the molecular assay may be the detection of Enteroviruses, but the primers and probes cross-react and amplify the Parechoviruses and hence result in false positive results. In most cases such as this the laboratory or manufacturer developing the assay would have suspected that there could be a possible cross-reaction. They would therefore have chosen their primers and probes carefully and made an initial database search before practice testing with known samples in order to ensure there was no cross-reaction. However, even if this approach is followed it is not always possible to obtain sufficient database information or enough known clinical samples in order to assess specificity. Hence, cross-reactivity may only be identified during the validation phase. For this reason it is important that analytical specificity is considered for taxonomically related organisms such as *N. meningitidis* and *N. lactamica* in case of *N. gonorrhoeae* molecular detection and also epidemiologically related organisms such as other respiratory agents in case of *Mycoplasma*, *Chlamydia*, when considering *M. tuberculosis* molecular testing.

Most of the aspects mentioned in the section above cover confirmation of the assays analytical performance characteristics. Prior to full implementation of the new technology or assay in the clinical microbiology laboratory it is necessary to show that it is fit for purpose within the clinical setting. This phase is referred to as *clinical validation* and it generally involves relating the test result to the presence of disease and refining the policies and procedures with which to manage and monitor the assay in the clinical setting. This involves establishing the clinical sensitivity and specificity of the assay as well as the positive predictive values (PPV) and negative predictive values (NPV) of the test. During clinical validation the laboratory may also have to establish the clinical utility of the new technology or assay.

It is often standard practice to consider the value of a new test in comparison with an existing assay/method, reference test or established “gold standard.” The problem with many molecular tests or new and improved methods is that they are typically more sensitive than the established test or gold standard which can complicate the subject when trying to establish equivalence through method comparison experiments. In order to resolve this issue discrepancy analysis is used where a third test method is employed to determine the validity of the results. Statistical techniques such as latent class analysis can also be used to evaluate and characterize the test results. In some cases the laboratory may also seek to use an “expanded gold standard” approach. This approach was used for *C. trachomatis* diagnosis where it was shown that the specificity and especially the PPV of non culture tests (including molecular) were underestimated when culture was used as the “gold standard.” By evaluating new test methods using an expanded gold standard based on multiple tests with proven specificity and sensitivity the new tests were more appropriately evaluated and represented (24).

In terms of the number of specimens that should be tested during the clinical validation phase this is largely dependent on the intended use and prevalence of the disease. Within the various guidelines, a 100 incoming routine “negative” specimens for the target are considered a good sample size for the evaluation of false positives rate in the negative population. In contrast, 50 clinical samples known to be positive for the target are considered accepted for evaluation of false positives (25, 26). Within the original NCCLS EP 12A guidelines now superseded by the CLSI EP12-A2 *User protocol for evaluation of qualitative test performance: approved guidelines.* The recommendation was for 50 positive and 50 negative specimens. However, it is sometimes difficult to obtain sufficiently characterized and suitable clinical specimens especially in the clinical matrix. For example low prevalence’ CSF positive for HSV and synovial fluid positive for *Borrelia burgdorferi*. In these situations the laboratory has to look for alternative evaluation approaches. These often include spiking suitable materials (including previous proficiency testing samples) into appropriate clinical matrices. In Europe, for qualitative assays a minimum of 20 known clinical positive samples and 50 negative have been proposed although this is not formalized with any guidelines. For quantitative assays the laboratory must also verify the upper and lower limits of detection, linearity across dynamic range using known clinical samples.

For many infectious disease targets it is necessary to perform the molecular diagnostic assay on samples from different clinical matrices. If a sample matrix is intended to be used for which the manufacture has not verified the molecular tests intended use. It is necessary for the clinical laboratory to verify and validate the molecular assay within that matrix. This is often done by running a series of experiments where know amounts of a reference material are spiked into the new matrix and the original matrix and a comparison made between the two matrices. The use of an internal control also helps to identify and monitor any potential matrix effects.

In addition to different matrix types with the increasing use of commercial multiplex assay systems and more laboratories developing in-house multiplex assays which focus on the detection and determination of multiple pathogen types, in distinct disease grops such as transplant patients and central nervous system diseases, the approach to clinical validation becomes more complex. There is an increased requirement for multiple sample types and strains. The requirement for highly characterized matrices which for some current molecular multiplexing system would need to be shown to be negative for over 20 target pathogens. In addition, the clinical laboratory has to assess the clinical performance and utility of the multiplex system for potential competition between amplification targets and cross-reactivity of the molecular reagents which could let to potential false positive results. This requires the clinical microbiology laboratory to establish suitable performance characteristics for each of the pathogen targets within the multiplex assay.

6. Standards and Controls

The requirement for exchangeable quantification among different molecular tests relies on the availability and acceptance of standards and defined reference materials. These materials help to facilitate method validation, intra laboratory comparisons, support the development of QC materials and also EQA or PT materials. In addition, they ultimately enhance the confidence of laboratory staff and management in the clinical decision-making process as well as support the laboratories regulatory requirements.

Acceptable Molecular standards in the microbiology area are somewhat limited. At present the International Standards (IS) provided by the World Health Organization (WHO) only cover six viruses (HCV-RNA, HBV-DNA, HIV-1-RNA, HAV-RNA, Parvovirus B19 DNA, HPV type 18 DNA) and one bacterium and *Plasmodium falciparum* DNA. In the mid-1990s Standardization of Genome Amplification Techniques (SoGAT) was established as a WHO working group in order to address safety testing within the blood banks. So the original focus was on blood product contamination. In June 2008 the SoGAT working group for clinical diagnostics was established (SoGAT-CD). With the aim to develop molecular primary reference materials and WHO International Standards for clinically relevant pathogens. The first WHO International Standards to be commissioned under SoGAT-CD will be HCMV and HEBV which were made available in 2010 and 2011 respectively.

Additional to the initiatives of WHO, other international standards organization have also recognized the need for more appropriate molecular reference materials within the microbiology area. One particular example is the development of a Standard Reference Material for HCMV by the National Institute for Standards and Technologies (NIST) in the United States. NIST aims to produce pure HCMV DNA reference materials which can be used as a calibrator and is traceable to the metric system/International system of Units (SI).

These primary reference materials can then be used by manufacturers of diagnostic assays, clinical laboratories, control companies, and EQA providers in the calibration of control material for routine day to day use.

7. Laboratory Regulation: Accreditation and Certification

A molecular test can be performed using either a protocol developed within the laboratory or a test kit developed by a commercial manufacturer of diagnostic kits or a combination of both. For example commercial extraction platform and kit, and in-house developed amplification kit or vice versa.

A commercial kit usually consists of all the reagents and instructions with which to perform the test and interpret the results. In the United States commercial assays are regulated by the FDA and in Europe through the In vitro Diagnostic Directive (IVDD) as in vitro diagnostics devices. Laboratory developed/in-house or home brew assays are developed within a laboratory using either FDA/IVD reagents or laboratories own reagents and analyte specific reagents. In-house assays are not actively regulated by the FDA although the FDA generally has a jurisdiction covering them. In Europe in-house developed assays are not covered by the IVDD and in-house assay performance would be covered under the laboratories own accreditation and certification requirements within that given country under standards such as ISO15189. However, the EU commission is currently undergoing a consultation process as regards to the revision of the IVDD. In particular the scope of the IVDD could potentially be widened, and this could lead to in-house assays either having to be CE marked or showing compliance with the IVDD essential technical requirements.

References

1. BS EN ISO 15189 (2003) Medical laboratories—particular requirements for quality and competence. British Standards Institute, Bristol, UK
2. Boone DJ, Hansen HJ, Hearn TL, Lewis DS, Dudley D (1982) Laboratory evaluation and assistance efforts: mailed, onsite and blind proficiency testing surveys conducted by the Centers for Disease Control. Am J Public Health 72:1364–1368
3. Bartlett RC, Mazens-Sullivan M, Tetreault JZ, Lobel S, Nivard J (1994) Evolving approaches to management of quality in clinical microbiology. Clin Microbiol Rev 7:55–87
4. Nutting PA, Main DS, Fischer PM, Stull TM, Pontious M, Seifert M, Boone DJ, Holcomb S (1996) Problems in laboratory testing in primary case. JAMA 275:635–639
5. Warford A (2000) Quality assurance in clinical virology. In: Richman DD, Whitley RJ, Hayden FG (eds) Clinical virology manual, 3rd edn. ASM Press, Washington, DC, pp 3–10
6. Niesters HGM (2002) Clinical virology in real time. J Clin Virol 25:S3–S12
7. Pasloske BL, Walkerpeach CR, Obermoeller RD, Winkler M, DuBois DB (1998) Armored RNA technology for production of ribonuclease-resistant viral RNA controls and standards. J Clin Microbiol 36(12):3590–3594, PMCID: PMC105245
8. van Doornum GJJ, Schutten M, Voermans J, Guldemeester GJJ, Niesters HGM (2007) Development and implementation of real-time nucleic acid amplification for the detection of enterovirus infections in comparison to rapid culture of various clinical specimens. J Med Virol 79:1868–1876
9. Libeer JC (2001) Role of external quality assurance schemes in assessing and improving quality in medical laboratories. Clin Chim Acta 309(2):173–177
10. Sunderman FW Sr (1992) The history of proficiency testing/quality control. Clin Chem 38(7):1205–1209
11. Belk WP, Sunderman FW (1947) A survey of the accuracy of chemical analyses in clinical laboratories. Am J Clin Pathol 17(11):853–861
12. Noordhoek GT, Kolk AH, Bjune G, Catty D, Dale JW, Fine PE, Godfrey-Faussett P, Cho SN, Shinnick T, Svenson SB (1994) Sensitivity and specificity of PCR for detection of *Mycobacterium tuberculosis*: a blind comparison study among seven laboratories. J Clin Microbiol 32(2):277–284
13. Valentine-Thon E (2002) Quality control in nucleic acid testing/where do we stand? J Clin Virol 25:S13–S21
14. Health Care Financing Administration (HCFA), Department of Health and Human Services (1992) Clinical Laboratory Improvement Amendments of 1988; final rule. Fed Regist. Feb 28:7146
15. ISO/IEC 17043: 2010(E). Conformity assessment—general requirements for proficiency testing. International Organisation for Standardization, Geneva, Switzerland
16. MM14-A. Proficiency testing (external quality assessment) for molecular methods; approved guideline to MM14-A2. Clinical and Laboratory Standards Institute, PA, USA
17. Sciacovelli L, Secchiero S, Zardo L, Plebani M (2001) External Quality Assessment Schemes: need for recognised requirements. Clin Chim Acta 309:183–199
18. Forbes BA (2003) Introducing a molecular test into the clinical microbiology laboratory: development, evaluation, and validation. Arch Pathol Lab Med 127(9):1106–1111
19. Pandit A, MacKay WG, Steel C, van Loon AM, Schuurman R (2008) HIV-1 drug resistance genotyping quality assessment: results of the ENVA7 Genotyping Proficiency Programme. J Clin Virol 43(4):401–406
20. Aberle SW, Puchhammer-Stöckl E (2002) Diagnosis of herpesvirus infections of the central nervous system. J Clin Virol 25: S79–S85
21. Staines HJ, Garcia-Fernandez L, Pogothata R, Wallace PS, MacKay WG, van Loon AM (2009) Monitoring performance of nucleic acid-based diagnostic measurement system users by EQA. Accred Qual Assur 14(5):243–252
22. College of American Pathologists (2007) Laboratory improvement: Laboratory Accreditation Program Checklists. CAP, Chicago
23. Bedson P, Sargent M (1996) The development and application of guidance on equipment qualification of analytical instruments. Accred Qual Assur 1:265–274
24. Thejls H, Gnarpe J, Gnarpe H, Larsson PG, Platz-Christensen JJ, Ostergaard L, Victor A (1994) Expanded gold standard in the diagnosis of *Chlamydia trachomatis* in a low prevalence population: diagnostic efficacy of tissue culture, direct immunofluorescence, enzyme immunoassay, PCR and serology. Genitourin Med 70(5):300–303
25. Green JM (1996) A practical guide to analytical method validation. Anal Chem News Features 68:305A–309A
26. Toney DM (2007) Validation of molecular methods in a Public Health Laboratory. In: APHL annual meeting, 5 June 2007

Chapter 4

Pre-analytical Sample Treatment and DNA Extraction Protocols for the Detection of Bacterial Pathogens from Whole Blood

Wendy L.J. Hansen, Cathrien A. Bruggeman, and Petra F.G. Wolffs

Abstract

Molecular diagnostics is an increasing popular approach for the direct detection and identification of pathogenic bacteria in clinical samples. Conventional culture techniques are time-consuming and therefore causing a delay in the diagnosis of the patient. Alternative techniques based on nucleic acid amplification offer a shorter turn-around-time and the ability to identify fastidious and non-cultivable organisms. However, molecular detection of bacteria in blood, by for example PCR, RT-PCR, or sequencing of the 16S rDNA genes is often complicated by the presence of PCR-inhibitory compounds. Here we describe several different methods for the extraction of bacterial DNA from whole blood samples. The methods differ regarding costs, hands-on time as well as regarding sensitivity. In combination with a model PCR the detection limits that can be reached using the different methods range from 1,000 to 50 cfu/ml.

Key words: Pathogen detection, Whole blood, Bacterial DNA extraction, Real-time PCR

1. Introduction

Blood cultures have become the gold standard in microbiological laboratories for the detection and identification of microorganisms causing bloodstream infections (1). Approximately 250,000 patients are diagnosed with bloodstream infections annually in the United States (2). Rapid detection and identification of the involved microorganisms is essential to start appropriate antimicrobial therapy as soon as possible. However, the time-to-results obtained by blood culture is relatively long (2–5 days) and blood culture may be difficult in the case of slow-growing or fastidious organisms. In addition, it may be impossible to recover bacteria if blood cultures are taken during antibiotic administration (3–5). For these

Mark Wilks (ed.), *PCR Detection of Microbial Pathogens: Second Edition*, Methods in Molecular Biology, vol. 943,
DOI 10.1007/978-1-60327-353-4_4, © Springer Science+Business Media, LLC 2013

reasons, molecular diagnostic methods are gaining increasing attention in this field. Various new approaches, using molecular technologies such as 16S rDNA or multiplex (real-time) PCR assays are being developed to improve the diagnosis of bloodstream infections (6–12). From that perspective, nucleic acid-based techniques can avoid the need for possible complex cultivation requirements and may offer a more rapid approach for the detection and identification of potentially pathogenic bacteria directly in blood. However, critical issues in the application of molecular methods are the sample treatment and nucleic acid isolation methods (13). The low concentration of pathogens and the presence of PCR-inhibitory compounds in blood are important challenges that are faced during the diagnosis of bloodstream infections (14–18). For these reasons, few studies in the past have focused on pathogen detection in whole blood samples. Instead, experiments were performed using bacterial suspensions or clinical sample materials such as pleural fluid, pus, synovial fluid, and pericardial fluid (7, 19). Zucol et al. evaluated different DNA extraction protocols for whole blood samples followed by broad-range real-time PCR, targeting the 16S rDNA gene in *Staphylococcus aureus* and *Escherichia coli.* They achieved the detection of bacterial concentrations of >10 cfu per PCR reaction, which was equivalent to 10^3 cfu per ml (20).

Recently, various new tools have been introduced to achieve the selective enrichment of bacterial DNA from total DNA. Here we describe the use of four different kits, i.e., QIAamp DNA Blood Mini Kit (Qiagen), the MagNA Pure LC Microbiology Kit M^{Grade} (Roche Diagnostics), the MolYsis Complete5 Kit (MolZym), and the MagSi-DNA Isolation Kit for Blood (MagnaMedics Diagnostics B.V). MolYsis Basic (MolZym) is a new pre-analytical sample treatment tool establishing the removal of human DNA and the enrichment of bacterial DNA from clinical samples such as whole blood. The method can be combined with a DNA extraction protocol of choice. The goal of this pre-analytic kit is to increase the diagnostic sensitivity and specificity for downstream applications. Bacterial DNA extraction is followed by amplification in a quantitative multiplex real-time PCR assay. Differences between the different methods are illustrated by using as a model a PCR targeting three different genes in MRSA (see Note 1).

2. Materials and Methods

2.1. Determination of Assay Sensitivity by Blood Spiking (see Note 2)

A clinical isolate of methicillin-resistant *Staphylococcus aureus* (MRSA) is grown overnight in brain heart infusion broth. Tenfold serial dilutions are made in sterile phosphate-buffered saline (PBS) to determine the viable count and to spike blood for the assay. The numbers of cfu/ml in the suspensions are calculated after overnight cultivation of 100 μl of the appropriate dilutions on blood agar plates. Each dilution series should also include a negative control, i.e., PBS without added bacteria. Blood from healthy blood donors is pooled and spiked with a tenfold dilution series of MRSA suspension.

2.2. Pre-analytical Sample Treatment

The pre-analytical sample treatment, MolYsis Basic, is used in combination with QIAamp DNA Blood Mini Kit, the MagNA Pure LC Microbiology Kit M^{Grade} and the MagSi-DNA Isolation Kit for Blood. MolYsis Basic is a three-step formula in which targeted DNA extraction from whole blood is preceded by the elimination of human DNA and other PCR inhibitors. Selective lysis of blood cells is followed by the degradation of the released human DNA. Finally, the bacteria are enriched and treated with a reagent optimized for the lysis of both Gram-positive and Gram-negative bacteria.

Product specifications:

1. Starting material: for optimal results, use only fresh blood (see Note 3): MolYsis Basic (0.2 ml blood) (see Notes 4 and 5).

Protocol for the use of MolYsis Basic 0.2 ml (perform before DNA isolation):

1. Add 50 μl buffer CM to 200 μl whole blood, vortex for 10 s, and let stand for 5 min.
2. Add 50 μl buffer DB1 and 10 μl MolDNase A to the lysate, vortex for 10 s and let stand for 15 min at room temperature.
3. Centrifuge for >12,000 × *g* for 5 min and remove the supernatant.
4. Add 1 ml buffer RS, vortex for 10 s, and centrifuge for 10 min (>12,000 × *g*). Carefully remove the supernatant.
5. Resuspend the pellet in 80 μl buffer RL. Vortex for 10 s.
6. Add 20 μl Buglysis solution, vortex for 10 s, and incubate at 37°C for 45 min.

2.3. DNA Extraction Protocols

Protocols for four commercially available DNA extraction kits, the QIAamp DNA Blood Mini Kit, the MagNA Pure LC Microbiology Kit M^{Grade}, the MolYsis Complete5 Kit, and the MagSi-DNA Isolation Kit for Blood are described. Results obtained by spiking blood with MRSA are shown to illustrate the differences between them. A schematic overview of the performed analytical processes for whole blood samples is given in Fig. 1.

2.3.1. QIAamp DNA Blood Mini Kit (QIAamp)

Targeted DNA isolation with the QIAamp DNA Blood Mini Kit is based on a fast spin-column procedure in which bacterial DNA binds to a silica-gel membrane. The Blood and Body Fluid Spin Protocol should be used according to the manufacturer's instructions

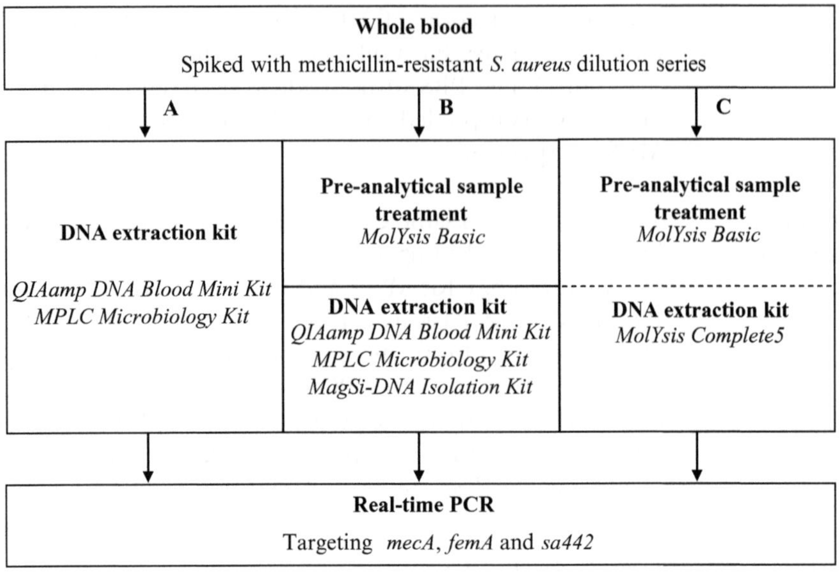

Fig. 1. Schematic overview of the analytical process for whole blood samples demonstrated by the extraction of bacterial DNA from blood spiked with MRSA. (A) DNA extraction with two conventional kits, without pre-analytical sample treatment. (B) DNA extraction with two conventional kits and one novel kit, following pre-analytical sample treatment with MolYsis Basic. (C) DNA extraction with a novel kit, which combines all the steps of the pre-analytical sample treatment (MolYsis Basic) with a complementary matching DNA extraction protocol. Each extraction was performed at least in three independent runs of serial dilutions in duplicate (modified from (22) with permission from the American Society for Microbiology).

with these modifications (incubation at 56°C for 1 h, followed by a short incubation step at 95°C for 10 min and elution in 60 μl of distilled water) (see Note 6).

2.3.2. MagNA Pure LC Microbiology Kit M^{Grade} (MPLC)

The MagNA Pure LC Microbiology Kit M^{Grade} binds target DNA using magnetic-bead technology. During processing on the MagNA Pure LC instrument a chaotropic buffer and magnetic glass particles are added. DNA binds to the surface of the magnetic glass particles, whereas unbound substances are removed by several washing steps. Finally, the purified DNA is eluted in 100 μl of elution buffer. Use according to the manufacturer's instructions.

2.3.3. MagSi-DNA Isolation Kit for Blood (MagSi-DNA)

The MagSi-DNA Isolation kit for blood is designed to be combined with the MolYsis Basic kit. The MagSi-DNA Isolation kit for Blood aims at optimal DNA binding by using specially coated, high paramagnetic particles (200–600 nm). In combination with non-chaotropic buffers bacterial DNA is eluted in 100 μl of elution buffer.

Protocol:

1. Add 200 μl lysis Buffer and 20 μl Proteinase K to the 50–200 μl sample (filled up to 200 μl with PBS). Vortex for 30 s. Incubate for 15 min at 60°C.

2. Add 400 μl binding buffer and vortex for 10 s. Add 20 μl MagSi-DNA beads and vortex for 1 min. Incubate at room temperature for 1 min.
3. Collect the beads to one side of the tube with the magnet. Wait for 30 s to attract all the beads and discard the supernatant.
4. Add 400 μl wash buffer 1, vortex for 10 s, and collect the beads with the magnet as described in step 3.
5. Wash beads with 800 μl wash buffer 2. Repeat this wash step.
6. After the supernatant is discarded, air-dry the beads for 5–10 min.
7. Elute the DNA by adding 50–200 μl preheated elution buffer. Incubate for 2 min at 60°C and mix 3 min. Collect beads and pipette the supernatant, containing the DNA, into a new tube.

2.3.4. MolYsis Complete5 (Complete)

MolYsis Complete5 combines the MolYsis Basic pre-analytical sample treatment with a complementary matching bacterial DNA isolation protocol in a spin-column format. Sample pretreatment involves the lysis of the blood cells followed by the degradation of the released nucleic acids. Next, bacterial cells are concentrated from the human blood cell lysate and bacterial DNA isolated by a bind–wash–elute procedure, using a spin-column format. The protocol for 1 ml blood samples is used according to the manufacturer's instructions. For optimal results, use only fresh blood.

Protocol:

Part I: pre-analytical sample treatment (see protocol MolYsis Basic, Subheading 2.2)
Part II: DNA extraction

1. Add 20 μl Proteinase K and 150 μl buffer RP. Vortex for 10 s and incubate at 56°C for 10 min.
2. Briefly centrifuge and add 250 μl binding buffer AB, vortex for 10 s.
3. Briefly centrifuge and transfer the lysates to a spin column. Centrifuge at ≥12,000 × *g* for 30 s. Discard the flow-through.
4. Add 400 μl buffer WB. Centrifuge at >12,000 × *g* or 30 s. Discard the flow-through.
5. Wash with 400 μl of 70% ethanol and centrifuge at ≥12,000 × *g* for 3 min.
6. Transfer the column to a 1.5 ml elution tube.
7. Place 100 μl water, preheated to 70°C, in the center of the column and incubate for 1 min. Centrifuge at ≥13,000 × *g* for 1 min to elute the DNA.

Table 1
Comparative analysis of the different DNA extraction procedures performed on whole blood spiked with a tenfold dilution series of MRSA. Adapted from (22) with permission from the American Society for Microbiology

DNA extraction method	V^a (ml)	Time[b] (min)	Detection limit[c], no. of cfu/ml							
			10^5	10^4	10^3	5×10^2	10^2	5×10^1	10^1	10^0
QIAamp	0.2	90	+++	+++	+++	++−	−−−	−−−	−−−	−−−
QIAamp + M.Basic	0.2	175	+++	+++	+++	+−−	−−−	−−−	−−−	−−−
MPLC	0.1	40	+++	+++	+++	−−−	−−−	−−−	−−−	−−−
MPLC + M.Basic	0.2	125	+++	+++	+++	+++	−−−	−−−	−−−	−−−
MagSi-DNA + M.Basic	0.2	130	+++	+++	+++	+−−	−−−	−−−	−−−	−−−
Complete	1.0	120	+++	+++	+++	+++	+++	+++	++−	−−−

[a]Blood volume in ml
[b]Hands-on time for eight samples
[c]Detection, represented by a Ct-value of <40, is indicated by +, no detection is indicated by −. Detection was determined for the three gene targets mecA, femA and sa442

2.4. Detection Limits after DNA Extraction and Real-Time PCR

The detection limit of the different DNA extraction procedures followed by a model real-time PCR assay is presented in Table 1 (see Note 7). Two conventional DNA extraction kits (QIAamp and MPLC) were tested with and without the MolYsis Basic pre-analytical sample treatment protocol. Pre-analytical sample treatment combined with the QIAamp extraction protocol, based on a spin-column format, does not result in an increase in the detection limit. Both procedures perform equally and are able to detect the three target genes at a minimal concentration of 10^3 cfu/ml. Here, the detection limit of 10^3 cfu/ml is equivalent to 16 cfu per PCR, taking the PCR sample volume and extraction input and output into account. The performance of the MPLC kit is roughly the same as the QIAamp kit, i.e., a detection limit of 10^3 cfu/ml. A 0.5 log unit increase of the lower detection limit can be achieved when the MPLC kit is combined with the MolYsis Basic pretreatment protocol. The minimal detectable amount of bacteria is 500 cfu/ml.

The MagSi-DNA kit, a novel combination of two sample preparation methods, showed results similar to the results obtained after conventional DNA extraction. A concentration of 10^3 cfu/ml can be detected after amplification of the isolated MRSA DNA. The second novel kit, MolYsis Complete5, is more sensitive allowing bacterial detection at a concentration of 50 cfu/ml. In this case, the detection of 50 cfu/ml was analogue to 4 cfu per PCR.

3. Discussion and Conclusions

More rapid diagnosis of bloodstream infections is still one of the most important challenges in the field of pathogen diagnostics. Implementation of rapid and accurate molecular testing directly from whole blood specimens could contribute to a faster diagnosis permitting an earlier administration of the appropriate antimicrobial therapy. Optimization of whole blood sample processing in combination with a sensitive real-time PCR assay could ameliorate the detection of pathogenic bacteria in patients with a suspected bloodstream infection. Here we describe four different DNA isolation procedures for the selective extraction of bacterial DNA from whole blood samples. The novel kit MolYsis Complete5 provides a combination of pre-analytical sample treatment and targeted DNA extraction containing all the buffers and reagents necessary for human DNA removal, bacteria enrichment, and bacterial DNA isolation. MolYsis Complete5 was the most successful kit for the detection of bacteria spiked in whole blood samples, suggesting that the novel combination of two complementary matching sample preparation procedures and the use of a larger volume of blood sample both contributed to a higher level of bacterial detection. The kit is superior to the other extraction kits, with a detection limit of less than 5 cfu per reaction (see Note 8).

Automation, ease of use, duration, and costs of the procedure are each important factors also contributing to the extent of implementation in diagnostic laboratories. Except for the MPLC Microbiology Kit M^{Grade}, which is performed on the automated MagNA Pure Instrument, all procedures are performed manually. It is clear that the degree of automation has its impact on the hands-on time needed for the different protocols: i.e., for MPLC extraction only 10 min of hands-on time is needed, whereas the combination of MolYsis Basic and QIAamp requires 170 min. The extraction methods are all considered as easy to perform. The data showed that there is no difference in performance between isolation using a spin-column format and isolation based on magnetic-bead technology.

The MolYsis Complete5 kit is superior compared to the other DNA extraction methods regarding detection limit. The combination of performing a pre-analytical sample treatment and using a larger sample volume allows the detection of only 50 cfu/ml of whole blood (<5 cfu per reaction), emphasizing that the rate of efficiency was attributed to more than one factor. The detection limit is improved by 1.5 log unit compared with the other DNA extraction methods. These results confirm that the efficiency of DNA extraction, especially for clinical samples such as whole blood, is a crucial element in the process of molecular pathogen detection.

Ultimately, a combination of optimal sample processing and molecular detection techniques will lead to rapid and accurate pathogen detection for diagnosis of bloodstream infections.

4. Notes

1. A model multiplex real-time PCR was used targeting three genes of MRSA. This PCR is chosen because the detection limit and the PCR efficiency of the three single PCRs within the multiplex vary and thus the impact of the DNA extraction in combination with varying PCRs can be studied.
2. The outcome of DNA extraction depends amongst others on the type of microorganism from which DNA is extracted. In this case, MRSA was used as a model organism. In general, DNA extraction from Gram-positive microorganisms (such as MRSA) is considered slightly more complex than extraction from Gram-negative microorganisms due to their cell wall composition. Similar experimental setups were used for DNA extraction of Gram-negative *E. coli* and similar results were obtained. It should be noted that results may be different when microorganisms with an even more rigid or complex cell wall are used.
3. It is essential to use fresh blood (which has not been frozen) when using MolYsis Basic or MolYsis Complete. These methods are based on sequential lysis of human blood cells, from which DNA is then degraded followed by bacterial lysis. When samples have been frozen, some bacterial cells are lysed prior to DNA extraction which results in premature loss of this bacterial DNA during the process.
4. MolYsis Basic is also available as MolYsis Basic5 for DNA extraction from 1 ml or 5 ml of blood.
5. The MolYsis Basis kit is compatible with any DNA isolation system, e.g., Qiagen DNeasy/QiaAmp, Roche High Pure PCR Template Preparation Kit, Epicentre Masterpure DNA Purification Kit. Also automated DNA isolation systems can be switched, e.g., Roche Magnapure, FujiFilm QuickGene-810, Chemagen Chemagic MSM I, PSS Magtration, Biomerieux easyMag.
6. For improved extraction, the incubation with proteinase K was extended to 1 h. To concentrate the DNA as much as possible, the elution volume was reduced from 200 to 60 μl.
7. Real-time PCR was set up in a final volume of 50 μl with 2× Absolute™ QPCR ROX (500 nM) Mix (Abgene, Epsom, UK). Final reactions contained 0.6 μM of *mecA* primer and 0.3 μM of *femA* and *sa442* primer, 100, 125, and 150 nM of

femA, *mecA*, and *sa442* probe respectively and 18.85 μl of template DNA. All tests were performed on the ABI PRISM® 7000 real-time PCR System (Applied Biosystems,) An initial denaturation step of 95°C for 15 min was followed by 42 cycles of denaturation at 95°C for 15 s and annealing at 60°C for 1 min.

8. Recently, a study by Horz et al. has investigated if MolYsis Basic could, indeed, eliminate human DNA in oral samples (21). They found that the use of MolYsis Basic prior to DNA isolation reduced the level of human DNA. However, this effect was accompanied with a partial loss of bacterial DNA. To date we have not determined the level of human DNA reduction. Instead, we observed the direct effect on the detection limit of the real-time PCR assay. Since the lower detection limit of our assay did not increase after MolYsis Basic pretreatment, loss of bacterial DNA could also have occurred during sample treatment with MolYsis Basic.

References

1. Peters RP, van Agtmael MA, Danner SA, Savelkoul PH, Vandenbroucke-Grauls CM (2004) New developments in the diagnosis of bloodstream infections. Lancet Infect Dis 4:751–760
2. Centers for Disease Control and Prevention (1992) Advance report of final mortality statistics. Centers for Disease Control and Prevention. Mon Vital Stat Rep 40:1–13
3. Glerant JC, Hellmuth D, Schmit JL, Ducroix JP, Jounieaux V (1999) Utility of blood cultures in community-acquired pneumonia requiring hospitalization: influence of antibiotic treatment before admission. Respir Med 93: 208–212
4. Grace CJ, Lieberman J, Pierce K, Littenberg B (2001) Usefulness of blood culture for hospitalized patients who are receiving antibiotic therapy. Clin Infect Dis 32:1651–1655
5. McKenzie R, Reimer LG (1987) Effect of antimicrobials on blood cultures in endocarditis. Diagn Microbiol Infect Dis 8:165–172
6. Jordan JA, Durso MB (2000) Comparison of 16S rRNA gene PCR and BACTEC 9240 for detection of neonatal bacteremia. J Clin Microbiol 38:2574–2578
7. Klaschik S, Lehmann LE, Raadts A et al (2004) Detection and differentiation of in vitro-spiked bacteria by real-time PCR and melting-curve analysis. J Clin Microbiol 42:512–517
8. Nadkarni MA, Martin FE, Jacques NA, Hunter N (2002) Determination of bacterial load by real-time PCR using a broad-range (universal) probe and primers set. Microbiology 148: 257–266
9. Rothman RE, Majmudar MD, Kelen GD et al (2002) Detection of bacteremia in emergency department patients at risk for infective endocarditis using universal 16S rRNA primers in a decontaminated polymerase chain reaction assay. J Infect Dis 186:1677–1681
10. Shafazand S, Weinacker AB (2002) Blood cultures in the critical care unit: improving utilization and yield. Chest 122:1727–1736
11. Kane TD, Alexander JW, Johannigman JA (1998) The detection of microbial DNA in the blood: a sensitive method for diagnosing bacteremia and/or bacterial translocation in surgical patients. Ann Surg 227:1–9
12. Klaschik S, Lehmann LE, Raadts A, Book M, Hoeft A, Stuber F (2002) Real-time PCR for detection and differentiation of gram-positive and gram-negative bacteria. J Clin Microbiol 40:4304–4307
13. Barken KB, Haagensen JA, Tolker-Nielsen T (2007) Advances in nucleic acid-based diagnostics of bacterial infections. Clin Chim Acta 384:1–11
14. Al-Soud WA, Radstrom P (2001) Purification and characterization of PCR-inhibitory components in blood cells. J Clin Microbiol 39:485–493
15. Kreader CA (1996) Relief of amplification inhibition in PCR with bovine serum albumin

or T4 gene 32 protein. Appl Environ Microbiol 62:1102–1106

16. Morata P, Queipo-Ortuno MI, de Dios Colmenero J (1998) Strategy for optimizing DNA amplification in a peripheral blood PCR assay used for diagnosis of human brucellosis. J Clin Microbiol 36:2443–2446
17. Smith K, Diggle MA, Clarke SC (2003) Comparison of commercial DNA extraction kits for extraction of bacterial genomic DNA from whole-blood samples. J Clin Microbiol 41:2440–2443
18. Wilson IG (1997) Inhibition and facilitation of nucleic acid amplification. Appl Environ Microbiol 63:3741–3751
19. Rantakokko-Jalava K, Jalava J (2002) Optimal DNA isolation method for detection of bacteria in clinical specimens by broad-range PCR. J Clin Microbiol 40:4211–4217
20. Zucol F, Ammann RA, Berger C et al (2006) Real-time quantitative broad-range PCR assay for detection of the 16S rRNA gene followed by sequencing for species identification. J Clin Microbiol 44:2750–2759
21. Horz HP, Scheer S, Huenger F, Vianna ME, Conrads G (2008) Selective isolation of bacterial DNA from human clinical specimens. J Microbiol Methods 72:98–102
22. Hansen WL, Bruggeman CA, Wolffs PF (2009) Evaluation of new preanalysis sample treatment tools and DNA isolation protocols to improve bacterial pathogen detection in whole blood. J Clin Microbiol 47: 2629–2631

Chapter 5

Detection of Bacterial Contamination in Platelet Concentrates Using Flow Cytometry and Real-Time PCR Methods

Tanja Vollmer, Knut Kleesiek, and Jens Dreier

Abstract

Despite considerable advances in the safety of blood components based on the application of highly sensitive and specific screening methods to minimize the viral infection risk, the prevention of transfusion-associated bacterial infection remains a major challenge in transfusion medicine. In particular, platelet concentrates represent the greatest infectious risk of transfusion-transmitted bacterial sepsis. The detection of bacterial contamination in platelet concentrates has been implemented in several blood services as a routine quality control testing. Although culture is likely to remain the gold standard method of detecting bacterial contamination, the use of rapid methods is likely to increase and play an important role in transfusion medicine in the future. In particular, flow cytometric methods and nucleic acid amplification techniques are powerful tools in bacterial screening assays. Compared to culture-based methods, the combination of high sensitivity and specificity, low contamination risk, ease of performance, and speed has made those technologies appealing alternatives to conventional culture-based testing methods.

Key words: Platelet concentrates, Bacterial contamination, Flow cytometry, Nucleic acid amplification techniques, Rapid screening methods

1. Introduction

Bacterial contamination of blood components is now recognized as the most frequent infectious complication in transfusion therapy (1, 2). In particular, bacterial contamination of platelet concentrates is a major problem as a result of the required storage conditions of PCs. Platelet concentrates are stored in gas-permeable containers at room temperature (22–24°C) with constant gentle agitation to preserve platelet function. The risk of receipt of a bacterially contaminated platelet unit is approximately 50- to 250-fold

Mark Wilks (ed.), *PCR Detection of Microbial Pathogens: Second Edition*, Methods in Molecular Biology, vol. 943,
DOI 10.1007/978-1-60327-353-4_5, © Springer Science+Business Media, LLC 2013

higher than the combined risk of potential viral infection (HIV, HCV, HBV, and HTLV-I/II (3)). Bacterial contaminants in blood components mostly originate from the donor's skin, or more uncommonly from donors with asymptomatic bacteremia (4). Immediately after donation, the initial bacterial contamination is of the order of approximately 10–100 CFU, which is equivalent to 0.03–0.30 CFU/mL assuming a typical blood product volume of 300 mL (5). While bacterial contamination may affect any blood component, the storage conditions of PCs support bacterial proliferation and small inocula of bacteria can easily grow to high numbers in a short period. Prospective testing of PCs reported a prevalence of bacterial contamination varying from 0.08 to 0.7%, depending on technology, testing protocols, and additional intervention methods (6, 7). Different screening methods have been developed for the detection of microbiological contamination in PCs which can be divided into the two methodological concepts: incubation or cultivation methods, and rapid-detection methods. Rapid methods include nucleic acid amplification techniques (NAT, (8–10)), fluorescence assorted cell sorting (FACS, (11–15)), or immunological detection methods (16, 17). General diagnostic difficulties include the limited diagnostic window, the wide range of possible organisms and the high demands on sensitivity to detect low initial bacterial amounts at the beginning of storage. Platelet bacteria screening comprises three different sampling strategies: (a) early-testing after pre-incubation for 24 h using culture methods with PCs release subject to a "negative-to-date" status, (b) early-testing after pre-incubation for 24 h using rapid methods (NAT, FACS), and (c) late-testing in combination with a rapid detection method (8). Incubation methods are dependent on bacterial growth kinetics, and the inherently long times required for detection necessitates early sampling (5), which may result in considerable sampling errors. Rapid methods, such as NAT, flow cytometry, and immunoassays, allow the possibility of sample collection at a later stage, thus overcoming the risk of sampling error due to initially low bacterial count or slow-growing bacterial species. Broad-range NAT assays should be attractive for the detection of bacteria in PCs because of their high sensitivity and specificity, as well as their rapidity. However, the prerequisite requirement to take advantage of PCR is an efficient protocol for nucleic acid extraction. The protocol described here uses a combination of a high-volume magnetic bead nucleic acid separation extraction method to minimize the sampling error (18) combined with sensitive 23S rRNA real-time PCR amplification for the detection of bacterial contamination in PCs. This approach is therefore applicable for both the early- and the late-sampling strategies. Compared with standard separation procedures, magnetic-separation techniques have several advantages, including simplicity in handling and a high automation potential (19). A second advantage is the usage of the 23S rRNA as a target for broad-range real-time PCR amplification combined with the used Tth polymerase. We experimentally determined that this combination showed no environmental contamination or cross reaction with human DNA, which is frequently observed using 16S rRNA (20, 21).

Flow cytometry for sterility testing of PCs has also been demonstrated to be a rapid and feasible approach (11–14). Early cytometer-based assays for bacterial screening used a one-step sample preparation procedure, i.e., simultaneous platelet lysis and bacteria labeling with the membrane-permeable dye thiazole orange (11, 12). This procedure is the most rapid and simple approach, but separation of bacteria population and platelet debris is complicated in the case of several bacterial species, due to an overlap of bacterial fluorescence signals and a high background from platelet debris. A two-step sample preparation with separate lysis and staining steps overcomes these disadvantages and leads to a better separation of bacteria population from platelet debris (11, 13). However, the comparatively low sensitivity still remains a disadvantage, even though pre-incubation methods at a higher temperature in culture medium (13) or in PC satellite bags (11) enhance sensitivity. The test procedure of the BactiFlow flow cytometric assay is based on fluorescent labeling of viable cells followed by flow cytometric enumeration of bacteria. The nonfluorescent fluorochrome passes the cell membrane of viable cells with intact membrane integrity and enzymatic activity, and is cleaved by intracellular esterases. The BactiFlow technology further facilitates the elimination of nontarget particles by specific enzymatic digestion of platelets and centrifiltration of platelet debris resulting in an enhanced sensitivity of the FACS method. Hence, the BactiFlow system encompasses the two major advantages of almost complete elimination of background signals combined with staining of metaboli cally active bacteria.

Here we describe the setup of nucleic acid extraction followed by 23S rDNA real-time PCR and the BactiFlow flow cytometric assay for the screening for bacterial contamination in PCs. NAT assays are the most sensitive rapid methods, but they are complex and the implementation requires a high degree of personnel qualification, careful considerations concerning work flow design. Additionally, their time-to-result of approximately 4 h is still relatively long. The detection limit of the BactiFlow flow cytometry is about ten times lower, but the time-to-result of approximately 1 h and the demanded personnel qualification are relatively low. The costs per single reaction are comparable for both assays, but NAT assays required a more extensive technical equipment. In our personnel evaluation, we suggest screening of PCs for bacterial contamination by flow-cytometry.

2. Materials

2.1. Bacterial Strains and Culture Media

1. Strain *Bacillus cereus* ATCC 11778 was purchased from the American Type Culture Collection (ATCC, LGC Promochem GmbH, Wesel, Germany). Strains *Staphylococcus epidermidis* PEI-DSM 3269, *Escherichia coli* PEI-B-19-03, *Klebsiella pneumoniae* PEI-B-08-08, and *Staphylococcus aureus* PEI-B-23-04 were obtained from the Paul-Ehrlich-Institute (PEI, Langen, Germany). However, the source of bacteria is irrelevant for the application of both assays and any other clinical isolates could also be used.
2. Bacteria cultivation media: 30 g/L tryptone soy broth from Oxoid (Oxoid Ltd., Basingstoke UK), autoclave prior to use, store at 4°C.
3. Bacteria plating media: 40 g/L tryptone soy agar from Oxoid, autoclave prior to use, store plates at 4°C.

2.2. Flow Cytometric Detection of Bacteria

1. Flow cytometer: BactiFlow (AES Chemunex GmbH, Bruchsal, Germany). The flow cytometer is equipped with a blue argon ion laser (488 nm) and fluorescence filter sets for FL1 (540 nm) and FL2 (590 nm). Fluorochromes cleaved by intracellular esterases of viable bacteria emit a specific FL1/FL2 ratio of 1.0 (range 0.8–1.2).
2. Reagents exclusively distributed from AES Chemunex GmbH: ChemSol S (liquid for flow cytometer operation), diluent M (phosphate buffered saline-like buffer), M1 (surfactant), M2 (enzyme), M3 (agent for protein denaturation), ChemSol A7 (phosphate buffered saline-like buffer including tenside), ChemSol B24 (buffer), ChemChrome V23 (fluorochrome), CS26A (reduction of background fluorescence, also includes validation beads), CRS6 powder (reduction of free fluorescein), standard G (flow cytometer calibration solution), cleaning solution 3 (rinse liquid flow cytometer), and ChemFilter 25 (centrifiltration of platelet debris). Store unused reagents at room temperature with the exception of M2 (–20°C), M3/ChemChrome V23/CS26A (4°C). Store all opened reagents at 4°C and warm up to room temperature prior to use. Mix reagent CS26A thoroughly prior to use to ensure homogenous distribution of validation beads.
3. Preparation of sterile M1 solution (for 24 samples): dissolve 1.1 g M1 in 11 mL of diluent M1 and remove particulate material by filter sterilization (use a 0.20 μm filter). The solution is stable at 4°C for 12 h or at –20°C for 2 weeks.
4. Enzyme mix (1 sample): mix 354 μL M1, 55.6 μL M2, and 22.4 μL M3. The solution is stable at 4°C for 12 h.

5. Staining solution (1 sample): mix 3 mL ChemSolB24 and 30 μL ChemChrome V23. The solution is stable at 4°C for 4 h, warm up to room temperature before use.

2.3. Nucleic Acid Extraction

1. Automated nucleic acid extraction system: Chemagen Magnetic Separation module I (PerkinElmer chemagen Technologie GmbH, Baesweiler, Germany). The Chemagen magnetic separation system features an electromagnet and magnetizable metal rods, which are immersed into the magnetic bead suspension and connected to a stirring motor.
2. Reagents (part of the Chemagen 2 K Viral DNA/RNA kit): protease, lysis buffer, binding buffer, wash buffer 3, wash buffer 4, magnetic beads (iron oxide core surrounded by a matrix that binds total nucleic acids). The reagents are stored at room temperature with the exception of magnetic beads (4°C) and protease (−20°C). Warm up binding buffer to 55°C prior to use.

2.4. Real-Time PCR Amplification

1. Amplification platform: Rotorgene 3000 (Corbett Life Sciences, Sydney, Australia).
2. Polymerase and buffer: Eurogentec *Tth* DNA polymerase (Eurogentec Deutschland, Köln, Germany), 10× *Tth* reaction buffer (Eurogentec).
3. Magnesium chloride: stock solution 25 mM
4. Deoxynucleotides triphosphates: prepare working solution including 5 mM dATP, dTTP, dCTP, and dGTP.
5. Primer and Probe sequences are shown in Table 1.

2.5. Blood Collection and Sterility Testing of PCs by Microbiological Cultivation

1. Platelet preparations (apharesis-derived single donor platelet preparations) were obtained with the Hemonetics MCS+system (Haemonetics GmbH, München, Germany).
2. PCs were analyzed with the BacT/Alert3D continuous monitoring system (bioMérieux, Nürtingen, Germany), culture bottles: anaerobic BacT/Alert BPN, aerobic BacT/Alert BPA (bioMérieux).

3. Methods

3.1. Preparation of Samples

1. Bacterial inoculation: PCs are spiked with low bacterial titers (<100 CFU/PC) of *K. pneumoniae*, *S. epidermidis*, *B. cereus* and *S. aureus*. For each inoculation, PC units were analyzed in parallel by the BactiFlow assay and 23S rRNA RT-PCR.
2. Sample drawing: PC samples are taken from the original platelet bag into an additional sterile sampling bag at 0, 6, 12, 24,

Table 1
Primer and probe sequences used

Primer or probe	Oligonucleotide sequence (5′-3′)	Nucleotide positions	Amplicon length (bp)	Reference
Amplification target: 23S rRNA				
23S-MGB-AG-forward	CTKCCAGGAAAAGCYTCTA		121	(18)
23S MGB-GA-forward	CTKCCGAGAAAAGCYTCTA	1560–1579[a]		
23S MGB-AA-forward	CTKCCAAGAAAAGCYTCTA			
1686w-reverse	CATTTTGCCKAGTTCCTT	1663–1682[a]		
23S-BL-LNA[b]-probe	[FAM][c]-ACTaCcTgTGtCGg TTT-[BHQ1][c]	1624–1607[a]		
Internal control (IC), amplification target ß2-microglobulin				
B2-MG-forward	TGAGTATGCCTGCCGTGTGA	343–362[d]	82	(22)
B2-MG-reverse	TGATGCTGCTTACATGTC TCGAT	1018–1040[d]		
B2-MG-probe	[JOE][c]-CCATGTGACTTTGTCA CAGC CCAAGATAG TT-[BHQ1][c]	364–394[d]		

[a]Positions according to the *E. coli* 23S rDNA sequence (GenBank accession no. AF053966)
[b]LNA (locked nucleic acid) bases are shown in lower case
[c][FAM], 6-carboxyfluorescein; [JOE], 2′,7′-dimethoxy-4′,5′-dichloro-6-carboxyfluorescein; [BHQ1], black hole quencher
[d]Positions according to the beta-2-microglobulin gene (GenBank accession no. M17987)

30, 48, 54, 72, and 144 h after bacterial inoculation. Therefore, the flexible tube of the satellite bag is connected/disconnected to the respective flexible tube of the original platelet bag using a sterile welding device. Samples can also be drawn using a needle and syringe.

3. To ensure baseline sterility of the PCs, microbiological sterility testing of PCs was performed prior to bacterial inoculation using the BacT/Alert3D continuous monitoring system (bioMérieux). 5 mL of samples were inoculated into the BacT/Alert BPA and the BacT/Alert BPN bottle. Bottles were incubated at 37°C until a positive signal was detected or for up to 7 days if they remained negative. Samples that did not react after 7-day storage were considered negative for bacterial contamination.

3.2. BactiFlow Flow Cytometric Assay for the Detection of Bacteria in PCs

1. All the reagents required are prepared and warmed up to room temperature.
2. The BactiFlow flow cytometer is switched on and rinsed with Chemsol S for 10 min. After warm up, the instrument is calibrated once with 600 μL Standard G and is then ready to use (see Note 2).

3. PC samples are collected using aseptic techniques as described above. Subsequently, 1 mL of PC sample is mixed with 432 μL enzyme mix and incubated at 37°C for 15 min.
4. The suspension was diluted with 7 mL ChemSol A7 buffer and PC cell debris was removed via ChemFilter 25 by centrifiltration (8 min, 2,000 × *g*).
5. The supernatant is decanted after centrifiltration and the pellet is completely resuspended by vortex mixing in 3 mL bacteria staining solution. Bacteria are stained at 30°C for 12 min in thermal block, followed by cooling to 4°C. It is critical that the supernatant is decanted immediately after centrifiltration to avoid premature resuspension. Samples were allowed to stand at 4°C for a maximum of 1 h before measurement.
6. For sample measurement, samples are individually warmed up at 30°C for 1 min. Afterwards, 140 μL of CS26A is added to the sample and mixed thoroughly (do not vortex). Using a spatula, CSR6 (0.03 g) is dissolved in the sample suspension (do not vortex), incubated at room temperature for 1 min and subsequently 600 μL of the sample suspension transferred to the BactiFlow tube for flow cytometric analysis. Each sample should be individually prepared directly prior to measurement and should be analyzed immediately after CSR6 incubation (see Note 3). Examples of the results (BactiFlow histograms) are given in Fig. 1. Results are displayed as counts/mL which corresponds to CFU/mL (15).
7. Negative controls (sterile water, unspiked PC) and positive controls (bacterial suspension with titers 30–50 × 10^3 CFU/mL) are included in each run (see Note 3).
8. After measurement, the BactiFlow flow cytometer should be rinsed for 10 min with Cleaning Solution 3 before the instrument is switched off.
9. A flowchart of the complete experimental procedure is given in Fig. 2.

3.3. Nucleic Acid Extraction Using the PCR Detection of Bacterial Contamination

1. Samples are collected as described above.
2. 2.4 mL of PC sample is mixed with 2.4 mL of lysis buffer.
3. Following the addition of 20 μL of protease and vortexing for 10 s, the sample suspension is incubated at 55°C for 10 min.
4. Subsequently, the lysate is mixed with 7.5 mL of binding buffer (see Note 5) containing 100 μL of magnetic beads.
5. Samples are transferred to the Chemagic Magnetic Separation Module I, which automatically performs the complete isolation process, including binding with continuous sample mixture (30 min), washing (two steps), and elution of the nucleic acids in 100 μL elution buffer. Nucleic acids were stored at –20°C until used for PCR amplification.

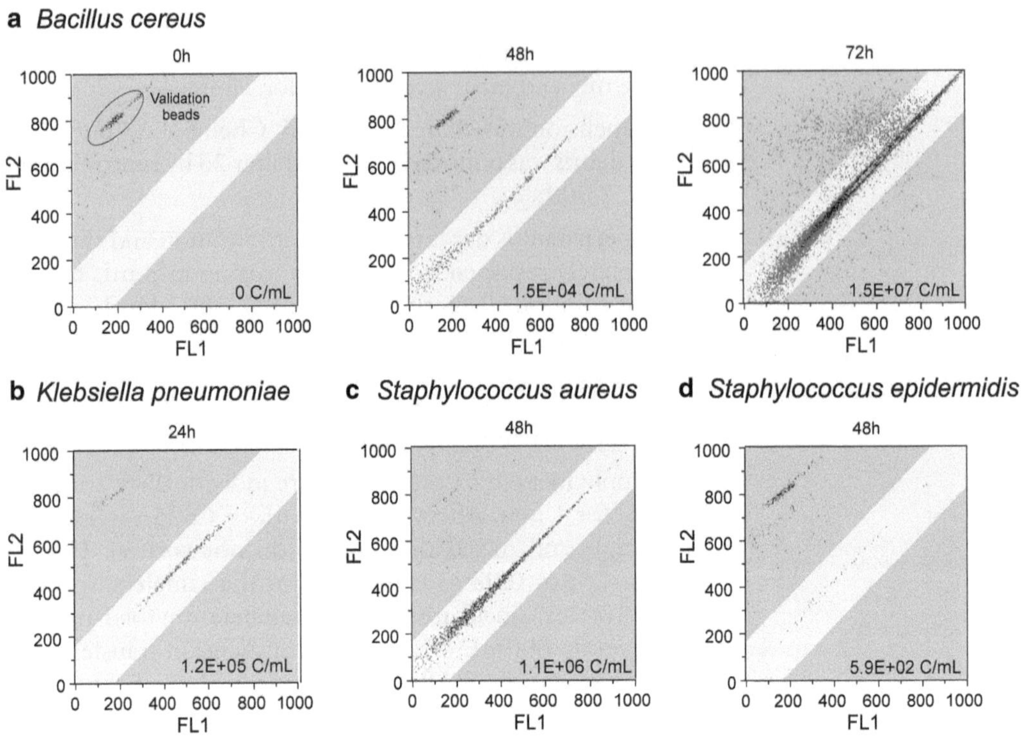

Fig. 1. Growth kinetics of bacteria detected by flow cytometry (15).
Samples from spiked PCs were analyzed using the BactiFlow assay: (**a**) Spiking with *Bacillus cereus* ATCC 11778, shown are the blank value at 0 h, the first positive result after 48 h and the subsequent sampling point (72 h, see Note 4). (**b**) Exemplarily the histograms of positive sampling points at are shown for PCs spiked with *Klebsiella pneumoniae* PEI-B-08-08, 24 h. (**c**) *Staphylococcus aureus* PEI-B-23-04, 48 h. (**d**) *Staphylococcus epidermidis* PEI-DSM 3269, 48 h. The *light shaded* area represents bacteria-specific counts (FL1/FL2 ratio of 1.0, range 0.8–1.2) and the *darker shaded* area represents background counts. The *upper left* area with dense counts is the validation bead area (exemplarily marked for *B. cereus* 0 h). The increase of the bacterial count in the bacteria specific area is definitely observable. Low numbers of PC background events were observed outside the area of interest and validation beads were clearly discriminable. Counts/mL (C/mL).

6. A flowchart of the complete experimental procedure is given in Fig. 2.

3.4. PCR

1. Extracted nucleic acid samples are analyzed by a one-step RT-PCR method incorporating *Tth* DNA polymerase (see Note 6). To detect false-negative results due to PCR inhibition, coextracted human β2-microglobulin (B2-MG) mRNA is coamplified in each PCR reaction.
2. RT-PCRs are carried out on the Rotor-Gene 3000 cycler system (see Note 7) in 0.2 mL tubes containing 40 μL of reaction mix and 10 μL of nucleic acid extract. The reaction mix consists of 1 × Eurogentec *Tth* buffer, 3.5 mmol/L $MgCl_2$, 400 μmol/L of each desoxynucleoside triphosphate,

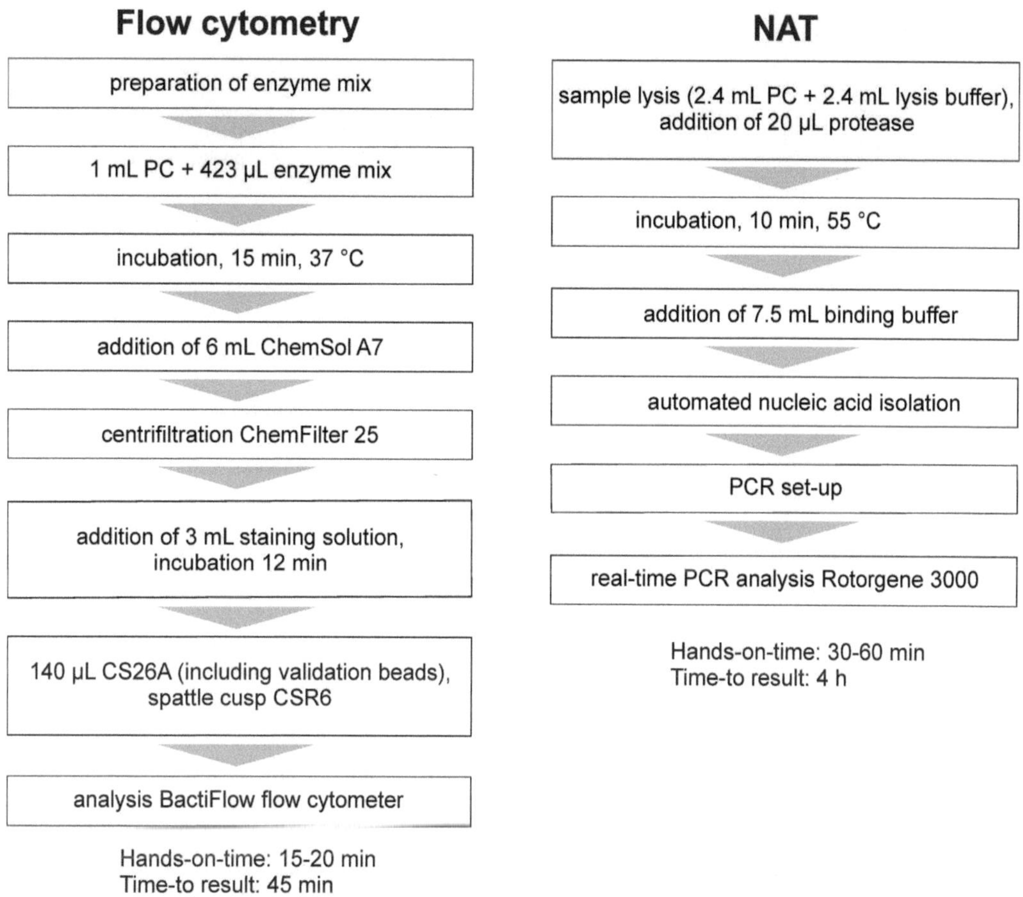

Fig. 2. Flow-chart of experimental procedures of BactiFlow flow cytometry and NAT.
The experimental procedures of BactiFlow flow cytometry and NAT are shown in a flow-chart. Hands-on time and time-to-result are calculated for the analysis of one sample at a time including operation of instruments.

200 nmol/L of each 23S forward primer, 600 nmol/L 23S reverse primer, 200 nmol/L of each IC primer, 200 nmol/L 23S fluorescent probe, 200 nmol/L IC fluorescent probe, and 2.5 U r*Tth* polymerase.

3. Cycling conditions: reverse transcription at 60°C for 20 min, denaturation at 95°C for 4 min, 40 cycles of denaturation at 95°C for 10 s, annealing at 60°C for 35 s, and extension at 72°C for 10 s.
4. Negative controls (sterile water, no template) and positive controls (PCs previously spiked with low bacterial titers ($30–50 \times 10^3$ CFU/L)) are included in each run.
5. An example of the 23S rRNA real-time amplification plot is shown in Fig. 3.

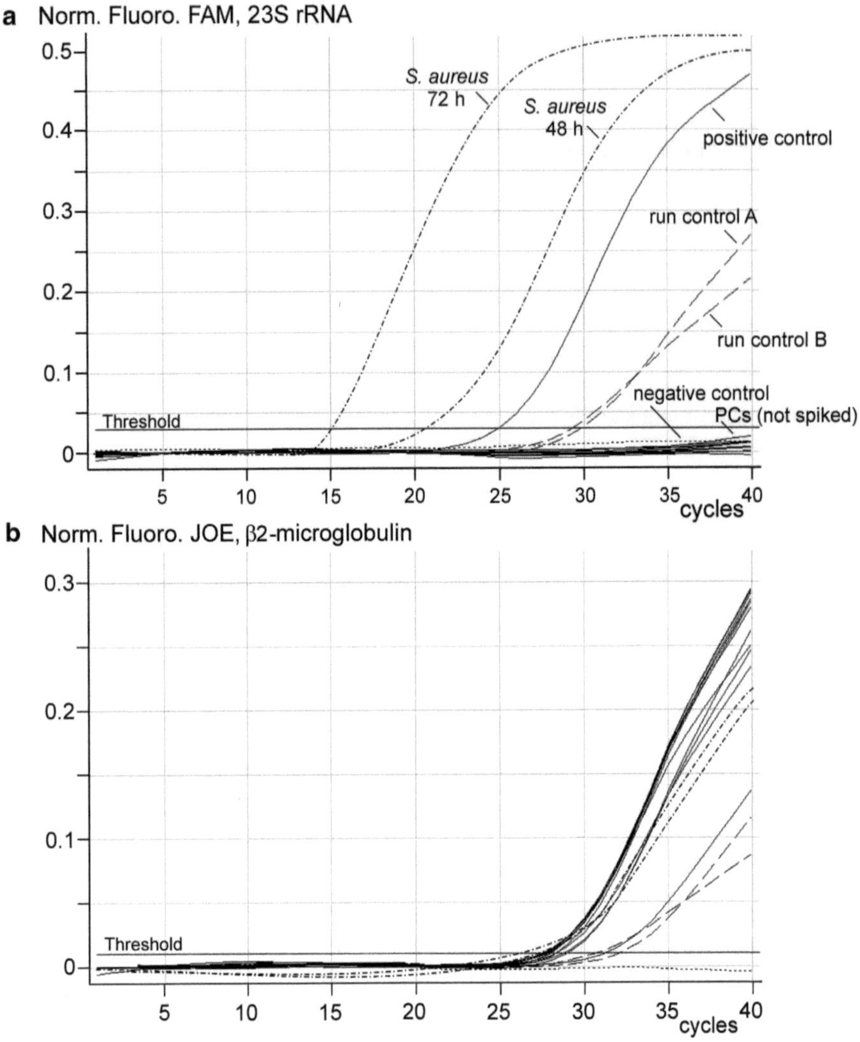

Fig. 3. Amplification plots of routine platelet screening using real-time RT-PCR (21).
Nucleic acids of sterile PCs were extracted from 2.4 mL PC using magnetic separation technology and analyzed by 23S rRNA real-time PCR. (**a**) Amplification plot of the 23S rRNA real-time PCR. For the positive control (*solid line*), a PC was spiked with *S. epidermidis* PEI-DSM 3269 (10^3 CFU/mL). For the low positive run controls (*dashed lines*) PCs were spiked with *S. epidermidis* PEI-DSM 3269 (50 CFU/mL) or *K. pneumoniae* PEI-B-08-08 (50 CFU/mL). The *dashed-dotted lines* show exemplarily the results of spiked PCs with *S. aureus* PEI-B-23-04 after incubation of 48 h and 72 h. (**b**) Amplification plot of the β2-mircoglobulin gene (internal control). Abbreviations: *Norm. Fluoro. FAM*: normalized fluorescence FAM, fluorescence of the amplification plots detecting the bacterial 23S rRNA in the FAM channel; *Norm. Fluoro. JOE*: fluorescence of the amplification plots detecting the internal control (IC) B2-MG mRNA in the JOE channel; negative control (H_2O, *dotted*).

4. Notes

1. A comparison of the two rapid detection methods BactiFlow flow cytometry and 23S rRNA real-time amplification in PCs inoculated with *S. epidermidis*, *K. pneumoniae*, *B. cereus* spores, and *S. aureus* is shown in Fig. 4. Both methods detected bacteria during the bacterial proliferation in the given range.

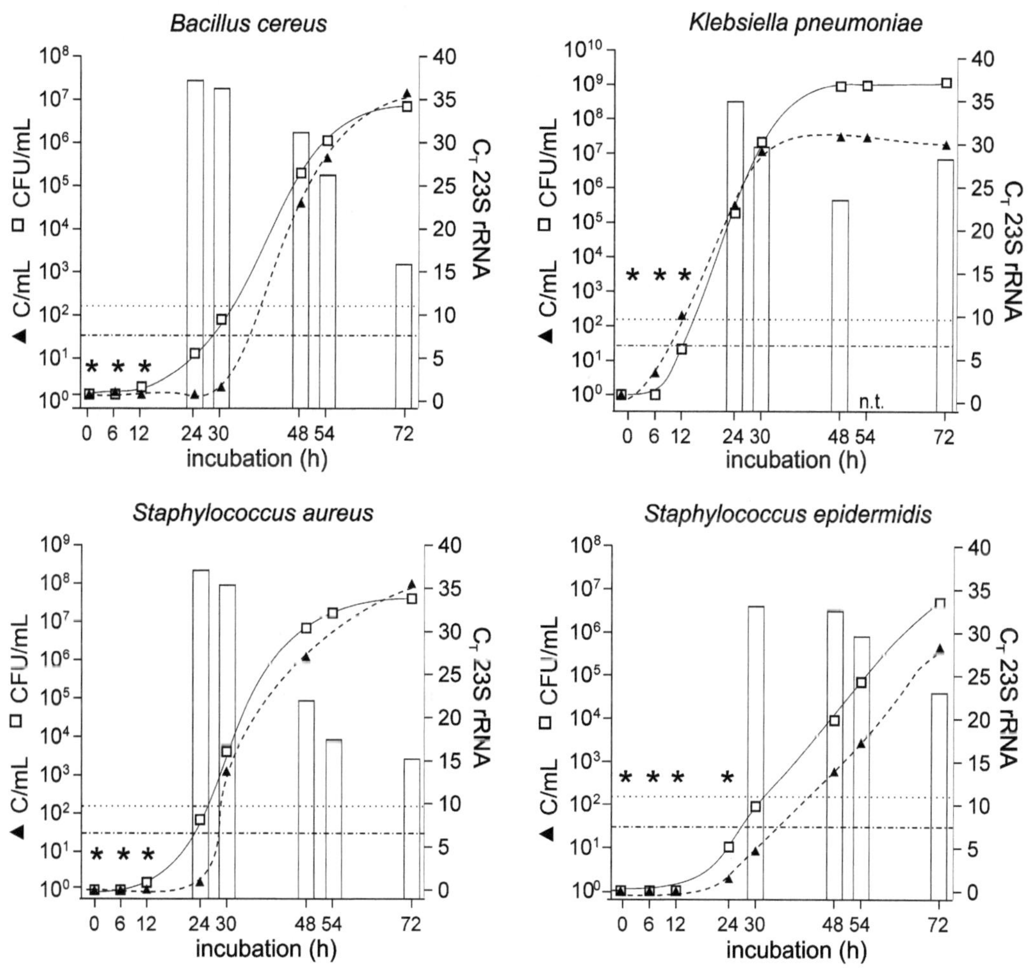

Fig. 4. Bacterial proliferation and detection by flow cytometry and real-time PCR (see Note 1).
One single apheresis-derived PC unit was spiked with <1 CFU/mL of bacteria (*B. cereus* ATCC 11778, *K. pneumoniae* (PEI-B-08-08), *S. aureus* (PEI-B-23-04), *S. epidermidis* (PEI-DSM 3269) and stored at 22°C with agitation. Samples were taken immediately (0 h) and 6, 12, 24, 48, 72 h after inoculation. Bacterial counts were performed by tenfold serial dilutions in PBS, surface spreading on TS agar and counting the number of colonies after incubation (*open square, solid lines*) and BactiFlow assay (*filled triangle, dashed lines*), left scale. Nucleic acids were extracted using magnetic separation technology and analyzed by real-time RT-PCR. C_T-values were used to interpret the results of the RT-PCR (C_T-values are displayed in bars, right scale). The *dotted horizontal lines* represent the lower detection limit of the BactiFlow assay (150 CFU/mL) and the *dashed-dotted horizontal line* represents the lower detection limit of the 23S rRNA assay (30 CFU/mL). *n.t.*: not tested, C_T: crossing point of threshold, *asterisk*: no detection of bacteria using the 23S rRNA assay (C_T > 40).

2. Failed calibration of the BactiFlow flow cytometer may be due to incorrect adjustment of the measuring cell. In this case, try to calibrate the measuring cell using standard C (supplemental reagent material) or contact technical support.
3. PC background in this flow cytometric assay is normally located outside the bacteria specific area (refer to Fig. 1). If negative controls show a strong background in the bacteria-specific area, there are two possible reasons: (a) sterile filtration of M1 solution is insufficient, or (b) reagents are contaminated with bacteria. The measurement has to be repeated with new reagents. If negative controls show a strong background within and/or in the left triangle of the histogram the amount of CSR 6 was insufficient.
4. If samples are contaminated with bacteria $>10^5$ counts/mL, the flow cytometer automatically reduces the analyzed sample volume to a tenth or less and the count for the complete sample volume is calculated by the software. In this case, a valid validation bead count is not necessary. If invalid validation bead count occurs in samples with $<10^5$ counts/mL, the analysis has to be repeated. This might have two reasons: (a) reagent CS26A was not mixed thoroughly prior to use, (b) the flow cytometer is blocked and has to be cleaned with cleaning 3.
5. Storage of binding buffer at room temperature could lead to precipitation of buffer components, warming to 55°C prior to use dissolves the precipitate.
6. Eurogentec *Tth* polymerase works well in this implementation. Use of other commercially available polymerases may lead to a considerable false-positive background-signal related to bacterial contamination of polymerases as a result of the production process in bacteria. Therefore, the application of alternative polymerases has to be validated prior to use with the DNA extract of unspiked PCs. Decontamination strategies (8-methoxypsoralen, UV irradiation, and ultracentrifugation) often result in a decreased sensitivity and are therefore not recommended (23).
7. The use of the Rotorgene 6000 or other thermal cyclers requires optimization.
8. The 23S rRNA protocol can also be adapted for the detection of bacteria in other clinical samples (e.g., heart valve tissue, whole blood) with modifications (change of internal control to avoid competitive amplification reactions).

References

1. Wagner SJ, Friedman LI, Dodd RY (1994) Transfusion-associated bacterial sepsis. Clin Microbiol Rev 7:290–302
2. Brecher ME, Hay SN (2005) Bacterial contamination of blood components. Clin Microbiol Rev 18:195–204
3. Reading FC, Brecher ME (2001) Transfusion-related bacterial sepsis. Curr Opin Hematol 8:380–386
4. Blajchman MA, Goldman M, Baeza F (2004) Improving the bacteriological safety of platelet transfusions. Transfus Med Rev 18:11–24
5. Montag T (2006) Perspectives and limitations in the bacterial screening of platelet concentrates. J Lab Med 30:60–65
6. te Boekhorst PA, Beckers EA, Vos MC, Vermeij H, van Rhenen DJ (2005) Clinical significance of bacteriologic screening in platelet concentrates. Transfusion 45:514–519
7. Walther-Wenke G (2006) Bacterial contamination of blood components-incidence and significance for homologous and autologous transfusion. J Lab Med 30:66–73
8. Dreier J, Störmer M, Pichl L, Schottstedt V, Grolle A, Bux J, Kleesiek K (2008) Sterility screening of platelet concentrates: questioning the optimal test strategy. Vox Sang 95: 181–188
9. Mohammadi T, Savelkoul PH, Pietersz RN, Reesink HW (2006) Applications of real-time PCR in the screening of platelet concentrates for bacterial contamination. Expert Rev Mol Diagn 6:865–872
10. Rood IG, Koppelman MH, Pettersson A, Savelkoul PH (2008) Development of an internally controlled PCR assay for broad range detection of bacteria in platelet concentrates. J Microbiol Methods 75:64–69
11. Karo O, Wahl A, Nicol SB, Brachert J, Lambrecht B, Spengler HP, Nauwelaers F, Schmidt M, Schneider CK, Muller TH, Montag T (2008) Bacteria detection by flow cytometry. Clin Chem Lab Med 46:947–953
12. Mohr H, Lambrecht B, Bayer A, Spengler HP, Nicol SB, Montag T, Muller TH (2006) Basics of flow cytometry-based sterility testing of platelet concentrates. Transfusion 46:41–49
13. Schmidt M, Hourfar MK, Nicol SB, Spengler HP, Montag T, Seifried E (2006) FACS technology used in a new rapid bacterial detection method. Transfus Med 16:355–361
14. Schmidt M, Karakassopoulos A, Burkhart J, Deitenbeck R, Asmus J, Muller TH, Weinauer F, Seifried E, Walther-Wenke G (2007) Comparison of three bacterial detection methods under routine conditions. Vox Sang 92:15–21
15. Dreier J, Vollmer T, Kleesiek K (2009) Novel flow cytometry-based screening for bacterial contamination of donor platelet preparations compared with other rapid screening methods. Clin Chem 55:1492–1502
16. Palavecino EL, Yomtovian RA, Jacobs MR (2006) Detecting bacterial contamination in platelet products. Clin Lab 52:443–456
17. Prowse C (2007) Zero tolerance. Transfusion 47:1106–1109
18. Störmer M, Kleesiek K, Dreier J (2007) High-volume extraction of nucleic acids by magnetic bead technology for ultrasensitive detection of bacteria in blood components. Clin Chem 53:104–110
19. Kleines M, Schellenberg K, Ritter K (2003) Efficient extraction of viral DNA and viral RNA by the Chemagic viral DNA/RNA kit allows sensitive detection of cytomegalovirus, hepatitis B virus, and hepatitis G virus by PCR. J Clin Microbiol 41:5273–5276
20. Marin M, Munoz P, Sanchez M, del Rosal M, Alcala L, Rodriguez-Creixems M, Bouza E (2007) Molecular diagnosis of infective endocarditis by real-time broad-range polymerase chain reaction (PCR) and sequencing directly from heart valve tissue. Medicine (Baltimore) 86:195–202
21. Dreier J, Störmer M, Kleesiek K (2007) Real-time polymerase chain reaction in transfusion medicine: applications for detection of bacterial contamination in blood products. Transfus Med Rev 21:237–254
22. Watzinger F, Ebner K, Lion T (2006) Detection and monitoring of virus infections by real-time PCR. Mol Aspects Med 27:254–298
23. Dreier J, Störmer M, Kleesiek K (2004) Two novel real-time reverse transcriptase PCR assays for rapid detection of bacterial contamination in platelet concentrates. J Clin Microbiol 42:4759–4764

Chapter 6

Multiplex Real-Time PCR Assay for the Detection of Meticillin-Resistant *Staphylococcus aureus* and Panton–Valentine Leukocidin from Clinical Samples

Lynne Renwick, Anne Holmes, and Kate Templeton

Abstract

The control and prevention of meticillin-resistant *Staphylococcus aureus* (MRSA) is a major challenge for healthcare establishments, especially as this pathogen continues to evolve. The emergence and spread of community associated MRSA producing Panton–Valentine leukocidin (PVL) causing severe, sometimes fatal, infections in otherwise healthy people is a significant cause of concern. Patient screening to detect MRSA is now widely used as part of an effective control program to limit the spread of this pathogen. Real-time PCR targeting specific MRSA markers offers a rapid alternative to conventional methods enabling earlier intervention, such as patient isolation and decolonization treatment. Herein we describe a multiplex real-time assay that combines primers and probes to detect MRSA and the genes for PVL to provide a rapid and informative assay.

Key words: Real-time PCR, MRSA, Staphylococcus cassette chromosome *mec*, Panton–Valentine leukocidin, Clinical samples

1. Introduction

Preventing meticillin-resistant *Staphylococcus aureus* (MRSA) infections by using effective control measures is a major public health priority. MRSA has long been associated with healthcare establishments; however, its emergence in the community setting (1) in recent years poses new challenges for microbiology laboratories to optimize their surveillance strategies.

An essential component of infection control measures to reduce the spread of this pathogen is to identify colonized patients, followed by their isolation to prevent cross-infection and decolonization treatment to eliminate carriage. PCR-based methods offer a

Mark Wilks (ed.), *PCR Detection of Microbial Pathogens: Second Edition*, Methods in Molecular Biology, vol. 943, DOI 10.1007/978-1-60327-353-4_6, © Springer Science+Business Media, LLC 2013

major advantage over conventional culture and identification by reducing the time to confirm results from 48–72 h to 1 day. Rapid detection of MRSA carriers may lead to early implementation of control measures, reduced MRSA transmission, improved patient outcomes, and reduced hospital bed stays (2–4).

Screening for MRSA carriage mostly involves swabbing superficial sites, such as the anterior nares (the most frequent site of MRSA colonization), axillae, and perineum. To identify MRSA in these non-sterile specimens it is necessary to use a PCR test that can differentiate MRSA from a mixture of *S. aureus* and meticillin-resistant coagulase-negative staphylococci. Meticillin resistance is mediated by the *mecA* gene, which is carried on a mobile genetic element called the Staphylococcal Chromosome Cassette *mec* (SCC*mec*). By targeting a region incorporating the SCC*mec* and the adjacent chromosomal *S. aureus*-specific *orfX* gene it is possible to directly detect MRSA from clinical specimens (5).

MRSA strains that have emerged in the community are genetically diverse and distinct from healthcare-associated MRSA (HA-MRSA). Often they carry the SCC*mec* IV cassette and the genes *lukf-PV* and *lukS-PV* (1), which encode Panton–Valentine leukocidin (PVL), a nonspecific pore forming toxin (6, 7). In the UK, there has been limited spread of community strains in hospitals (8); however, the emergence of CA-MRSA USA300 as a major cause of healthcare associated infections in the United States highlights the need for active surveillance procedures to closely monitor these evolving strains (9, 10).

This protocol describes a qualitative, multiplex real-time PCR assay for the detection of MRSA and the PVL genes from clinical samples. Five forward primers targeting various SCC*mec* types combined with an *S. aureus orfX*-specific reverse primer are used to amplify a region of MRSA DNA (5), which is detected in real-time using a generic, *orfX*-specific fluorogenic probe (11). Multiplexed with the MRSA oligonucleotides are primers that target a conserved area across the *lukF* and *lukS* genes, and a PVL probe labelled with a different fluorophore to the SCC*mec* probe to allow multiplexing (11, 12). The three main procedures involved, namely, sample preparation, nucleic acid extraction, and real-time PCR, are described in detail below.

2. Materials

2.1. Sample Preparation

1. Swabs in Stuart's medium.
2. Overnight culture of positive control, e.g., *S. aureus* ATCC-BAA-1752, a PVL producing strain of MRSA.
3. 3 ml saline (0.85%).

Table 1
Primers and probes used in the MRSA PCR

Assay	Primer	Conc.	Sequence (5′–3′)
MRSA	Mecii	100 μM	GTCAAAAATCATGAACCTCATTACTTAGT
	Meciii	100 μM	ATTTCATATATGTAATTCCTCCACATCTC
	Meciv	100 μM	CAAATATTATCTCGTAATTTACCTTGTTC
	Mecv	100 μM	CTCTGCTTTATATTATAAAATTACGGCTG
	Mecvii	100 μM	CACTTTTTATTCTTCAAAGATTTGAGC
	OrfX	100 μM	GGATCAAACGGCCTGCACA
	Probe	25 μM	**FAM**-CRTAGTTACTRCGTTGTAAGACGTC-**BHQ**
PVL	PVL F	50 μM	ACACACTATGGCAATAGTTATTT
	PVL R	50 μM	AAAGCAATGCAATTGATGTA
	Probe	5 μM	**HEX**-ATTTGTAAACAGAAATTACACAGTTAAATATGA-**BHQ**

HEX and FAM = fluorophores, BHQ = black hole quencher

2.2. Nucleic Acid Extraction

1. NucliSens® easyMAG™ (bioMérieux, Basingstoke, UK) incl. all buffers and plastic consumables.
2. 1.5 ml microcentrifuge tubes.
3. Nuclease free water.

2.3. Mastermix

1. Make primers and probes (Eurogentec, Southampton, UK) up to required stock concentrations in nuclease free water (see Table 1).
2. HotStar Taq DNA Polymerase (QIAGEN, Crawley, UK), which contains 15 mM $MgCl_2$.
3. dNTP mix made to 2 mM in nuclease free water.
4. $MgCl_2$ 25 mM.
5. 1.5 ml microcentrifuge tubes.
6. Axygen PCR microplates.
7. Optical adhesive film to seal plates.

2.4. Real-Time PCR

1. Applied Biosystems 7500 Real-Time PCR System.

3. Methods

3.1. Sample Preparation

This assay is performed directly from skin/wound swabs.

1. Use a sterile loop to pick a single colony of the positive control and suspend in saline (3 ml).
2. Place the swabs in saline and agitate for 10 s to suspend bacteria.

3.2. Nucleic Acid Extraction

Extraction is carried out using the NucliSens® easyMAG™ according to the manufacturer's instructions. This is an automated platform based on the Boom method (13) that utilizes magnetic silica to increase washing efficiency (see Note 1).

1. Switch on the apparatus and define the extraction requests as follows:

Sample ID	Enter the sample number
Matrix	Set as "other"
Protocol	Keep as default, Generic 2.0.1
Volume (ml)	Set to 0.2
Eluate (μl)	Set to 100
Type	Set as "primary" for onboard lysis
Priority	Set as "normal"

2. Load 200 μl of each sample in to the appropriate well in the sample strip and click to dispense lysis buffer.
3. Following the 10 min lysis incubation, for each 8-well sample strip take a vial of magnetic silica and add 550 μl of nuclease free water to a 1.5 ml microcentrifuge tube and 550 μl of magnetic silica vortexing thoroughly.
4. Add 100 μl of 1:1 silica and water to each sample well ensuring adequate mixing and start the run.
5. At the end of the run, pipette the extracts into labelled microcentrifuge tubes.

3.3. Mastermix

1. Thaw reagents as required and keep on ice.
2. Make enough mastermix for samples and controls plus 10% extra (see Note 2). Primers and probe sequences and the volumes of reagents required are described in Tables 1 and 2.
3. Dispense 15 μl mastermix in the appropriate wells of the PCR microplate.
4. Add 10 μl extract to each appropriate tube and seal with the adhesive film.

3.4. Real-Time PCR

The assay uses primers and TaqMan probes to detect MRSA and the PVL genes. The probes are 5′ labelled with different fluorescent reporter dyes (FAM and HEX), and 3′ labelled with BHQ-1, a nonfluorescent quencher dye. In the absence of target, fluorescence is quenched by fluorescence resonance energy transfer (FRET). In the presence of target, during amplification the probes anneal to their complementary single strand DNA sequences and are subsequently degraded by the 5′–3′ exonuclease activity of *Taq* DNA polymerase enzyme, separating the quencher from the reporter.

Table 2
Volumes of reagents required per reaction and the final concentrations used

	Volume	Final concentration
10× Buffer	2.5 ml	1×
dNTP (2 mM)	2.5 ml	200 μM
$MgCl_2$ (25 mM)	3.0 ml	4.5 mM
Taq (5 U/μl)	0.25 ml	1.25 U
mecii	0.125 ml	0.5 μM
meciii	0.125 ml	0.5 μM
meciv	0.125 ml	0.5 μM
mecv	0.125 ml	0.5 μM
mecvii	0.125 ml	0.5 μM
orfX	0.125 ml	0.5 μM
MRSAprobe	0.35 ml	0.35 μM
PVLF	0.2 ml	0.4 μM
PVLR	0.4 ml	0.8 μM
PVLprobe	0.5 ml	0.1 μM
dH_2O	4.55 ml	

Due to the release of the quenching effect on the reporter, the fluorescence intensity of the reporter dye increases and is measured during each cycle by the real-time PCR machine.

1. The following protocol is intended for use with the 7500 Real-Time PCR System from Applied Biosystems (see Note 3).
2. Before performing a run, create an SDS file template with the appropriate detectors and thermal cycling conditions (see Note 4). Switch on the PCR machine and open the Applied Biosystems 7500 System SDS software. Select File > New to open the New Document Wizard. In the Assay drop-down list select Standard Curve (Absolute Quantitation) (see Note 5). Click Next, followed by New Detector. Enter MRSA in the Name field and select FAM and BHQ-1 from the Reporter Dye and Quencher Dye drop-down lists, respectively (see Note 6). Click OK to save the detector information and repeat for the PVL target, except select HEX and BHQ-1 for the Reporter Dye and Quencher Dye respectively. Select both detectors from the list and click Add. Select Next. Highlight all plate wells and select the detectors and tasks. Click Finish to create the plate

document. To specify thermal cycling conditions select the Instrument Tab and 95°C for 15 min, followed by 50 cycles of 95°C 20 s, 60°C 40 s, and 72°C 20 s. Set sample volume to 25 μl and Stage 2, step 2 (60°C at 0:40) for Data Collection. Save the file as a SDS Template using an appropriate name such as MRSA_PVL template.

3. To run samples select File > New > Browse and select the MRSA_PVL template. A plate should appear with two detectors in each well. Load samples into the machine and enter the sample names and controls in the appropriate wells on the plate setup. Save the plate document (see Note 7), select the Instrument tab, and click Start.
4. Once the run has finished the data can be analyzed (see Note 8). The analysis settings can be set automatically; however, to obtain optimal results it is often best to manually set the baseline and threshold values, which must be done for each detector.
5. Select all test and control wells, and then click on the Results tab, followed by the Amplification Plot tab. Select Delta Rn vs Cycle from the Data drop-down list. To set the baseline for the MRSA detector, select MRSA from the Detector drop-down list and then enter the Start cycle and End cycle values (see Note 9). Set the threshold value by clicking on the red line and dragging it to the middle of the exponential region on the graph. Click Analyse.
6. Repeat this process for the PVL detector. Once both detectors have been analyzed, select All from the Detector drop-down list on the right hand side of the screen. Select the Report tab to view results, which may be printed or exported by using the Report Settings tool.
7. To ensure the test is valid check that the controls have worked. Positive and negative controls are included in each run to monitor the assay performance (see Note 10). An invalid control invalidates the assay run. If the controls have worked the results may be interpreted as shown in Table 3. Figure 1 shows amplification plots of a typical run with positive and negative samples (see Note 11). The graph shows the delta Rn against cycle number. Delta Rn is calculated by the software and represents the normalized reporter minus the baseline. The amplification curve should be sigmoidal in shape. If irregular amplification curves are observed this may be due to incorrect baseline and threshold settings, which can be modified manually as mentioned earlier. Amplification efficiency can also be checked by testing a tenfold serial dilution of the positive control in triplicate; the reaction is 100% efficient when the slope of the standard curve is −3.32 (±10% of this value is typically acceptable).

Table 3
Real-time PCR result interpretation

MRSA C_t value[a]	MRSA Result	PVL C_t value[a]	PVL Result	Interpretation
≤40	Positive	≤40	Positive	PVL-positive MRSA present
≤40	Positive	Not Detected	Negative	PVL-negative MRSA present
Not detected	Negative	Not detected	Negative	No MRSA present
Not detected	Negative	≤40	Positive	PVL genes detected but no MRSA present[b]

[a]Samples with C_t values >40 are classified as equivocal and the test should be repeated or a repeat swab should be requested

[b]PVL genes can be found in meticillin-sensitive *Staphylococcus aureus*, which may be present in the clinical samples

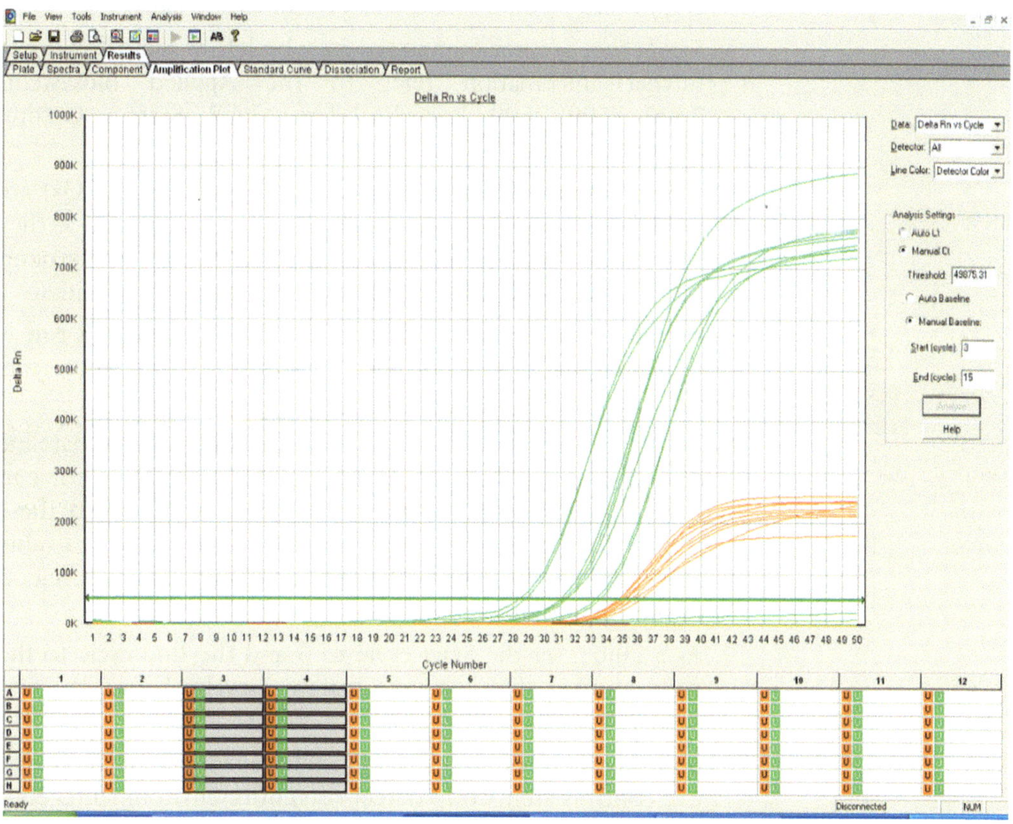

Fig. 1. Typical amplification plot of MRSA and PVL multiplex assay. The *green* and *orange lines* are fluorescent signals from the FAM-labelled MRSA probe and the HEX-labelled PVL probe, respectively. Positive samples show increases in fluorescent above the threshold (*green horizontal line*).

4. Notes

1. It is possible to use other automated or manual extraction methods that are suitable for bacteria.
2. To save time and reduce pipetting errors a 10× primer/probe mastermix can be prepared.
3. This protocol is intended for use with Applied Biosystems Real-time PCR machines; however, other platforms such as the Roche Applied Science LightCycler 2.0 or 480, Bio-Rad Laboratories iQ5, Stratagene Multiplex qPCR Systems, the Corbett Research Rotor-Gene Systems, and the Cepheid SmartCycler TD System may be used.
4. When programming the real-time PCR machine it is essential that the correct detectors are selected for the reporter dyes. By creating a SDS template, the appropriate detectors and amplification conditions are stored for future runs.
5. Fields that are not mentioned may be left as the default. For further information refer to the Applied Biosystems 7300/7500/7500 Fast Real-Time PCR System Getting Started Guide.
6. It is recommended that Black Hole Quenchers (BHQs) are used rather than fluorescent quenchers (such as Tamra) as they emit heat rather than fluorescence so lower background fluorescence signals are produced improving assay sensitivity.
7. A plate document stores data collected from each PCR run as well as all run information such as the sample names, detectors and amplification conditions.
8. To provide accurate data, it is vital that the optimal baseline settings and threshold values are used for each reporter dye in every run. The baseline is the initial cycles of PCR where there is little change in fluorescence, whilst the threshold is a value above the baseline and within the exponential growth region of the amplification curve.
9. As a guide, set the Start cycle to 3 and the End cycle to the cycle number that is approximately 3 cycles before the first positive appears.
10. A positive control is used to monitor reagent failure, and negative controls are used to detect reagent or environmental contamination or carryover of MRSA DNA or MRSA amplicons.
11. The assay will detect MRSA carrying SCC*mec* types I–IV. Also, *S. aureus* isolates that have lost the *mec*A gene but still carry SCC*mec* elements may be detected using this assay (14).

References

1. Vandenesch F, Naimi T, Enright MC, Lina G, Nimmo GR, Heffernan H, Liassine N, Bes M, Greenland T, Reverdy ME, Etienne J (2003) Community-acquired methicillin-resistant *Staphylococcus aureus* carrying Panton-Valentine leukocidin genes: worldwide emergence. Emerg Infect Dis 9:978–984
2. Hardy K, Price C, Szczepura A, Gossain S, Davies R, Stallard N, Shabir S, McMurray C, Bradbury A, Hawkey PM (2010) Reduction in the rate of methicillin-resistant *Staphylococcus aureus* acquisition in surgical wards by rapid screening for colonisation: a prospective, cross-over study. Clin Microbiol Infect 16(4): 333–339
3. Jog S, Cunningham R, Cooper S, Wallis M, Marchbank A, Vasco-Knight P, Jenks PJ (2008) Impact of preoperative screening for methicillin-resistant *Staphylococcus aureus* by real-time polymerase chain reaction in patients undergoing cardiac surgery. J Hosp Infect 69:124–130
4. Keshtgar MR, Khalilli A, Coen PG, Carder C, Macrae B, Jeanes A, Folan P, Baker D, Wren M, Wilson AP (2008) Impact of rapid molecular screening for methicillin-resistant *Staphylococcus aureus* in surgical wards. Br J Surg 95: 381–386
5. Huletsky A, Giroux R, Rossbach V, Gagnon M, Vaillancourt M, Bernier M, Gagnon F, Truchon K, Bastien M, Picard FJ, van Belkum A, Ouelle M, Roy PH, Bergeron MG (2004) New real-time PCR assay for the rapid detection of methicillin resistant *Staphylococcus aureus* directly from specimens containing a mixture of staphylococci. J Clin Microbiol 42:1875–1884
6. Prevost G, Cribier B, Couppie P, Petiau P, Supersac G, Finck-Barbancon V, Monteil H, Piemont Y (1995) Panton-Valentine leucocidin and gamma-hemolysin from *Staphylococcus aureus* ATCC 49775 are encoded by distinct genetic loci and have different biological activities. Infect Immun 63:4121–4129
7. Prevost GL, Mourey L, Colin DA, Menestrina G (2001) Staphylococcal pore-forming toxins. Curr Top Microbiol Immunol 257:53–83
8. Ellington MJ, Perry C, Ganner M, Warner M, McCormick Smith I, Hill RL, Shallcross L, Khatri K, Sabersheikh S, Holmes A, Cookson BD, Kearns AM (2009) Clinical and molecular epidemiology of ciprofloxacin-susceptible MRSA encoding PVL in England and Wales. Eur J Clin Microbiol Infect Dis 28: 1113–1121
9. Seybold U, Kourbatova EV, Johnson JG, Halvosa SJ, Wang YF, King MD, Ray SM, Blumberg HM (2006) Emergence of community-associated methicillin-resistant *Staphylococcus aureus* USA 300 genotype as a major cause of health care—associated blood stream infections. Clin Infect Dis 42:647–666
10. Gonzalez BE, Rueda AM, Shelburne SA, Musher DM, Hamill RJ, Hulten KG (2006) Community-associated strains of methicillin-resistant *Staphylococcus aureus* as the cause of healthcare-associated infection. Infect Control Hosp Epidemiol 27:1051–1056
11. McDonald RR, Antonishyn NA, Hansen T, Snook LA, Nagle E, Mulvey MR, Levett PN, Horsman GB (2005) Development of a triplex real-time PCR assay for detection of Panton-Valentine leukocidin toxin genes in clinical isolates of methicillin-resistant *Staphylococus aureus*. J Clin Microbiol 43:6147–6149
12. Renwick L, Hardie A, Girvan EK, Smith M, Leadbetter G, Claas E, Morrison D, Gibb AP, Dave J, Templeton KE (2008) Detection of meticillin-resistant *Staphylococcus aureus* and Panton-Valentine leukocidin directly from clinical samples and the development of a multiplex assay using real-time polymerase chain reaction. Eur J Clin Microbiol Infect Dis 27:791–796
13. Boom R, Sol CJ, Salimans MM, Jansen CL, Wertheim-van Dillen PM, van der Noordaa J (1990) Rapid and simple method for extraction of nucleic acids. J Clin Microbiol 28: 495–503
14. Shore AC, Rossney AS, O'Connell B, Herra CM, Sullivan DJ, Humphreys H, Coleman DC (2008) Detection of staphylococcal cassette chromosome *mec*-associated DNA elements in multiresistant methicillin-susceptible *Staphylococcus aureus* (MSSA) and identification of *Staphylococcus epidermidis* ccrAB4 in both methcillin-resistant *S. aureus* and MSSA. Antimicrob Agents Chemother 52:4407–4419

Chapter 7

PCR Detection of *Haemophilus influenzae* from Respiratory Specimens

Guma M.K. Abdeldaim and Björn Herrmann

Abstract

The detection of *Haemophilus influenzae* by conventional methods like culture is time-consuming and may give false-negative results, especially during ongoing antibiotic treatment. Therefore, non-culture based methods that are sensitive, specific, and rapid are valuable for early diagnosis and effective therapy. Here we describe a quantitative real-time PCR assay based on the outer membrane P6 gene *omp6*, to detect *H. influenzae* and its application on respiratory tract specimens.

Key words: *Haemophilus influenzae*, Outer membrane P6, Quantitative PCR, Real-time PCR, Pneumonia

1. Introduction

Haemophilus influenzae is a common pathogen causing a variety of infections, including respiratory tract infections and meningitis. Detection of *H. influenzae* by conventional methods like culture is cheap and facilitates antibiotic resistance detection; however, it is time-consuming and may give false-negative results especially during ongoing antibiotic treatment. Therefore, there is a need for non-culture methods that are sensitive, specific and rapid for early diagnosis and effective therapy. In the last decade an abundance of nucleic acid-based assays have been developed for the detection of *H. influenzae*.

The genus *Haemophilus* is characterized by a promiscuous ability to exchange genetic material by transformation and recombination, both between strains within species (1–3) and between species (4). This genetic exchange can lead to high strain to strain variation

Mark Wilks (ed.), *PCR Detection of Microbial Pathogens: Second Edition*, Methods in Molecular Biology, vol. 943,
DOI 10.1007/978-1-60327-353-4_7,

in the gene content within the same species. Thus, delineation of *H. influenzae* is still reported as an unresolved challenge when MLST, phenotyping and detection of marker genes was used (5). As the specificity of PCR depends on the selection of target genes, it is therefore a challenge to design specific PCR for the detection of *H. influenzae*. The other challenge is to distinguish the disease causing *H. influenzae* infections from colonization, especially in the case of respiratory tract infection, since *H. influenzae* can be found as normal flora in the upper respiratory tract. As has been shown for *Streptococcus pneumoniae*, the concentration of *H. influenzae* in the respiratory tract will probably be higher during infection causing disease than during carriage, and therefore, quantitative methods will probably improve the diagnostic quality (6).

We have developed a sensitive and fairly specific quantitative real time PCR based on the outer membrane protein P6 (*ompP6*) (7) (see Note 1). The specificity of the assay was determined by analysis of 29 strains of 11 different *Haemophilus* spp. and was compared with PCR assays having other target genes: *rnpB*, *16S rRNA*, and *bexA*. The analytical sensitivity of the assay is 3–30 genome copies per reaction tube. The method was evaluated on nasopharyngeal aspirates from 166 adult patients with community-acquired pneumonia and from 84 adult controls without respiratory symptoms. When 10^4 DNA copies/mL was used as cutoff limit for the method, *ompP6* PCR had a sensitivity of 97.5% and a specificity of 96.0% compared with the culture. Furthermore, the method was combined with quantitative PCR assays for the detection of *S. pneumoniae* targeted against the Spn9802 DNA fragment (6) and *Neisseria meningitidis* targeting the *ctrA* gene (8) in a multiplex PCR format. Evaluation was performed on 156 bronchoalveolar lavage (BAL) samples from patients with lower respiratory tract infection (heavy smokers and/or under antibiotic treatment) and 31 BAL from control patients. When 10^5 DNA copies/mL was used as cutoff limit for the method, the sensitivity of the *ompP6* PCR decreased from 93 to 63% while the specificity increased from 96 to 100% compared with the culture and *fucK* PCR (data not yet published). The use of this QPCR on respiratory specimens should assist other workers in the differentiation of carriage from disease.

2. Materials

2.1. P6 Primers and Probe

One of the most important factors in successful PCR is proper primer design. Primers that only amplify one product will provide the best assay sensitivity and specificity. In this study, the chosen target genes for *H. influenzae* were first tested by *in silico* exploration using the online software BLAST http://blast.ncbi.nlm.nih.gov/Blast.cgi.

Primers and probes were evaluated using the online software Oligo Analyzer 3.0 http://www.idtdna.com/analyzer/Applications/OligoAnalyzer/. Probes containing locked nucleic acids (LNAs) were evaluated using the online software http://oligo.lnatools.com/expression/ (see Note 2).

From a previously sequence determined *omp* P6 gene (*H. influenzae*, GenBank accession no. M19391), the following oligonucleotides were chosen:

1. Oligonucleotide primers: Forward primer (Hi P6 F) 5′-CCA GCT GCT AAA GTA TTA GTA GAA G-3′ (position 302–326) and reverse primer (Hi P6 R) 5′-TTC ACC GTA AGA TAC TGT GCC-3′ (position 477–457) defined an amplicon of 156 bp.
2. For detection, a JOE-labeled probe with black hole quencher was designed: Hi P6 pro 5′-JOE-CAg ATg CAg TTg AAg GTt Att tAG-BHQ1-3′.

To increase the Tm of the probe, we used a locked nucleic acid probe (9), here denoted as positions with lower case letters (see Note 3). Primers and probe were purchased from Eurogentec (Seraing, Belgium) (see Note 4).

2.2. Master Mix Reagents

1. HotStarTaq DNA Polymerase kit Cat (QIAGEN).
2. Deoxyribonucleotide triphosphate (dNTP) Mix. dNTP Mix is a premixed solution containing sodium salts of dATP, dCTP, dGTP, and dTTP (Roche). Store all reagents at −20°C. Avoid exposure to frequent temperature changes.
3. Nuclease-Free Water Cat (Ambion). Store at room temperature.

2.3. DNA Extraction

BAL: Centrifuge at 1,000×*g* for 10 min and remove the supernatant, but leave about 1 mL to resuspend the pellet. Dispense 200 μL sample to two micro-tubes, of which one will be spiked with a positive control (see Subheading 2.4.1).

Sputum: Treat the sample with sputolysin solution (Calbiochem). Dithiothreitol (DTT) should be diluted in 0.1 M phosphate buffer pH 6.8 to a final concentration of 0.9%. The solution is stable for 2 days at 4°C. Mix by vortexing, wait for 5 min, and vortex again.

QIAamp® DNA Mini Kit Cat (see Note 5):

1. QIAamp spin columns
2. Collection tubes (2 mL)
3. Buffer AL
4. Buffer AW1 (concentrate)
5. Buffer AW2 (concentrate)

6. Buffer AE
7. Proteinase K

2.4. Controls and DNA Standard

2.4.1. Positive Controls

Positive control is important for quality assurance of the PCR assay. There are two types of positive controls:

(a) Extraction control in which a suspension of target bacterial cells is used in parallel with the clinical samples during the DNA preparation and the subsequent PCR detection step. The function of the extraction control is to monitor efficiency of the whole procedure (DNA extraction and PCR amplification).

(b) DNA amplification control in which purified DNA of target bacteria is added directly to PCR reaction tubes. This functions as a technical control of PCR reaction by monitoring DNA amplification. Target DNA may also be used to spike a duplicate reaction tube for each clinical sample, and function as a monitor of inhibition.

2.4.2. Negative Controls

A negative control is also important for quality assurance of the PCR assay. There are two types of negative controls:

(a) Extraction control in which sterile water (non-template) is used in parallel with the clinical samples during the DNA preparation and the subsequent PCR detection step. The function of the extraction control is to check if there is a contamination during the DNA extraction procedure.

(b) DNA amplification control in which sterile water is added directly to PCR reaction tubes. The function is to check if the primers and/or master mix reagents are contaminated.

2.4.3. DNA Standard for Quantification

To obtain a reliable standard curve for quantification at least three measuring points (concentrations) in duplicate is needed. To obtain reproducible quantifications concentrations of more than 1,000 target gene copies per reaction is recommended, e.g., 10^3, 10^4, and 10^5.

3. Methods

3.1. Primers and Probe Preparation

Primers and probe must be prepared (rehydrated) according to manufacturer's description, and there are some recommendations that must be considered during the preparation (see Note 6).

3.2. PCR Amplification and Quantification

To avoid contamination, the PCR master mix should be prepared in a clean room (i.e., room that is not used for DNA preparation or culture) (see Note 7). Sample and control DNA is added to tubes with dispensed master mix in a separate room with a UV equipped PCR cabinet.

The optimized real-time PCR amplifications are performed in 25-μl reactions containing:

2.5 μL of 10× PCR Buffer (final concentration 1×).

2.5 μL of 2 mM dNTP (final concentration 0.2 mM).

2 μL of 25 mM $MgCl_2$ (final concentration 3.5 mM including $MgCl_2$ in PCR buffer).

0.2 μL of HotStar Taq polymerase (final concentration 1 U).

0.75 μL of each primer (final concentration 0.3 μM).

0.25 μL of the probe (final concentration 0.2 μM).

11.8 μL of nuclease-free H_2O (Ambion®).

A total of 5 μL of target DNA is used in the assay.

The PCR cycling has the following program: 15 min of enzyme activation at 95°C, followed by 45 cycles at 95°C for 15 s and, 60°C for 40 s. The assay was performed using Qiagen/Corbett Rotor-Gene® 3000 (Qiagen, Hilden, Germany), and the time taken for the assay was 90 min.

Standard curves for quantification were based on duplicates of three measuring points containing 500, 2,000, and 10,000 genome copies of the P6 target.

3.3. Clinical Samples

Different types of samples can be used for the detection of *H. influenzae* from the respiratory tract. Sputum is a noninvasive lower respiratory tract sample. Sputum samples are often contaminated by the oropharyngeal flora. In order to reduce the risk of false positive sputum culture results, a cutoff limit of 10^5 colony-forming units (CFU/mL) has been applied. In cases when no sputum sample can be obtained, nasopharyngeal aspirates (NPA) or swabs from patients with pneumonia can be used for detection of *H. influenzae* in populations with expectedly low carriage rates of these bacteria. If available, bronchoalveolar lavage (BAL) is useful and is a specimen type with minor contamination problems. However, as the bronchoscope passes the pharynx the oropharyngeal flora may contaminate the lower respiratory tract. Thus, to differentiate between infection and colonization, a cutoff limit of 10^4 CFU/mL may be used for BAL culture. In order to completely exclude the risk of contamination from the pharynx, transtracheal aspiration or transthoracic needle aspiration can be performed, but this technique has been associated with complications. It should be remembered that CFU/mL values may not be equivalent to copy/mL figures, as bacteria may aggregate into clumps so that one colony is derived from a large number of bacteria, thus underestimating the number of bacteria present.

Robust cutoff limits are desirable, but difficult to define. As described above several factors influence the sensitivity and specificity, and for a specific assay the specimen type, antibiotic treatment, and patient population must be considered when elaborating cutoff levels. In addition it is not easy to determine a correct quantitative DNA-standard due to limitations in the accuracy of DNA concentration determination.

3.4. DNA Extraction from Clinical Samples

DNA preparation is a crucial step when performing quantitative PCR. The presence of inhibiting substances in the sample may lead to a complete absence of amplification or affect the quantification of PCR products. Numerous methods are available for DNA preparation, either manual like Qiamp DNA mini kit (Qiagen, Hilden, Germany) or automatic systems like the QIAsymphony (Qiagen), the MagNa Pure LC DNA-Isolation System (Roche Diagnostics), and the NucliSens easyMAG instrument (Biomérieux).

The manual Qiamp DNA mini Kit works well and is described below:

1. Pipette 20 μL Proteinase K into the bottom of a 1.5-mL microcentrifuge tube.
2. Add 200 μL sample to the microcentrifuge tube (see Note 8).
3. Add 200 μL Buffer AL to the sample and mix by vortexing for 15 s (see Note 9).
4. Incubate at 56°C for 10 min.
5. Briefly centrifuge the 1.5-mL microcentrifuge tube to remove drops from the inside of the lid.
6. Add 200 μL ethanol (96–100%) to the sample and mix by vortexing for 15 s. After mixing, briefly centrifuge to remove drops from the inside of the lid.
7. Carefully apply the mixture from step 6 to a QIAamp spin column (in a clean 2-mL collection tube) without wetting the rim, close the cap, and centrifuge at 6,000 × *g* for 1 min. Place the QIAamp spin column in a clean 2-mL collection tube and discard the tube containing the filtrate.
8. Carefully open the QIAamp spin column and add 500 μL Buffer AW1 without wetting the rim. Close the cap and centrifuge at 6,000 × *g* for 1 min. Place the QIAamp spin column in a clean 2-mL collection tube and discard the tube containing the filtrate.
9. Carefully open the QIAamp spin column and add 500 μL Buffer AW2 without wetting the rim. Close the cap and centrifuge at full speed (20,000 × *g*; 14,000 rpm) for 3 min.
10. Place the QIAamp spin column in a clean 1.5-mL microcentrifuge tube and discard the tube containing the filtrate. Carefully open the QIAamp spin column and add 200 μL Buffer AE or distilled water. Incubate at room temperature for 1 min and centrifuge at 6,000 × *g* for 1 (see Note 10).

3.5. DNA Concentration Determination

The concentration of purified DNA used as standard for quantification was measured in a NanoDrop spectrophotometer (NanoDrop Technologies, Wilmington, DE), and the number of genome copies was calculated according to the following formula: Number of copies = ($C \times 6{,}023 \times 10^{23}$)/(Genome weight $\times 10^{-9}$).

C = concentration in ng/μL

4. Notes

1. The specificity of the P6 PCR and other PCR assays having other target genes (*rnpB*, *16S rRNA* and *bexA*) was tested on DNA preparation from the following *Haemophilus* strains: *H. influenzae* type a (CCUG 6881), *H. influenzae* type b (CCUG 23946), *H. influenzae* type b (CCUG 15195), *H. influenzae* type c (CCUG 6879), *H. influenzae* type d (CCUG 6878), *H. influenzae* type e (CCUG 6877), *H. influenzae* type f (CCUG 15435), *H. influenzae biovar I* (CCUG 45442), *H. influenzae biovar II* (CCUG 23945^T), *H. influenzae biovar II* (CCUG 45156), *H. influenzae biovar III* (CCUG 35407), *H. influenzae biovar V* (CCUG 36704), *H. parainfluenzae* (CCUG 8259), *H. parainfluenzae* (CCUG 12836^T), *H. parainfluenzae* (CCUG 44486), *H. parainfluenzae* (CCUG 44743), *H. parainfluenzae* (CCUG 45191), *H. parainfluenzae* (CCUG 7596), *Haemophilus intermedius* subsp. *intermedius* (CCUG 11096), *H. intermedius* subsp. *intermedius* (CCUG 32367), *H. haemolyticus* (CCUG 12834^T), *H. parahaemolyticus* (CCUG 3716^T), *H. parahaemolyticus* (CCUG 51599), *H. aegyptius* (CCUG 25716^T), *Haemophilus* cryptic genospecies (CCUG 31340), *H. pittmaniae* (CCUG 48703^T), *Aggregatibacter aphrophilus* (CCUG 3715^T), *Aggregatibacter segnis* (CCUG 10787^T), *Aggregatibacter segnis* (CCUG 46700), and 7 clinical isolates of *H. parainfluenzae.*

   Our results show that all the target genes have limitations and the P6 PCR gave the best results, since it detected all 12 strains of *H. influenzae* even though it also detected three other species (*H. aegyptius*, *Haemophilus* cryptic genospecies, and *H. haemolyticus*) (7). In that analysis *H. intermedius subsp. intermedius* was also detected by the P6 PCR, upon retesting the two strains with fresh isolates obtained directly from CCUG, this species was found to be negative in the P6 PCR.

2. When designing primer and probe, the length should normally be between 18 and 30 nucleotides, while the G-C content should be between 20 and 80%. Primer and probe self- and inter-complementation should be avoided. The melting temperature (T_m) of the probe should be several degrees above the

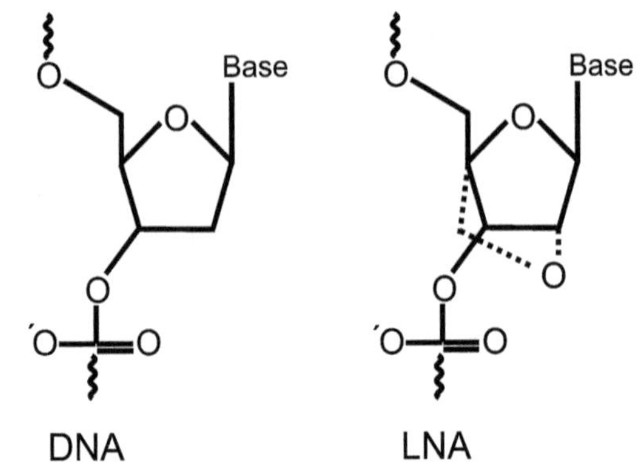

Fig. 1. Chemical structure of 2′-O,4′-C-methylene linked LNA residues.

T_m of the primers. Gs on the 5′ end of the probe should be avoided and a strand should be selected that gives the probe more Cs then Gs

3. Locked nucleic acid (LNA) is a class of nucleic acid containing nucleosides whose major distinguishing characteristic is the presence of a methylene bridge that connects the 2′-oxygen of ribose with the 4′-carbon of the ribose ring (Fig. 1). Usually, LNA/DNA duplexes have increased thermal stability (3–8°C per modified base in the oligonucleotide) compared with similar duplexes formed by DNA alone. Therefore, even short conserved stretches could be considered for probe or primer design (Fig. 1).

4. It is often difficult to find manufacturers who can reproducibly synthesize oligonucleotides of high quality. This is especially important for probes.

5. Buffer AL and Buffer AW1 are irritant; Buffer AW2 is highly toxic; QIAamp spin columns and buffers can be stored dry at room temperature (15–25°C); ready to use Proteinase K is stable for up to 1 year when stored at room temperature. To prolong the lifetime of Proteinase K, storage in aliquots at 2–8°C is recommended.

6. Preparing stock solutions of the primers and probe at 100 μM concentration and aliquoting them into several tubes is strongly recommended.

   From the 100 μM primers and probe stock solutions, make working solutions with concentration of 10 μM and aliquot in many tubes, e.g., 30 μL per tube.

   Stock solutions must be stored at −70°C, while working solutions can be stored at −20°C.

   Each tube must not be exposed to more than three cycles of freezing and thawing.

7. Before starting, leave primers, probe and dNTP at room temperature for a few minutes to thaw them.

   Keep the probe away from direct sunlight.

   The HotStar *Taq* polymerase must be kept at −20°C and since it contains glycerol which keeps it liquid even at −20°C, it can be removed from the freezer, used immediately, and returned to the freezer promptly.

8. If the sample volume is less than 200 μL, make up to 200 μL with PBS.

9. Do not add Proteinase K directly to Buffer AL.

10. To increase the yield of DNA, incubate the QIAamp spin column loaded with Buffer AE for 5 min at room temperature.

    Elution in 50 μL Buffer AE is recommended if samples contain less than 1 μg of DNA. In normal clinical specimens as NPAs and BAL the amount of DNA is supposed to be <1 μg, sputum samples may be diffferent, but 50 μL buffer works well according to our experience.

References

1. Kroll JS, Moxon ER (1990) Capsulation in distantly related strains of *Haemophilus influenzae* type b: genetic drift and gene transfer at the capsulation locus. J Bacteriol 172: 1374–1379
2. Kroll JS, Moxon ER, Loynds BM (1994) Natural genetic transfer of a putative virulence-enhancing mutation to *Haemophilus influenzae* type a. J Infect Dis 169:676–679
3. Mukundan D, Ecevit Z, Patel M, Marrs CF, Gilsdorf JR (2007) Pharyngeal colonization dynamics of *Haemophilus influenzae* and *Haemophilus haemolyticus* in healthy adult carriers. J Clin Microbiol 45:3207–3217
4. Kroll JS, Wilks KE, Farrant JL, Langford PR (1998) Natural genetic exchange between Haemophilus and Neisseria: intergeneric transfer of chromosomal genes between major human pathogens. Proc Natl Acad Sci USA 95:12381–12385
5. Norskov-Lauritsen N, Overballe MD, Kilian M (2009) Delineation of the species *Haemophilus influenzae* by phenotype, multilocus sequence phylogeny, and detection of marker genes. J Bacteriol 191:822–831
6. Abdeldaim GM, Stralin K, Olcen P, Blomberg J, Herrmann B (2008) Toward a quantitative DNA-based definition of pneumococcal pneumonia: a comparison of *Streptococcus pneumoniae* target genes, with special reference to the Spn9802 fragment. Diagn Microbiol Infect Dis 60:143–150
7. Abdeldaim GM, Stralin K, Kirsebom LA, Olcen P, Blomberg J, Herrmann B (2009) Detection of *Haemophilus influenzae* in respiratory secretions from pneumonia patients by quantitative real-time polymerase chain reaction. Diagn Microbiol Infect Dis 64:366–373
8. Taha MK, Alonso JM, Cafferkey M, Caugant DA, Clarke SC, Diggle MA, Fox A, Frosch M, Gray SJ, Guiver M, Heuberger S, Kalmusova J, Kesanopoulos K, Klem AM, Kriz P, Marsh J, Molling P, Murphy K, Olcen P, Sanou O, Tzanakaki G, Vogel U (2005) Interlaboratory comparison of PCR-based identification and genogrouping of *Neisseria meningitidis*. J Clin Microbiol 43:144–149
9. Braasch DA, Corey DR (2001) Locked nucleic acid (LNA): fine-tuning the recognition of DNA and RNA. Chem Biol 8:1–7

Chapter 8

Rapid Detection of Atypical Respiratory Bacterial Pathogens by Real-Time PCR

Clare L. Ling and Timothy D. McHugh

Abstract

The accurate diagnosis of lower respiratory tract infection remains a challenge, with conventional diagnostic methods often failing to identify a causative agent. Here we describe a multiplex real-time PCR assay that has been validated for the detection of the rarely identified atypical bacterial pathogens: *Legionella pneumophila*, *Mycoplasma pneumoniae*, and *Chlamydophila pneumoniae*. Due to the complexity and poor performance of the existing cultural and serological methods for the detection of these pathogens, investigation for them is rarely initiated. This is likely to result in underdetection and empirical treatment regardless of a microbiological diagnosis. The assay described here is designed to address this need.

Key words: *Legionella pneumophila*, *Mycoplasma pneumoniae*, *Chlamydophila pneumoniae*, Real-time PCR, Multiplex, Dual-labeled probes

1. Introduction

Respiratory infection is the most common reason for patients to present for a primary care consultation and approximately one-third of these consultations relate to lower respiratory tract infection (LRTI) (1–3). Despite the clinical imperative for accurate diagnosis, the etiology of LRTI is rarely established and the detection of causative organisms limited (detection rate 16–55%; (4–6)). We previously showed that, using a combination of molecular tools and conventional microbiology, the yield of potential pathogens was improved to 69%, demonstrating the value of nucleic acid amplification assays in closing the "diagnostic gap" (2). In this study in a primary care setting, the "atypical"

Mark Wilks (ed.), *PCR Detection of Microbial Pathogens: Second Edition*, Methods in Molecular Biology, vol. 943,
DOI 10.1007/978-1-60327-353-4_8,

respiratory bacterial pathogens were rarely identified. However, as they cannot be distinguished clinically from other infections (2, 7), it is essential that they are tested for in a timely fashion in order to enable appropriate patient management and reduce morbidity. In the cases of *Legionella pneumophila* and *Mycoplasma pneumoniae* infections, it is important to rapidly identify potential outbreaks and implement the required public health measures.

L. pneumophila, *M. pneumoniae*, and *Chlamydophila pneumoniae* represent a challenge to conventional microbiological diagnosis, usually requiring a specific request from the reporting physician or the clinical microbiologist. They are all fastidious organisms and culture is only routinely used to diagnose *L. pneumophila*; however, it is slow (3–7 days) and insensitive. Urinary antigen detection tests are more sensitive than culture for *L. pneumophila* detection, but they only detect serogroup 1. Serological methods are relied upon for the diagnosis of *M. pneumoniae* and *C. pneumoniae*; however the demonstration of the presence or the absence of IgM can be misleading and the requirement for paired sera to demonstrate a rising IgG titer enables only a retrospective diagnosis (8–11).

Due to the complexity and poor reliability of the existing methods for the microbiological diagnosis of these pathogens, clinicians are often reluctant to initiate these investigations. This is likely to result in underdetection and, where atypical infection is suspected, empirical treatment regardless of microbiological data. It is in this context that we established a multiplex real-time PCR protocol for their timely detection. The multiplex assay utilizes dual-labeled probes, known as Taqman® probes. These are fluorescently labeled with different fluorophores (dyes) allowing multiple targets to be tested in one assay, resulting in reduced hands-on time and costs. Due to the nature of the assay it can easily be performed in any laboratory with a real-time PCR platform. The multiplex PCR described here targets the *L. pneumophila mip* gene, *M. pneumoniae P1* adhesin gene, and *C. pneumoniae Pst*-1 fragment. The product sizes are 102, 117, and 82 bp, respectively. To test for inhibitors, an extra reaction is set up for each sample which is spiked with *L. pneumophila* DNA; this reaction needs to be positive for a negative result to be accepted. The analytical sensitivity of the assay is 0.1 pg for each target and the assay was found to be 100% specific when challenged with 1 ng DNA from a range of organisms. This methodology can be adapted to detect other etiologies of respiratory tract infections including bacterial and viral, resulting in a single and comprehensive approach to diagnosing respiratory tract infections.

2. Materials

2.1. Specimen Preparation

1. Dithiothreitol, e.g., Sputasol (Oxoid, UK).

2.2. DNA Extraction

1. Phosphate-buffered saline (PBS, 1×).
2. Chelex 100 (sterile, 10%).
3. Molecular grade water.
4. Water bath or hot block, required temperature range 56–95°C.

2.3. PCR

1. Platinum® qPCR Supermix-UDG (Invitrogen).
2. $MgCl_2$.
3. Primers, salt purified grade (Table 1, see Note 1).
4. Probes, HPLC purified grade (Table 1, see Note 1).
5. Molecular grade water.
6. *L. pneumophila* control DNA (5 ng/μl).
7. *M. pneumoniae* control DNA (1 ng/μl).
8. *C. pneumoniae* control DNA (10^4 DNA copies/μl).
9. Rotor-Disc 72 (Qiagen).
10. Rotor-Disc sealing film (Qiagen).
11. Rotor-Disc heat sealer (Qiagen).
12. QIAgility (Qiagen).
13. Rotor-Gene Q 5plex or 6plex platform and consumables (Qiagen).

Table 1
Primer sets for multiplex PCR

Organism	Target	Primer/ probe name	Sequence 5′–3′	Reference
L. pneumophila	Mip	Lp-F	AAAGGCATGCAAGACGCTAT	Developed in-house (see Note 4.2)
		Lp-R	GTACGYTTTGCCATCAAATCTT	
		Lp-P	ROX-TGTTAAGAACGTCTTTCA TTTGCTGTTCGG-BHQ2	
M. pneumoniae	P1	Mp-F	GGTCAACACATCAACCTTTTGGT	Developed in-house (see Note 4.2)
		Mp-R	TGTGATTGTGCTCAGTGTTACCT	
		Mp-P	YAK-ACCCAGCCTTCAAGGCCTG TTTGTTCTTGT-Dabcyl	
C. pneumoniae	*Pst*-1 fragment	CLPM1	CATGGTGTCATTCGCCAAGT	(9)
		CLPM2	CGTGTCGTCCAGCCATTTTA	
		CLP-P	6FAM-TCTACGTTGCCTCTAAGA GAAAACTTCAAGTTGGA-Dabcyl	

3. Methods

3.1. Specimen Preparation (see Notes 3 and 4)

NB: All processes should be performed under Containment Level 3 conditions until the specimen has been heat killed (see Note 5).

1. Add an equal volume of Dithiothreitol (1:10) to each sputum sample and homogenize by vortexing.
2. Transfer 1 ml of each homogenized sputum sample into labeled 2 ml screw-cap tubes.
3. Place the 1 ml aliquots in an 80°C water bath for 30 min to heat kill.

3.2. DNA Extraction (12)

1. Centrifuge the heat-killed samples for 10 min at 10,000 ×*g*.
2. Remove the supernatants, resuspend the pellets with 1 ml sterile PBS (1×), and centrifuge at 10,000 ×*g* for 10 min.
3. Remove the supernatant, resuspend the pellet with 1 ml PBS (1×), and centrifuge at 10,000 ×*g* for 10 min.
4. Remove the supernatant and resuspend the pellet with molecular grade water using a volume 5× that of the pellet by vortexing.
5. Using a wide-tipped pastette double the volume with sterile 10% Chelex 100.
6. Vortex for 20 s, and then incubate at 56°C for 20 min in a water bath.
7. Vortex briefly, and then incubate at 95°C for 5 min in a water bath (or hot block).
8. Vortex briefly, then cool on ice, and centrifuge at 13,000 ×*g* for 10 min.
9. Remove the supernatant and place in a sterile labeled 1.5 ml tube (discard the pellet).

3.3. Multiplex Real-Time PCR

1. If using the QIAgility, program it to set up the reactions described in Tables 2 and 3 following the manufacturer's instructions. This can be achieved by programming the QIAgility to make two master mixes: one without *L. pneumophila* DNA and one with *L. pneumophila* DNA (for the spiked reactions) (see Notes 6–8).
2. Set up the reactions with the QIAgility using the Rotor-Disc 72 loading block, Rotor-Disc 72, the master mix reagents, and the extracts/control DNA.
3. Once the QIAgility has completed setting up the reactions, seal the Rotor-Disc with the Rotor-Disc sealing film and heat sealer as per manufacturer's instructions.

Table 2
Master mix reagents, their starting concentrations, final concentrations, and volumes per reaction

Reagent	Starting conc.	Final conc.	Volume/reaction
Platinum supermix	–	–	12.5
$MgCl_2$	50 mM	4.5 μM	1.5
Lp-F	50 μM	0.3 μM	0.15
Mp-F	50 μM	0.3 μM	0.15
Cpn-F	50 μM	0.3 μM	0.15
Lp-R	50 μM	0.5 μM	0.25
Mp-R	50 μM	0.5 μM	0.25
Cpn-R	50 μM	0.5 μM	0.25
Lp-P	50 μM	0.25 μM	0.125
Mp-P	50 μM	0.25 μM	0.125
Cpn-P	50 μM	0.25 μM	0.125
Water	–	–	4.425
Extract/control	–	–	5
Final volume			25 μl

Table 3
Reactions required

	Sample (μl)	Water (μl)	Positive control (μl)
Samples neat	5	–	–
Samples spiked	4	–	1[a]
Negative extraction	5	–	–
Negative extraction spiked	4	–	1[a]
Negative control	–	5	–
L. pneumophila positive control	–	4	1[a]
M. pneumoniae positive control	–	–	5[b]
C. pneumoniae positive control	–	–	5[c]

[a] *L. pneumophila* DNA (5 ng/μl)
[b] *M. pneumoniae* control DNA (1 ng/μl)
[c] *C. pneumoniae* control DNA (10^4 DNA copies/μl)

Table 4
Parameters for analysis of Rotor-Gene Q data

Channel	Target	Threshold
Green	*C. pneumoniae*	0.05
Yellow	*M. pneumoniae*	0.05
Orange	*L. pneumophila*	0.05

4. Place the Rotor-Disc 72 containing the reactions into the Rotor-Gene Q and perform the following program: 3 min at 95°C, 40 cycles of 15 s 95°C + 45 s 60°C. Data is acquired in the Green, Yellow, and Orange channels at the end of each cycle. The gains for these channels should be auto-set (see Note 9).

3.4. Interpretation of Data

1. Data from the Rotor-Gene Q is analyzed using the appropriate thresholds and cutoffs. The raw data from each channel is analyzed in conjunction with the quantitation analysis. For the quantitation analysis the dynamic tube and slope correct functions are used. A positive identification is indicated by an increase above the set threshold for the quantitation analysis coupled with an increase in fluorescence in the raw channel (Table 4 and see Notes 10 and 11).
2. The assay is validated by review of the controls: All negative controls should be negative and all positive controls and spikes should be positive. If any of the controls fail, the assay needs to be repeated.
3. A sample is deemed negative if it does not have a C_t value for any of the targets, but does have a C_t value for the spiked sample in the orange channel; positive if it has a C_t value <35 for one or more of the targets; equivocal if it has a C_t value ≥35 for one or more of the targets; and inhibited if a C_t value is not obtained for the spiked sample in the orange channel (see Note 12).

4. Notes

1. Stock solutions of primers and probes (100 μM) should be stored at −20°C and working stocks (50 μM) can be stored at 4°C. Aliquot probes in small volumes in amber tubes to avoid repeated freezing/thawing and to reduce exposure to light (they are light sensitive). For ease of use, primers and probes

can be stored pooled as working stocks. If stored correctly we have found that primers and probes can be used for at least 1 year.

2. Following the alignment of available sequences the primers and probes were designed with the aid of Primer 3 and Integrated DNA technology OligoAnalyser Web sites, http://frodo.wi.mit.edu/primer3/ and http://eu.idtdna.com/analyzer/Applications/OligoAnalyzer/, respectively. The melt temperature, GC content, and occurrence of secondary structures were determined. NCBI BLAST searches were also performed (http://www.ncbi.nlm.nih.gov/blast/).
3. This method describes testing sputum samples using a manual Chelex-based extraction method. Other extraction methods could be employed, e.g., an automated method and different respiratory samples tested; however, these would have to be validated.
4. It is essential that sterile reagents and plasticware are used and that standard precautions are taken to avoid contamination including dedicated areas, pipettes, plugged tips, gloves, and gowns. A negative extraction should be set up with each batch of extractions.
5. Due to the risk of respiratory samples containing *Mycobacterium tuberculosis* samples are treated as Containment level 3 specimens until heat killing is completed, after which the samples are processed in a Containment level 2 laboratory.
6. The QIAgility is a liquid handler used to set up the PCR reactions; it, along with Rotor-Discs, is used to increase accuracy and to reduce hands-on time. However, reactions could be set up manually or using an alternative liquid handling system.
7. It is important to ensure that a negative result is due to the absence of target DNA and not due to the presence of inhibitors preventing amplification. Although the inhibition rate has been found to be low using the extraction method described, a spiked sample is used to check for inhibition. If inhibition is found to be a problem samples can be tested diluted to reduce the effect. We have found that diluting the samples 1 in 2 is sufficient to reduce inhibition with minimal effect on sensitivity. This assay could be adapted to include an extraction/internal control.
8. To ensure that the assay is capable of detecting the relevant and circulating strains, it is important to continually check the suitability of the primer/probe sets. This can be done by sequence analysis and testing of isolates (if available). This is especially important for *L. pneumophila* where the target gene is known to be polymorphic and is used for genotyping (13).

9. This assay has been validated for use on the Rotor-Gene Q; however, it could be adapted for other real-time platforms.
10. The threshold and cutoffs described here were established when evaluating the sensitivity and specificity of the assay. It is important to ensure that these thresholds and cutoffs are appropriate for the platform being used and the samples being tested.
11. It is important to thoroughly inspect the amplification curves; if they are not sigmoidal in the linear scale check for amplification in the raw data. Note any anomalies and treat the data with caution.
12. It is essential that the results from this assay are interpreted in the context of the clinical picture and results from other diagnostic methods. We also recommend that positive results are confirmed.

Acknowledgments

The assay described here was developed with the input of several trainee Clinical Scientists; the authors would like to acknowledge the specific input of Holly Ciesielczuk, Marcus Pond, and Kerry Williams together with the staff of the Department for Medical Microbiology, Royal Free NHS Trust.

References

1. Office of Population Censuses and Surveys (1995) Morbidity statistics from general practice. Fourth National Study, 1991–1992, vol Series MB5, 3. HMSO, London
2. Creer DD, Dilworth JP, Gillespie SH, Johnston AR, Johnston SL, Ling CL, Sanderson G, Wallace PG, McHugh TD (2006) Aetiological role of viral and bacterial infections in acute adult lower respiratory tract infection (LRTI) in primary care. Thorax 61:75–79
3. Huchon G, Woodhead M, Gialdroni-Grassi G et al (1998) Management of adult community-acquired lower respiratory tract infections. Eur Respir Rev 8(61):391–426
4. Macfarlane J, Holmes W, Gard P, Macfarlane R, Rose D, Weston V, Leinonen M, Saikku P, Myint S (2001) Prospective study of the incidence, aetiology and outcome of adult lower respiratory tract illness in the community. Thorax 56:109–114
5. Boldy DA, Skidmore SJ, Ayres JG (1990) Acute bronchitis in the community: clinical features, infective factors, changes in pulmonary function and bronchial reactivity to histamine. Respir Med 84(5):377–385
6. Jonsson JS, Sigurdsson JA, Kristinsson KG et al (1997) Acute bronchitis in adults. How close do we come to its aetiology in general practice? Scand J Prim Health Care 15(3):156–160
7. Waites KB, Talkington DF (2004) *Mycoplasma pneumoniae* and its role as a human pathogen. Clin Microbiol Rev 17:697–728
8. Dowell SF, Peeling RW, Boman J, Carlone GM, Fields BS, Guarner MR et al (2001) Standardising *Chlamydia pneumoniae* assays: recommendations from the Centers for Disease Control and Prevention (USA) and the Laboratory Centre for Disease Control (Canada). Clin Infect Dis 33:492–503
9. Welti M, Jaton K, Altwegg M, Sahli R, Wenger A, Bille J (2003) Development of a multiplex real-time quantitative PCR assay to detect *Chlamydia pneumoniae*, *Legionella pneumophila* and *Mycoplasma pneumoniae* in respiratory tract secretions. Diagn Microbiol Infect Dis 45:85–95

10. Waites KB (2003) New concepts of *Mycoplasma pneumoniae* infections in children. Pediatr Pulmonol 36:267–278
11. Plouffe JF, File TM, Breiman RF, Hackman BA, Salstrom SJ, Marston BJ et al (1995) Re-evaluation of the definition of Legionnaires disease: use of the urinary antigen assay. Clin Infect Dis 20:1286–1291
12. Pitcher D HPA Centre for Infections, Colindale, UK (Personal communication)
13. Fry NK, Afshar B, Bellamy W, Underwood AP, Ratcliff RM, Harrison TG, European Working Group for Legionella Infections (2007) Identification of *Legionella* spp. by 19 European reference laboratories: results of the European Working Group for Legionella Infections External Quality Assessment Scheme using DNA sequencing of the macrophage infectivity potentiator gene and dedicated online tools. Clin Microbiol Infect 13(11):1119–1124

Chapter 9

Utilization of Multiple Real-Time PCR Assays for the Diagnosis of *Bordetella* spp. in Clinical Specimens

Kathleen M. Tatti and Maria Lucia Tondella

Abstract

Bordetella pertussis causes an upper respiratory infection in infants, adolescents, and adults. Diagnosis of pertussis, a vaccine-preventable disease, can be difficult, but recent implementation of real-time PCR assays in laboratories has hastened the ability of clinicians to make an accurate diagnosis. In this paper we describe the method of nasopharyngeal specimen collection, extraction of DNA, and real-time PCR assays that will allow the detection and identification of *Bordetella* spp. in clinical specimens.

1. Introduction

Pertussis, more commonly known as whooping cough, is an acute respiratory infection caused by *Bordetella pertussis*, a Gram-negative bacterium. The most common *Bordetella* human upper respiratory pathogens are *B. pertussis*, *B. parapertussis*, *B. holmesii*, and *B. bronchiseptica*, an animal pathogen, which rarely infects healthy humans (1, 2).

Laboratory diagnosis of pertussis is challenging. Culture is considered the primary diagnostic test for pertussis, but due to its limited sensitivity and lengthy incubation period of 4–14 days, it has been gradually replaced by real-time PCR assays (R-PCR) in most clinical and public health laboratories. The Centers for Disease Control and Prevention (CDC) recommends that culture capacity should be maintained by laboratories as the gold standard test and that PCR is complementary. In addition, serology is useful as a diagnostic, especially in the later stages of pertussis disease.

The R-PCR algorithm for *B. pertussis* detection used at the CDC consists of positive amplification of two targets, IS*481* and *ptxS*1. The insertion sequence IS*481* is present in 50–238 copies in

Mark Wilks (ed.), *PCR Detection of Microbial Pathogens: Second Edition*, Methods in Molecular Biology, vol. 943, DOI 10.1007/978-1-60327-353-4_9,

B. pertussis and 8–10 copies in *B. holmesii*, while the pertussis toxin gene (*ptxA*), also called *ptxS1*, is present in a single copy in *B. pertussis*, *B. parapertussis*, and *B. bronchiseptica* (3). The high cycle threshold (C_t) values obtained with the multi-copy target IS*481* can be indicative of false positive PCR results leading to pseudo-outbreaks (4). The occurrence of these pseudo-outbreaks emphasizes the importance of rigorous quality assurance and quality control, the need for linkage between epidemiology and laboratory results, and the necessity for the clinical evaluation of laboratory findings.

The R-PCR assays described herein are intended for the qualitative detection of *B. pertussis*, *B. parapertussis*, and *B. holmesii* DNA extracted from clinical specimens, formalin-fixed paraffin-embedded tissues (FFPET) (5, 6), or culture isolates. They are used to confirm identification of *B. pertussis*, *B. parapertussis*, and *B. holmesii* culture isolates and to determine infection with *B. pertussis*, *B. parapertussis*, or *B. holmesii* in clinical specimens or FFPET collected from patients presenting with signs or symptoms that lead to suspicion of pertussis.

Each specimen, whether it is an isolate, clinical specimen from a nasopharyngeal (NP) aspirate, NP swab, or FFPET, is extracted for DNA either by lysis with proteinase K treatment followed by purification on a silica membrane (7) or by lysis with proteinase K treatment followed by purification with magnetic glass particles (MGPs) (8). Detection of *Bordetella* DNA is performed by an R-PCR assay which uses a TaqMan® probe, consisting of an oligonucleotide with a reporter dye attached to the 5′ end and a quencher dye attached to the 3′ end (Table 1) (9). The extracted DNA is reacted separately with each of the two primer/probe sets referred to as IS*481* and *ptxS*1.

2. Materials

2.1. Specimen Collection

1. Regan-Lowe medium containing charcoal, agar (Oxoid, Basingstoke, UK), and 10% defibrinated horse blood (storage at 2–8°C).
2. Polyester (such as Dacron), rayon, or nylon-flocked tipped NP swabs.
3. Npak—Nasopharygneal aspiration kit (M-Pro).
4. 10% buffered formalin pH 6.8–7.2 (Richard Allan Scientific).

2.2. DNA Extraction (see Notes 1 and 2)

1. Lid locks (ISC Bioexpress).
2. Sterile 0.85% NaCl (storage at 15–25°C).
3. QIAamp DNA Mini Kit (Qiagen, Valencia, CA) (storage at 15–25°C).

Table 1
Sequences and optimal concentrations of primers and probes used in the real-time PCR assays

Target	Primer/probe	Sequence 5′→3′	Amplicon length (bp)	Optimal concentration per reaction (nM)
IS*481*[a]	852U18	CAAGGCCGAACGCTTCAT	66 bp	300
	894 L24	GAGTTCTGGTAGGTGTGAGCGTAA		300
	871U22P[b]	CAGTCGGCCTTGCGTGAGTGGG		300
*ptxS*1[c]	402U16	CGCCAGCTCGTACTTC	55 bp	700
	442 L15	GATACGGCCGGCATT		700
	419U22P[b]	AATACGTCGACACTTATGGCGA		300
rnaseP[d]	*rnaseP*forward	CCAAGTGTGAGGGCTGAAAAG	80 bp	400
	*rnaseP*reverse	TGTTGTGGCTGATGAACTATAAAAGG		400
	*rnaseP*probe[b]	CCCCAGTCTCTGTCAGCACTCCCTTC		100

[a]Accession no. M28220
[b]Probe 5′ end labeled with 6-carboxyfluorescein (FAM) and 3′ end labeled with Black Hole Quencher 1 (BHQ1)
[c]Accession no. M14378
[d]Accession no.NM_006413

4. Ethanol (96–100%) (storage at 15–25°C).
5. Xylene.
6. Teflon pestles for tissue homogenization (Fisher).
7. MagNA Pure LC Instrument (Roche).
8. MagNA Pure LC Plastic Consumables needed for the MagNA Pure LC (Roche).
9. MagNA Pure LC Total Nucleic Acid Isolation Kit (Roche).
10. Dithiothreitol (DTT) (storage at 2–8°C). Stock solutions (storage at −15 to −25°C).

2.3. R-PCR

1. Reagents
   1. IS*481*, *ptxS*1, and *rnaseP* primer/probe sets; 10× concentrations of primers and probes using sterile PCR grade water; 1× working concentrations of primers and probes prepared before running the assay (Table 1) (storage at −15 to −25°C up to 6 months in non-frost-free freezers; cover all tubes of probes using foil or use dark colored tubes; do not refreeze thawed aliquots).
   2. TaqMan®Gene Expression Master Mix for IS*481*, *ptxS*1, and *rnaseP* (Applied Biosystems) (storage at 2–8°C).
   3. PCR grade water (storage at 2–8°C and −15 to −25°C).

4. *B. pertussis* positive control DNA (for IS*481* and *ptxS*1 assays; DNA extracted from CDC isolate A639 or any *B. pertussis* isolate) (storage at −15 to −25°C).
5. Human genomic positive control DNA (10 ng/μl for *rnaseP* assay; Applied Biosystems) (storage at −15 to −25°C).
6. 10% bleach (made fresh weekly).
7. Ethanol (70 and 100%).
8. DNA AWAY or similar type of reagent (ISC Bioexpress).

2. Equipment
   1. Eppendorf Centrifuge 5810 for 96-well plates.
   2. ABI7500 thermal cycler and software (Applied Biosystems).
   3. 96-well optical plates (Applied Biosystems).
   4. Optical caps (Applied Biosystems).

3. Methods

3.1. Specimen Collection (see Note 3)

1. *Clinical isolates*: *Bordetella* spp. isolates are grown for 4 days at 37°C under high humidity on modified Regan-Lowe medium, containing charcoal, agar, and 10% defibrinated horse blood.
2. *FFPET*: Clinical isolates of *Bordetella* spp. are grown as described in Subheading 3.1, step 1, fixed in 10% buffered formalin, minced with uninfected tissues, and embedded in paraffin. Lung and/or upper airway tissues are obtained from patients, fixed in formalin and paraffin embedded. Embedding and processing of tissue specimens from patients is a different scope of material that can be reviewed elsewhere (10).
3. *Clinical specimens using NP aspirates*: Acceptable clinical specimens for R-PCR should be taken from patients with suspected pertussis or who are epidemiologically linked to someone who meets the case definition for pertussis (see Note 4). The neck of patients must be extended which is *very important* as this allows pooling of the aspirate in the nasopharynx (Fig. 1). Disposable gloves should be worn to collect the aspirates and should be changed between each aspirate. For the NPak kit (www.n-pak.com), the luer catheter is attached to the syringe and the catheter is generously lubricated with jelly. The syringe plunger is pushed and pulled quickly in a smooth motion without moving the catheter out of position. The recovered aspirate sample should be approximately 2 ml in volume. The aspirate specimen is transported to the clinical specimen lab in the capped syringe using cold packs to maintain

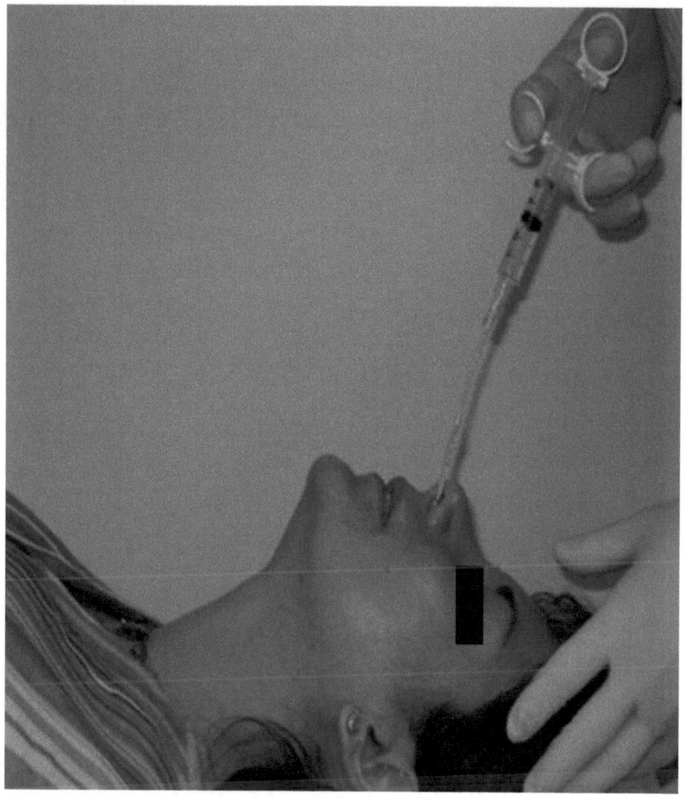

Fig. 1. Collection of a nasopharyngeal aspirate.

the specimen at 4–8°C. The specimen must not be frozen in the syringe. NP aspirates are dispensed into aliquots and can be stored at 2–8°C for less than 24 h or −15 to −25°C for storage longer than 24 h.

4. *Clinical specimens using NP swabs*: A polyester, rayon, or nylon-flocked tipped NP swab with an aluminum or plastic shaft must be used for specimen collection (http://video.cdc.gov/asx-gen/nip/isd/swabdemo.wmv) (see Note 5).

   Swab specimens should be obtained from the posterior nasopharynx (see Note 6). Disposable gloves should be worn to collect the swabs and should be changed between each swab. Tubes with NP swabs are stored at 4°C and transported to the lab within 24 h of collection with ice packs to maintain 4°C. Alternatively, the NP swab could be placed in a Regan-Lowe transport media for PCR and bacterial cultivation. Culture is highly recommended for confirmation in outbreak situations.

3.2. DNA Extraction

DNA from culture isolates with colony morphology consistent with *Bordetella* spp. and from clinical specimens or tissues that meet the criteria described in Subheading 3.1, step 3, can be extracted by any of the following methods. The extraction method that we found most suitable for a particular specimen type is indicated. All DNA extracts are stored at –20 or –80°C until ready for PCR. Freezing and thawing DNA extracts should be avoided.

1. Using silica-based membrane for FFPET—QIAamp DNA mini kit (7) (see Notes 1 and 7)

   DNA is extracted from one 10-µm section of control FFPET containing *Bordetella* spp. isolates or from the lung tissue (<25 mg) of patients placed in 2 ml centrifuge tube and processed in a clinical specimen biological safety cabinet.

   1,200 µl of xylene is added to the tissue, vortexed, and centrifuged at 20,000×*g* for 5 min, and the supernatant is removed by pipetting.

   The procedure is repeated for each specimen, including a sterile (blank) specimen. Gloves are changed after every blank or whenever they become contaminated, throughout the procedure.

   On the lab bench, specimens are loaded into a centrifuge rotor so that specimens are not side by side for all centrifugations.

   1,200 µl ethanol (96–100%) is added to the pellet to remove residual xylene, vortexed gently, and centrifuged at 20,000×g for 5 min, and ethanol removed. The ethanol extraction is repeated one more time.

   The tube containing the pellet is incubated at 37°C for 10–15 min or until the ethanol has evaporated.

   The pellet is washed with 200 µl 0.85% NaCl to help remove formalin.

   The pellet is resuspended with a pestle in 180 µl SDS-containing lysis buffer (Qiagen), 20 µl proteinase K (15 mg/ml) is added, and the procedure of the QIAamp DNA Mini kit (Qiagen) for FFPET is followed (7). The DNA is bound to a silica membrane and is eluted in 100 µl of Buffer AE (10 mM Tris–Cl, 0.5 mM EDTA pH 9.0). From 100 µl, only 4 µl of the DNA is used per R-PCR reaction to test for *Bordetella* spp.

2. Using MGPs for NP aspirates, and NP swabs—Roche MagNA Pure Total Nucleic Acid Kit (8) (Notes 2 and 7).

   NP specimens are processed in a clinical specimen hood.

   NP Aspirates may be viscous or mucous and liquefaction with DTT may be necessary (11, 12). A DTT stock solution (e.g., 5× conc.=0.75%) is prepared and the final concentration of DTT in the sample is adjusted to 0.15% by adding appropriate amounts of DTT stock solution. 200 µl of aspirate sample is incubated with the appropriate amount of DTT at 37°C, with shaking until the sample can be pipetted easily. The remaining NP aspirate is stored at −80°C.

NP Swabs are eluted in 400 μl of saline, extensively vortexed, the swab tip is removed, and 200 μl is retained for storage.

200 μl of the liquefied NP aspirate or 200 μl of resuspended NP swab is transferred to the MagNA Pure LC sample cartridge.

The quantities of reagents, including the lysis buffer and proteinase K, are based on the number of samples being extracted and are dispensed according to the program for total nucleic acid extraction (Roche).

The MGP suspension is vortexed extensively to ensure resuspension, and is loaded immediately before the extraction starts. The nucleic acid is eluted from the beads in 100 μl of elution buffer (60–70 mM Tris–Cl pH 8.3) and divided into two aliquots. One 50 μl aliquot is stored at −80°C.

The other 50 μl aliquot is placed at 4°C and used within 48 h or stored at −80°C until ready to perform R-PCR. 4 μl of the total nucleic acid is added per R-PCR reaction.

3.3. Real-Time PCR (Notes 8 and 9)

The R-PCR is performed in four separate rooms: the clean room where the master mix is prepared and no DNA is allowed, the template addition room where DNA extracted from clinical specimens is added, the PCR workstation where DNA from *B. pertussis* isolates (positive control DNA) is added, and, finally, an area that contains the R-PCR instruments and where any post-PCR processing occurs. DNA from any of the methods in Subheading 3.2 is used in R-PCR.

1. *Master Mix preparation*: A master mix (*N*+1) is prepared for each plate, where N is the number of reactions being performed. For each reaction the following reagents are added: 2.5 μl of the working concentration (1×) of each primer and probe (Table 1) which have been thoroughly mixed after thawing; 1 μl of water; 12.5 μl of the 2× master mix (TaqMan®Gene expression master mix containing the ampliTaqGold® DNA polymerase; passive reference dye Rox; dNTPs with UTP; AmpErase® UNG; and proprietary optimized buffer system for 5′ nuclease assay). The master mix is gently resuspended and centrifuged, and 21 μl of this master mix is dispensed to each well.

2. *Addition of clinical specimen DNA*: 4 μl of sterile PCR grade water is added to the non-template control (NTC) and the well is closed with an optical cap.

   4 μl of clinical specimen DNA, NP specimen, or FFPET, from any of the aforementioned methods, is added to the appropriate well and the well is closed with an optical cap. All specimens should be tested in duplicate. A third reaction should be performed with 1:5 dilution of the DNA extract.

Table 2
Real-time PCR cycling conditions

Thermal cycling conditions

Target	Stage 1	Stage 2	Stage 3	
	Thermal (1 cycle)	Denature (1 cycle)	Amplification (45 cycles)	
IS*481*	50°C, 2 min	95°C, 10 min	95°C, 15 s	60°C, 1 min
*ptxS*1	50°C, 2 min	95°C, 10 min	95°C, 15 s	57°C, 1 min
rnaseP	50°C, 2 min	95°C, 10 min	95°C, 15 s	60°C, 1 min

3. *Positive control of isolate DNA* (Note 10): 4 μl of each positive control is added to the final wells and wells are resealed with the optical caps.

4. *Prepare Plate for Run*: 96-well plate is centrifuged for 3 min at 1,800 × *g* to confirm that the reaction mixes are at the bottom of the tubes and no bubbles are present. The centrifugation is repeated if necessary. Thermal cycling conditions are specified according to Table 2 with the sample volume set to 25 μl and the run is started.

5. *Analysis of controls for DNA extraction and R-PCR*: Extracts from sterile swabs or sterile water (blanks) should be tested to ensure that there was no cross-contamination during the DNA extraction (Note 7).

Assays that are positive for the extraction blanks indicate that there was contamination during the DNA extraction process and specimens must be re-extracted for confirmation. Any positive specimens that were re-extracted and are still positive should be regarded with suspicion. Epidemiological and clinical corroboration are required before regarding these specimens as true positives.

NTC, using sterile water in the place of DNA, should be included as a negative control in each assay to ensure that there was no cross-contamination during PCR setup. A positive NTC indicates that there was contamination during the PCR setup and the assay must be repeated.

For the IS*481* and *ptxS*1 assays, a negative reaction of PCR positive control indicates either that the PCR did not work or that the positive control is no longer usable. The assay is repeated using a fresh positive control extract before reporting results. If there is still a problem, test extracts again with primer/probe set dilutions prepared before running the assay.

When DNA from clinical specimens is tested, include the *rnaseP* set to serve as an external positive control for the assay. All

Table 3
Lower limit of detection of the PCR assay dual-target on AB7500

	B. pertussis	
	IS*481*	*ptxS*1
Genomic equivalent[a]	C_t[b] values	C_t values
5,000	18.4	28.7
500	22.0	31.3
50	25.7	34.9
10	27.0	35.8
5	29.2	36.7
1	31.2	39.2
0.1	35.8	Negative

[a]*B. pertussis* isolate A639
[b]C_t = cycle threshold

clinical specimens and human genomic DNA positive controls should be positive for *rnaseP*. For the *rnaseP* assay, negative or high C_t positive *rnaseP* in undiluted DNA extracts indicates that the specimen extraction contains inhibitors to the PCR assay, a low DNA yield from the DNA extraction process occurred, or the specimen was not collected correctly. To determine if inhibitors are present, check the DNA extracts at 1:5 dilution. If the *rnaseP* results improve with the diluted extracts, the IS*481* and *ptxS*1 assays must be repeated using the diluted extracts. If the *rnaseP* results do not improve, this may indicate that the NP aspirate or swab was not collected or stored correctly.

DNA extracted from a *B. pertussis* strain at low DNA concentration should be included as a positive control for the IS*481* and *ptxS*1 assays (Note 11). The different C_t values obtained for each genomic equivalent of DNA with IS*481* and *ptxS*1 demonstrate the multi-copy nature of IS*481* (Table 3).

6. *Interpretation/results according to algorithm* (Table 4) (Note 12): When the above controls are met, any clinical specimen that is positive for two of the three replicates (two reactions with undiluted DNA extract and one reaction with DNA diluted 1:5) in both the IS*481* and *ptxS*1 assays is considered positive for the presence of *B. pertussis*; any specimen that is positive in only one of the replicates should be repeated. If the repeat results are positive in at least one of the three replicates for *ptxS*1, the test is considered positive. If a specimen is negative in the IS*481* assay and positive in

Table 4
PCR algorithm for different species

IS*481* PCR result	*ptxS*1 PCR result[a]	Interpretation
C_t < 35 cycles	Positive	*Bordetella pertussis*
$C_t \geq$ 35 cycles	Positive	*Bordetella pertussis*
$C_t \geq$ 35 cycles	Negative	Indeterminate—Requires confirmation by other means (culture, serology, or epi linkage)
C_t < 35 cycles	Negative	*Bordetella* spp.—Possibly *B. holmesii* or *B. bronchiseptica*[b]
Negative	Positive	*Bordetella* spp.—Possibly *B. parapertussis* or *B. bronchiseptica*[b]

[a] C_t < 40 cycles is considered a positive result
[b] *B. bronchiseptica* is an animal pathogen, rarely found in humans. *B. bronchiseptica* cannot be differentiated from other *Bordetella* species by this algorithm

the *ptxS*1 assay the combined result is considered positive for *Bordetella* spp., possibly *B. parapertussis* (Table 4).

Any specimen positive in the IS*481* assay before cycle 35 in at least two of the three replicates and negative in the ptxS1 assay is suggestive of the presence of *Bordetella* spp., possibly *B. holmesii* (Table 4). Confirmation with a specific PCR target for the presence of *B. holmesii* is suggested.

In outbreak settings, when many specimens are handled in a short period of time, low levels of *B. pertussis* DNA detected by high C_t IS*481* positive and *ptxS*1 negative may be due to DNA contamination during specimen collection, DNA extraction, or PCR setup. In this case the PCR results should be interpreted as indeterminate if not confirmed by another laboratory test such as culture, serology, or immunohistochemistry for FFPET (6). Epidemiological and clinical relevance of high C_t IS*481* positive PCR tests should be taken into consideration. It is also important to keep a record of the C_t values of all reactions in a database in order to facilitate result interpretation (Note 13).

PCR tests are repeated for extracts where the C_t values for the two undiluted extract reactions are not similar or when only the diluted extract reaction has a C_t value. If only one duplicate of a DNA extract has a C_t value after repeated PCR, the result for this specimen is considered negative. If only the diluted extract has a C_t value after repeated PCR, the specimen is noted as having an inhibitor present. If inhibition of the *rnaseP* assay is noted for a clinical specimen, extracted DNA should be tested at two or more dilutions (e.g., 1:5 and 1:10) to verify the result. Results from the diluted extract reaction are interpreted according to Table 4.

7. *Limitations:* If inhibitors are present in a DNA extraction, PCR assays may produce a false negative result.

Data suggest that clinical specimens collected subsequent to initiation of antimicrobial treatment may not be positive for *Bordetella* spp. due to reduction in the number of organisms. Whenever possible, specimens collected prior to administration of antimicrobial agents should be used to determine infection with *Bordetella* spp.

4. Notes

1. DNA extraction using silica-based membrane can be performed with several commercial kits. Here we described the method using the Qiagen column kit.
2. DNA extraction using MGPs can be performed with several commercial kits. Here we described the method using the Roche MagNA Pure LC Total Nucleic Acid Isolation Kit; however, the MagNA Pure LC DNA Isolation Kit III (Bacterial, Fungi) and Compact Roche kit for bacterial extraction on the Compact MagNA Pure work as well as the MagNA Pure LC Total Nucleic Acid kit for extraction of DNA from *Bordetella* spp. The MagNA Pure LC DNA Isolation Kit III (Bacteria, Fungi) uses 100 μl of specimen with external lysis by proteinase K (15–20 mg/ml) and heating for 10 min at 65°C.
3. Specimen collection for pertussis diagnostic testing must be performed in a separate room far from where pertussis vaccines are administered (13, 14).
4. The preferred specimen type is an NP aspirate, if it is obtained correctly, because the overall yield and recovery of organisms are optimized.
5. Calcium alginate- or cotton-tipped swabs are not acceptable for specimen collection due to possible inhibition of the PCR reaction.
6. Throat and anterior nasal swabs have unacceptably low rates of recovery.
7. Because of the sensitivity of R-PCR, special precautions should be taken to prevent contamination during DNA extraction. The following steps are recommended:

   Maintain a biological safety cabinet (BSC) and lab dedicated to working with clinical specimens only, not bacterial isolates.

   Maintain separate equipment (pipettes, microcentrifuges, etc.) and supplies for DNA extraction from clinical specimens.

   Extract NP aspirates in batches of no more than 28 aspirates or swabs at one time.

Extract sterile water (blank) after every seventh specimen to ensure that there is no cross-contamination during DNA extraction.

Change gloves frequently.

Decontaminate BSC, lab bench, centrifuge rotor and lid, pipettes, scissors, forceps, and tube rack with 10% bleach (prepared weekly), 70% ethanol which inactivates the bleach, DNA AWAY, and UV light after each DNA extraction batch.

Do not extract isolates when you extract clinical specimens.

If you have used the instrument for extraction of isolates, decontaminate and perform a run with water only.

Multiple centrifugation steps and extensive hands-on manipulations during extraction processes can lead to contamination of specimens and equipment; try to minimize these steps.

Do not extract positive controls with clinical specimens to reduce the risk of contamination.

8. Maintain separate rooms for PCR and culture. PCR should never be performed in a room where bacterial cultivation occurs.
9. Because of the sensitivity of R-PCR, special precautions should be taken to prevent false positive amplifications. The following steps are recommended:

   Maintain separate rooms for master mix setup and DNA addition.

   Do not bring DNA into the master mix setup room, and maintain separate equipment (using aerosol barrier pipettes, microcentrifuges, etc.) and supplies for assay setup and DNA addition.

   Wear a clean lab coat and gloves when setting up assay.

   Change gloves frequently.

   PCR workstations or BSC should be treated with UV light for at least 1 h prior to setting up assay.

   After setting up assays in the rooms, clean equipment and bench or cabinet with 10% bleach, 70% ethanol, DNA AWAY, and UV light.
10. Physically separating the addition of the positive control DNA from the addition of specimen DNA is an additional step to prevent contamination. A separate pipette should be maintained for use with positive control DNA.
11. Positive control should contain an average C_t value (C_t 25–30) and a lower limit control ($C_t \geq 35$), never high concentrations of DNA with low C_t values.
12. This algorithm is used for laboratory diagnosis in public health settings during pertussis outbreaks where specificity is needed. The clinical decisions regarding treatment for pertussis are

driven not only by the PCR results but also by the clinical and epidemiological data.

13. Care should be taken when using this algorithm to interpret high C_t value (35 cycles or greater) IS*481* positive/*ptxS1* negative clinical specimens. Since IS*481* is present in many copies and *ptxS1* is only present in one copy, a clinical specimen that is high C_t IS*481* positive and *ptxS1* negative may indicate the presence of low levels of *B. pertussis* due to infection or false positive due to the multi-copy nature of IS*481*. Pseudo outbreaks of pertussis have been described due to the high C_t values of IS*481* from contamination (4).

References

1. Mattoo S, Cherry JD (2005) Molecular pathogenesis, epidemiology, and clinical manifestations of respiratory infections due to *Bordetella pertussis* and other Bordetella subspecies. Clin Microbiol Rev 18:326–382
2. Goebel EM, Zhang X, Harvill ET (2009) *Bordetella pertussis* infection or vaccination substantially protects mice against *B. bronchiseptica* infection. PLoS One 4:e6778
3. Tatti K, Wu K, Tondella L et al (2008) Development and evaluation of dual-target real-time polymerase chain reaction assays to detect *Bordetella* species. Diagn Microbiol Infect Dis 61:264–272
4. CDC (2007) Outbreaks of respiratory illness mistakenly attributed to pertussis—New Hampshire, Massachusetts, and Tennessee, 2004–2006. MMWR 56:837–842
5. Tatti KM, Wu KH, Sanden GN et al (2006) Molecular diagnosis of *Bordetella pertussis* infection by evaluation of formalin-fixed tissue specimens. J Clin Microbiol 44:1074–1076
6. Paddock CD, Sanden GN, Cherry JD et al (2008) Pathology and pathogenesis of fatal *Bordetella pertussis* infection in infants. Clin Infect Dis 47:28–38
7. Qiagen (2007) QIAamp®DNA Mini and Blood Mini Handbook. Qiagen, Valencia, CA
8. Roche Applied Science (2008) MagNA Pure LC Total Nucleic Acid Isolation Kit Handbook. Roche Applied Science, Mannheim
9. Espy MJ, Uhl JR, Sloan LM et al (2006) Real-time PCR in clinical microbiology: applications for routine laboratory testing. Clin Microbiol Rev 19:165–256
10. Prophet EB, Mills B, Arrington JB, Sobin LH (1992) Laboratory methods in histotechnology. American Registry of Pathology, Washington, DC
11. Dundas N, Leos NK, Mitui M et al (2008) Comparison of automated nucleic acid extraction methods with manual extraction. J Mol Diagn 10:311–316
12. Riffelman M, Schmetz J, Bock S et al (2007) Preparation of *Bordetella pertussis* DNA from respiratory samples for real-time PCR by commercial kits. Eur J Clin Microbiol Infect Dis 27:145–148
13. Taranger J, Trollfors B, Lind L et al (1994) Environmental contamination leading to false-positive polymerase chain reaction for pertussis. Pediatr Infect Dis J 13:936–937
14. Tatti KM, Slade B, Patel M et al (2008) Real-time polymerase chain reaction detection of *Bordetella pertussis* DNA in acellular pertussis vaccines. Pediatr Infect Dis J 27:73–74

Chapter 10

Detection of *Mycoplasma pneumoniae* by Real-Time PCR

Jonas M. Winchell and Stephanie L. Mitchell

Abstract

Mycoplasma pneumoniae is a significant cause of respiratory disease, accounting for approximately 20% of cases of community-acquired pneumonia. Although several diagnostic methods exist to detect *M. pneumoniae* in respiratory specimens, real-time PCR has emerged as a significant improvement for the rapid diagnosis of this pathogen. The method described herein details the procedure for the detection of *M. pneumoniae* by real-time PCR (qPCR). The qPCR assay described can be performed with three targets specific for *M. pneumoniae* (Mp181, Mp3, and Mp7) and one marker for the detection of the *RNaseP* gene found in human nucleic acid as an internal control reaction. Recent studies have demonstrated the ability of this procedure to reliably identify this agent and facilitate the timely recognition of an outbreak.

Key words: *Mycoplasma pneumoniae*, Community-acquired pneumonia, Real-time PCR, qPCR

1. Introduction

Mycoplasma pneumoniae is a significant cause of community-acquired pneumonia (CAP) in both children and adults. Outbreaks of *M. pneumoniae* most often occur in closed settings such as schools, prisons, and hospitals (1–4). Although many infections result in mild, asymptomatic, and often self-limiting cases, up to 25% of those infected with *M. pneumoniae* may experience extra-pulmonary complication and/or require hospitalization due to severe pneumonia. A comprehensive overview of *M. pneumoniae* and its pathogenicity is provided elsewhere (5–7). A reliable and rapid detection method is needed for an effective public health response due to the high prevalence and significance of this pathogen.

Historically, *M. pneumoniae* infection has been difficult to diagnose, primarily because of the lack of available standardized

Mark Wilks (ed.), *PCR Detection of Microbial Pathogens: Second Edition*, Methods in Molecular Biology, vol. 943, DOI 10.1007/978-1-60327-353-4_10, © Springer Science+Business Media, LLC 2013

and specific diagnostic tests. Culture, complement fixation, and serological tests have been used to identify *M. pneumoniae* infections, but each has substantial drawbacks and limitations. These methods are labor intensive and have questionable sensitivity and specificity. Serologic tests require paired serum specimens to distinguish between active and previous infections and do not provide timely results. Similarly, isolation of *M. pneumoniae* is slow and requires specialized expertise.

The use of real-time polymerase chain reaction (qPCR) greatly improved the diagnostic capabilities for this agent. This technique provides greater sensitivity, specificity, and speed to that of traditional tests. In the past several years, studies have reported the development of real-time PCR assays for the detection of *M. pneumoniae* (8–10). Recently, it has been shown that real-time PCR may be more useful during the early stages of infection, probably due to the presence of a higher pathogen burden and specimen collection occurring prior to the patient receiving antibiotic treatment (11). Therefore, specimen collection and testing should be performed as early as possible (see Note 7). Due to the overall enhanced sensitivity, ease of use, and widespread availability, real-time PCR is becoming the primary means of detecting *M. pneumoniae* infections.

As with any nucleic acid amplification procedure, optimal laboratory setup, practices, and proper technique are paramount. Maintaining separate and dedicated procedural rooms, equipment, and consumables is essential for decreasing contamination. A "clean room" should house primer/probe dilutions, enzyme reagents, all the necessary consumables, and dedicated equipment to properly set up a PCR reaction plate. A template or "dirty" room should only be used for the addition of nucleic acid and centrifugation of capped plates prior to transfer into the PCR instrument room. Nucleic acid isolation and any post-PCR analysis that may occur, such as gels or other amplicon analysis, should be conducted in a separate "Nucleic Acid Isolation" and "Post-PCR Analysis" rooms, respectively. These practices provide a safeguard to prevent contamination of the laboratory and should be strictly adhered to.

The currently described assay uses hydrolysis-based probes (Taqman) and primers that are specific for *M. pneumoniae*. Two markers (Mp3 and Mp7) target the *ATPase* gene, whilst the third, Mp181, targets the newly described CARDS toxin gene (12). These markers have been previously evaluated and are reported elsewhere (8). The widely used RNaseP (RP) target is included as an internal control to ensure template quality and proper PCR setup and execution. By using all four markers on a 96-well platform, 22 specimens can be tested along with a no-template control (NTC) and a positive control (PC). If higher throughput is required for an initial screening of specimens, the Mp181 and RP can be used to screen specimens followed by retesting any positive samples with the full panel. Mp181 is recommended as the primary marker for screening due to its slightly enhanced sensitivity. This procedure should enhance the detection rates for this agent and provide a convenient method to rapidly test patients suspected of having an *M. pneumoniae* infection.

2. Materials

2.1. Reagents

1. TaqMan (TM) primers/probes (Table 1).

   Primers should optimally be synthesized as "sequencing grade" (see Note 1). Working dilutions of each primer are suggested to be at 50 μM (diluted in nuclease-free molecular-grade water).

   Probes should be made with a 5′ FAM label and a 3′ black hole quencher 1 (BHQ1). Working dilution for the probe is recommended at 10 μM (diluted in nuclease-free molecular-grade water). The probe should be protected from light.

   Working dilutions of primers and probes should be kept at 4°C. For long-term storage, aliquots may be frozen at −20°C (see Note 2).

2. Platinum Quantitative PCR SuperMix-UDG and 50 mM $MgCl_2$ (Invitrogen). Store at −20°C (non-frost-free freezer).

Table 1
***M. pneumoniae* real-time PCR primer/probe sequences**

Primer/probe	Sequence 5′→3′	Target
Mp181-F	tttggtagctggttacgggaat	CARDS Tx
Mp181-R	ggtcggcacgaatttcatataag	
Mp181-P	tgtaccagagcaccccagaagggct	
Mp3-F	cgatctatgtgccagctgatga	ATPase
Mp3-R	agcatccaggtgggtaaaggt	
Mp3-P	ttgactgaccccgctccggc	
Mp7-F	actaacaattaccgtgcttacaatgaa	ATPase
Mp7-R	ccacacctttgtcttggatcac	
Mp7-P	actctt(t)gccaaccaacaaaacgagtcct	
RP-F	agatttggacctgcgagcg	RNaseP
RP-R	gagcggctgtctccacaagt	
RP-P	ttctgacctgaaggctctgcgcg	

Probes are labeled with 5′ FAM and 3′ black hole quencher (BHQ) 1. Mp7-P is internally quenched with BHQ1, represented by "(t)." "CARDS Tx" refers to community-acquired respiratory distress syndrome toxin. RNaseP (RP) is used as an internal positive control for the detection of human nucleic acid

3. Platinum *Taq* DNA Polymerase (5 U/μl) (Invitrogen). Store at −20°C (non-frost-free freezer).
4. 10 mM PCR Nucleotide Mix (Promega). Store at −20°C (non-frost-free freezer).
5. Nuclease-free water. Store at room temperature. Separate water bottles are recommended for master mix use and dilution of primer/probe working stocks.

2.2. Consumables

1. MicroAmp Optical 96-well reaction plates and MicroAmp Optical 8-cap strips (Applied Biosystems).
2. Nuclease-free 1.5 mL Eppendorf tubes.
3. Aerosol barrier pipette tips.
4. Pipettes: P20, P100, P200, P1000.
5. Disposable gowns and gloves.
6. RNase Away (Molecular BioProducts, cat # 7005) or freshly prepared 10% bleach.

2.3. Instruments

1. ABI 7500 (Standard or FAST real-time PCR thermocycler by Applied Biosytems) or a comparable instrument (i.e., Stratagene Mx 3000/3005, Rotorgene 6000, iCycler, etc.).

3. Methods

The steps outlined below describe the setup, execution, and analysis of the *M. pneumoniae* real-time PCR assay.

3.1. "Clean" Room Procedures

1. The PCR master mix should be prepared in a designated "clean room," as far away as possible from any areas that may generate potential nucleic acid contamination. Disposable gown and gloves should be donned and all surfaces and shafts of each pipette should be thoroughly wiped clean with RNase Away or an equivalent decontaminant (i.e., 10% freshly prepared bleach) after applying with a disposable wipe. The master mix should be used immediately after preparation and not be stored any longer than 1 h at 4°C after being formulated. Longer storage after formulation may result in inferior performance, including false positives/negatives.
2. The master mix(es) contains the following reagents per reaction:

   12.5 μl of Platinum® Quantitative PCR SuperMix-UDG.

   1.5 μl of 50 mM $MgCl_2$.

   0.5 μM final concentrations of each primer (0.25 μl per reaction when a 50 μM stock is used).

   0.1 μM final concentration of the probe (0.25 μl per reaction when a 10 μM stock is used).

		1	2	3	4	5	6	7	8	9	10	11	12
Mp181	A	NTC	S2	S4	S6	S8	S10	S12	S14	S16	S18	S20	S22
Mp3	B	NTC	S2	S4	S6	S8	S10	S12	S14	S16	S18	S20	S22
Mp7	C	NTC	S2	S4	S6	S8	S10	S12	S14	S16	S18	S20	S22
RP	D	NTC	S2	S4	S6	S8	S10	S12	S14	S16	S18	S20	S22
Mp181	E	S1	S3	S5	S7	S9	S11	S13	S15	S17	S19	S21	PC
Mp3	F	S1	S3	S5	S7	S9	S11	S13	S15	S17	S19	S21	PC
Mp7	G	S1	S3	S5	S7	S9	S11	S13	S15	S17	S19	S21	PC
RP	H	S1	S3	S5	S7	S9	S11	S13	S15	S17	S19	S21	PC

Fig. 1. Plate setup. For a 96-well plate, individual primer/probe master mixes are added in rows (A–H). For use of all markers, Mp181 (rows A and E), Mp3 (rows B and F), Mp7 (rows C and G), and RP (rows D and H) are shown. Specimens (S1–S22) are added in columns. The NTCs (A1–D1) are added first, followed by each specimen, and finally, the Positive Control (PC, wells E9–H12).

1.25 U Platinum® Taq DNA Polymerase.

1 μL of 10 mM PCR Nucleotide Mix.

Nuclease-free, molecular-grade water to 20 μl final volume.

The total volume of the PCR reaction mixture + 5 μl of extracted nucleic acid will equal 25 μl.

3. Plate setup and dispensing of master mix are critical to minimize contamination potential. For best results, master mix(es) should be added in rows, while specimens are added in columns (Fig. 1).
4. Prior to dispensing into the plate, the master mix(es) should be mixed by gently pipetting the mix 5–10 times. Dispense 20 μl of the master mix into the corresponding wells of a MicroAmp Optical 96-well reaction plate.
5. Add 5 μl of nuclease-free water to wells that will be used as "NTC"; cap all NTC wells in the clean room before moving to the template room (see Note 3).

3.2. Template Room Procedures

1. Transport the plate to a designated template ("dirty") room (see Note 4).
2. Extracted nucleic acid should be maintained on ice or cold block throughout this procedure and added to the plate within a dedicated BSC or contained setup/isolation cabinet (see Note 5).
3. Add 5 μl of extracted nucleic acid from specimens for a final volume of 25 μl per well. Gloves should be changed if contaminated with specimen or defective (see Note 6).

4. A strip of four caps should be placed on each column of wells after the specimen has been added before proceeding to the next group of wells.
5. After all specimens have been added and capped, a positive control of *M. pneumoniae* with a Ct value between 25 and 30 should be added and capped.
6. The plate should be centrifuged for 1 min at ~500–750 × *g* to ensure that all reagents and template are at the bottom of each well.
7. Prior to leaving the "template" room, all caps should be examined to ensure that they are flushed with the plate (no uneven caps). Some manufacturers provide a "capping" tool to assist. Carefully examine the wells to ensure that no large air bubbles exist. If so, repeat the centrifugation in step 6.
8. Gowns and gloves should be disposed of before exiting the "template" room and eventually autoclaved. Gowns should not be reused.
9. Transport the prepared PCR plate to the designated PCR instrument room.

3.3. PCR Instrument Room Procedures

1. The ABI 7500 (or an equivalent instrument) should be programmed using the following cycling conditions and be run in "standard" mode with the 25 μl reaction volume condition:

   95 °C 2 min, followed by 45 cycles of 95 °C 10 s and 60 °C 30 s (data acquisition for FAM)
2. The passive reference dye (if using the ABI 7500) should be set to "none."
3. For the ABI 7500 instrument, when the run has completed (~1.5 h), select "Analyze" to set a threshold value (the threshold line will turn from red to green).
4. The default displays the data in a logarithmic format but it may be more easily visualized on a linear scale. To change from logarithmic to linear, double click on the *y*-axis, check the box that indicates linear view on the "post-run settings" block, click "apply," and view the data as a sigmoidal curve.
5. Adjust the threshold value so that the line crosses the curves at the beginning of the exponential phase. All controls should be analyzed first to validate the experiment. Amplification growth curves that demonstrate an exponential increase and cross the threshold are considered a "positive" PCR result for that marker; those that do not possess an exponential growth curve within 45 cycles are considered a "negative" result for those specimens for that marker (Fig. 2).
6. A report file may be generated using the software that may be programmed to suit the investigator's needs, including well number, specimen information, threshold data, and C_t value.

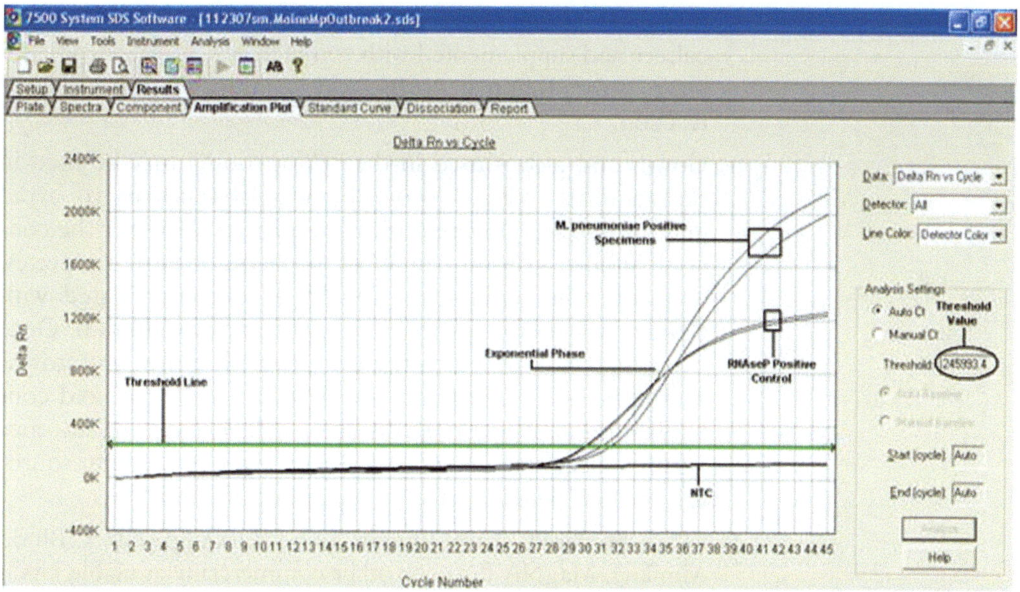

Fig. 2. Real-time PCR Data File. Detail view of data analysis on the ABI 7500 instrument. Typical amplification curves are shown (in duplicate), for a clinical sample tested with Mp181 and the RNaseP control. Non-template controls (NTC) are also shown (see Notes 8–12).

4. Notes

1. Primer purity is important. Many manufacturers offer different "purity grades." We recommend the use of "sequencing grade," as this provides superior performance due to higher purity. Primers and probes may be synthesized by many reputable vendors or may be ordered internally at the Biotechnology Core Facility at the investigator's institute.
2. Repeated freeze–thaw cycles for primers/probes are not recommended for stability purposes. Working dilutions should be stable for up to 6 months and may be quality control tested (using known template) after this period to verify performance. The stocks from which the working dilutions are made should be stable at 4°C for longer periods (up to a year) but should also be quality control tested if used past this period.
3. Plate setup in Fig. 1 shows the furthest possible separation of NTC from the Positive Control wells. This should be standard practice to ensure the quality of the assay setup and execution. The NTC should be added in the "clean" room and capped. The PC should be added last to the plate, after all other specimens have been added and capped. The PC can be a clinical specimen known to be positive for *M. pneumoniae* or can be

prepared by using extracted nucleic acid from an *M. pneumoniae* culture and supplemented with commercially available (or laboratory isolated) human nucleic acid to ensure a positive RNaseP reaction.

4. Gowns and gloves used in the "clean room" may be used in the template "dirty" room by the laboratorian to avoid de-gowning/re-gloving during this process; however the converse should *never* occur. Work surfaces, dedicated pipettes, and any other equipment should be decontaminated with RNase Away before proceeding with template addition. Once the template is added to the test plate, all gloves and gowns should be disposed of and autoclaved in order to avoid contamination and pipette shafts and work surfaces should once again be decontaminated with RNase away. A UV light source can be used on all surfaces to aid in decontamination.
5. DNA prepared from Qiagen (e.g., QiaAmp DNA Blood Minikit), MagNA Pure, or other commercially available kits is acceptable. Both DNA and total nucleic acid (TNA) are acceptable.

   Specimens should be extracted and tested as soon as possible after receipt to ensure optimal results.
6. Use 5 μl of sample to ensure accurate pipetting and specimen addition. If the sample volume is limited, less than 5 μl may be added and the volume of water should be adjusted to total 25 μl.
7. Specimens should be taken from patients during the early phase of infection or first onset of symptoms of respiratory infection. Acceptable specimens include oropharyngeal and/or nasopharyngeal swabs, transported and stored in ~2 ml of universal viral transport media (VTM, Becton Dickinson or equivalent) in a screw-top sterile falcon tube. Specimens transported within 72 h can be shipped at 4°C. If over 72 h, specimens should be stored between −20 and −70°C and shipped on dry ice. BAL specimens are acceptable.
8. Positive specimens should show an exponential curve when crossing the threshold value. Depending on the specimen quality, time of acquisition, and other factors (i.e., antibiotic use), C_t values may range dramatically (i.e., from ~20 to 40 or greater). In our experience, late positive C_t values have yielded culture isolates. Specimens resulting in a late curve (C_t value of 40–45) or displaying a curve that is not exponential should be retested again by either qPCR or a traditional method. It is critical to note the shape of the curves, especially for those which have late C_t values. It should be emphasized that specimens that result in a C_t value >40 need to be examined closely. These curves may, in fact, be accurate in detecting a true posi-

tive specimen. They also may be an artifact. It is critical to examine the suspect curve(s) in the context of the NTCs, note the shape of the curve(s), and inspect the "Spectra" and "Component" tabs (if using the ABI 7500) to help with analysis. Ultimately, the specimen may require concentration to be retested.

9. Since different specific markers (i.e., Mp181, Mp3, and Mp7) can be used on the same specimen, they each may contain different background fluorescent levels and require separate analysis in order to properly assess the specimen's status. The three markers can also serve to increase the confidence in interpretation of results (8).
10. The RNaseP value can be helpful in assessing the quality of the extracted nucleic acid. A nonexistent or poor (i.e., non-exponential) amplification curve signifies poor-quality template and the specimen should be re-extracted. Lack of amplification on any specimen, including the positive control, may indicate improper master mix preparation.
11. Because the RNaseP marker detects human DNA, it is quite easy to contaminate stocks, PCR consumables, and common equipment if strict PCR and proper laboratory practices are not followed. If stocks become contaminated, they must be discarded and reordered. Occasionally, one may observe a "positive" curve in an RNaseP NTC. These should be interpreted with care and indicate that contamination has occurred within the laboratory. These "positives" underscore the sensitivity of the procedure and the constant need for routine decontamination of surfaces and equipment with RNase Away or equivalent.
12. If post-PCR analysis of the reaction mixture is needed but cannot be performed immediately, plates should be stored at -20°C to ensure DNA amplicon stability.

References

1. Walter ND, Grant GB, Bandy U, Alexander NE, Winchell JM, Jordan HT, Sejvar JJ, Hicks LA, Gifford DR, Alexander NT, Thurman KA, Schwartz SB, Dennehy PH, Nino K, Fields BS, Dillon MT, Erdman DD, Whitney CG, Moore MR (2008) Community outbreak of *Mycoplasma pneumoniae*: school-based cluster of neurologic disease associated with household transmission of respiratory illness. J Infect Dis 198:1365–1374
2. Klement E, Talkington DF, Wasserzug O, Kayouf R, Davidovitch N, Dumke R, Bar-Zeev Y, Merav R, Boxman J, Thacker WL, Wolf D, Lazarovich T, Shemer-Avni Y, Glikman D, Jacobs E, Grotto I, Block C, Nir-Paz R (2006) Identification of risk factors for infection in an outbreak of *Mycoplasma pneumoniae* respiratory tract disease. Clin Infect Dis 43: 1239–1245
3. Mezarina KB, Huffmire A, Downing J, Core N, Gershman K, Hoffman R (2001) Outbreak of community-acquired pneumonia caused by *Mycoplasma pneumoniae*—Colorado 2000. MMWR 50:227–230
4. Waring AL, Halse TA, Csiza CK, Carlyn CJ, Musser KA, Limberger RJ (2001) Development of a genomics-based PCR assay for detection of *Mycoplasma pneumoniae* in a large outbreak in New York State. J Clin Microbiol 39: 1385–1390

5. Waites KB, Talkington DF (2004) *Mycoplasma pneumoniae* and its role as a human pathogen. Clin Microbiol Rev 17:697–728
6. Waites KB, Balish MF, Atkinson TP (2008) New insights into the pathogenesis and detection of *Mycoplasma pneumoniae* infections. Future Microbiol 3:635–648
7. Hammerschlag MR (2001) *Mycoplasma pneumoniae* infections. Curr Opin Infect Dis 14:181–186
8. Winchell JM, Thurman KA, Mitchell SL, Thacker WL, Fields BS (2008) Evaluation of three real-time PCR assays for the detection of *Mycoplasma pneumoniae* in an outbreak investigation. J Clin Microbiol 46: 3116–3118
9. Ursi D, Dirven K, Loens K, Ieven M, Goossens H (2003) Detection of *Mycoplasma pneumoniae* in respiratory samples by real-time PCR using an inhibition control. J Microbiol Methods 55:149–153
10. Loens K, Leven M, Ursi D, Beck T, Overdijk M, Sillekens P, Goossens H (2003) Detection of *Mycoplasma pneumoniae* by real-time nucleic acid sequence-based amplification. J Clin Microbiol 41:4448–4450
11. Thurman KA, Walter ND, Schwartz SB, Mitchell SL, Dillon MT, Deutscher M, Fulton JP, Tongren JE, Hicks LA, Winchell JM (2008) Comparison of laboratory diagnostic procedures for detection of *Mycoplasma pneumoniae* in community outbreaks. Clin Infect Dis 48:1244–1249
12. Kannan TR, Baseman JB (2006) ADP-ribosylating and vacuolating cytotoxin of *Mycoplasma pneumoniae* represents unique virulence determinant among bacterial pathogens. Proc Natl Acad Sci USA 103:6724–6729

Chapter 11

Real-Time PCR Assay for the Diagnosis of *Pneumocystis jirovecii* Pneumonia

Judith Fillaux and Antoine Berry

Abstract

Pneumocystis jirovecii is a common cause of life-threatening pneumonia among immunocompromised patients. Since *P. jirovecii* cannot be cultured, specific identification of it depends on examining respiratory specimens. In the last decade, PCR has been developed which allows the detection of very low levels of *P. jirovecii* not detectable by routine histochemical staining. We have shown that the direct immunofluorescence assay can be replaced by a real-time PCR assay given its feasibility, sensitivity, and specificity, for the detection of *P. jirovecii*. A negative PCR, performed on a LightCycler System®, enables a diagnosis of *Pneumocystis jirovecii* pneumonia (PjP) to be excluded, and the semiquantitative result with the application of some cutoff values can have a role in distinguishing between colonized or subclinically infected patients and PjP patients.

Key words: *Pneumocystis jirovecii*, Bronchoalveolar lavage, Diagnosis, Real-time PCR

1. Introduction

Pneumocystis jirovecii, an opportunistic fungus, causes pneumonia (*Pneumocystis jirovecii* pneumonia (PjP)) that is an important cause of morbidity and mortality among immunocompromised patients (1–4). In HIV-infected individuals, PjP presents with fever, cough, and dyspnea on exertion, and upon radiology, symmetric interstitial or granular opacities are observed (5, 6). The presentation of PjP in other immunocompromised patients differs from this classical acute presentation, causing diagnostic difficulties (7, 8). Since *P. jirovecii* cannot be cultured, specific identification of it depends on examining respiratory specimens. Bronchoscopy with bronchoalveolar lavage (BAL) is the preferred diagnostic procedure with reported sensitivities ranging from 89% to greater than 98%, using

Mark Wilks (ed.), *PCR Detection of Microbial Pathogens: Second Edition*, Methods in Molecular Biology, vol. 943, DOI 10.1007/978-1-60327-353-4_11, © Springer Science+Business Media, LLC 2013

any microscopy-based diagnostic technique (9, 10). In the last decade, PCR has been developed which allows the detection of very low levels of *P. jirovecii* not detectable by routine histochemical staining. These approaches have shown that *P. jirovecii* is also carried in asymptomatic individuals with only mild immunosuppression (long-term glucocorticoid therapy), in immunocompetent individuals with chronic pulmonary diseases, and in immunocompetent health-care workers (11–15).

The direct immunofluorescence assay (DFA), our "gold standard" method, can be replaced by the real-time PCR assay, given its feasibility, sensitivity, and specificity (16). A negative PCR enables a diagnosis of PjP to be excluded, and the semiquantitative result with the application of some cutoff values can have a role in distinguishing between colonized or subclinically infected patients and PjP patients.

2. Materials

The real-time PCR assay is performed according to a modification of the Larsen et al. (17) protocol. Patient samples should be tested, once, in duplicate.

2.1. Nonspecific Materials

1. Nuclease-free, aerosol-resistant pipette tips.
2. Absolute ethanol.
3. Isopropanol.

2.2. Patients' Specimens

1. Fresh BAL (see Note 1).
2. Mucolytic agent such as Digest-EUR® (Eurobio, Les Ulis, France).

2.3. Controls

1. External control: *P. jirovecii*-positive and -negative BAL samples, as determined by previous DFA and PCR, are included in each PCR run.
2. Contamination control: PCR-grade water.
3. Human internal control: Human beta-globin. From each extract, amplification of the human beta-globin gene is performed to assess the correct progress of DNA extraction and to underline the absence of PCR inhibitors.

2.4. DNA Extraction and Amplification

1. High Pure PCR Template Preparation Kit (Roche Diagnostics).
2. LightCycler® 2.0 System (Roche Diagnostics).
3. LightCycler® capillaries (Roche Diagnostics).

4. *P. jirovecii*: Primers and probes (17) are derived from the *P. jirovecii* major surface glycoprotein (MSG) gene (GenBank accession number: AF372980).
   - LightCycler® FastStart DNA Master HybProbe (Roche Diagnostics).
   - Primers: JKK14/15 (5′-GAA TGC AAA TCY TTA CAG ACA ACA G-3′) and JKK17 (5′-AAA TCA TGA ACG AAA TAA CCA TTG C-3′) (Eurogentec, Angers, France) amplify a 250-bp segment of the multicopy MSG gene family. Primers are used at a final concentration of 0.5 μM.
   - Probes labeled with fluorescein and Red 640: PCMSGFRET1U (5′-CAA AAA TAA CAY TSA CAT CAA CRA GGC G-fluorescein-3′) and PCMSGFRET1D (5′-Red 640-TGC AAA CCA ACC AAG TGT ACG ACA GG-3′) (Tib Molbiol, Berlin, Germany). The final concentrations of the fluorescein-labeled probes and LC-Red 640-labeled probes are 0.1 μM and 0.5 μM, respectively.
5. Human beta-globin: This PCR assay is performed separately.
   - LightCycler® FastStart DNA Master SYBR Green (Roche Diagnostics).
   - Primers: BG07 (5′-GGT TGG CCA ATC TAC TCC CAG G-3′) and BG08 (5′-TGG TCT CCT TAA ACC TGT CTT G-3′) (18). Primers are used at a final concentration of 0.1 μM.

3. Methods

3.1. Processing of Patients' Specimens

These should be handled under the laminar flow hood of the extraction room (see Note 2).

1. Vortex the patient's sample.
2. To a 2 ml sterile screw microcentrifuge tube, add 1.5 ml of the patient's sample. If the sample is too slimy, pretreat it with a mucolytic agent at a 1:1 v/v ratio. Leave at room temperature, mixing occasionally until it becomes liquified.
3. Centrifuge for 20 min at 900 × *g*.
4. Remove the supernatant, retain a 200-μl pellet, and perform the DNA extraction as described below.

3.2. DNA Extraction

This step should be performed in the DNA extraction room.

1. Turn on the incubator to 72°C.
2. Take 40 μl of proteinase K per patient sample from the freezer.

3. Fill a microcentrifuge tube with 200 μl elution buffer per patient sample and incubate at 72°C. The next three (4–6) steps are carried out under the laminar flow hood.
4. To the 200-μl pellet of the sample material, add 200 μl Binding Buffer and vortex.
5. Add 40 μl proteinase K and vortex.
6. Incubate at 72°C for 10 min. The next 18 steps (7–24) are carried out on the laboratory bench.
7. Mix briefly to avoid aerosol creation.
8. Add 100 μl isopropanol in order to inactivate Proteinase K and vortex.
9. Mix briefly to avoid aerosol creation.
10. Insert one filter tube in one collection tube.
11. Pipette the sample into the upper buffer reservoir and close.
12. Centrifuge for 90 s at 8,000 × *g*.
13. Remove the filter tube from the collection tube, and combine the filter tube with a new collection tube.
14. Add 500 μl "Inhibitor Removal Buffer" to the upper reservoir and close.
15. Centrifuge for 90 s at 8,000 × *g*.
16. Remove the filter tube from the collection tube, and combine the filter tube with a new collection tube.
17. Add 500 μl "Wash Buffer" to the upper reservoir and close.
18. Centrifuge for 90 s at 8,000 × *g*, and then 90 s at 13,000 × *g*.
19. Remove the filter tube from the collection tube, and combine the filter tube with a new collection tube.
20. Add 500 μl prewarmed "Elution Buffer" (see step 3 above) to the upper reservoir and close.
21. Centrifuge for 90 s at 8,000 × *g*.
22. Remove the filter tube from the collection tube, and combine the filter tube with a new collection tube.
23. The collection tube now contains eluted DNA. Remove the filter tube. Transfer the eluted DNA in a 2 ml microcentrifuge tube.
24. Store the microcentrifuge tube at +4°C until amplification.

3.3. *P. jirovecii* amplification (see Note 3)

These steps should take place in the DNA-free room; work on a small sheet of aluminum foil.

3.3.1. Preparation of the Master Mix

1. Thaw one vial of "Reaction Mix."
2. Briefly centrifuge one vial "Enzyme" and the thawed vial of "Reaction Mix."

3. Pipette 60 μl from the "Reaction Mix" vial into the "Enzyme" vial.
4. Mix by pipetting up and down.
5. Label the "Enzyme" vial, "Pneumocystis Taq Mix," and keep refrigerated for a maximum of 1 week. Do not freeze.

3.3.2. Preparation of the PCR Mix

This step should take place in the DNA-free room; work on a small sheet of aluminum foil:

1. Thaw the solutions of primers and probes. Do not vortex or refreeze. Keep refrigerated until 10 days after thawing and then discard the remaining reagent.
2. Prepare a 10× conc. solution of v/v pooled primers and a 1× conc. solution of probes.
3. Mix the $MgCl_2$ stock solution.
4. In a 2 ml reaction tube, prepare the PCR mix by adding the following components in the order shown in Table 1.
5. Check no residue remains in the tip after each pipetting step.
6. Do not vortex the PCR mix.
7. Take a new box of LightCycler capillary tubes and remove the top.
8. Transfer the capillaries to an empty sterilized box.
9. Carefully mix the components of the PCR mix by pipetting up and down.

Table 1
Preparation and final concentration of components of the test PCR

Component	Volume (μl)	Final concentration
Water	3	
$MgCl_2$	1.2	4 mM[a]
JKK 14/15, JKK 17 Primer mix	2	1 μM/primer
Red Probe mix	0.25	0.5 μM
Fluo Probe mix	0.05	0.1 μM
Pneumocystis Taq Mix	1	1×
Total volume	7.5	

[a]The $MgCl_2$ final concentration takes into account the $MgCl_2$ contained in the Pneumocystis Taq Mix

10. Pipette 7.5 μl PCR mix into each precooled capillary with the same tip without sealing (see Note 4).
11. Close the sterilized box and store the PCR mix at −20°C (max 1 month).

3.3.3. Addition of the DNA Template

This is performed in the DNA extraction room on the bench.

1. Take the required number of capillaries, prepared with the PCR mix, from the freezer and place them in precooled centrifuge adapters wrapped up in aluminum foil.
2. Take the DNA samples from the fridge and thaw the controls.
3. Mix and then briefly centrifuge the DNA samples and the controls in a standard benchtop microcentrifuge.
4. Add 2.5 μl of the DNA templates and seal with a stopper after filling each capillary (see Note 4).
5. Transfer the capillaries into the sample carousel of the LightCycler®.

3.3.4. Program the LightCycler®

Use the conditions shown in Table 2

3.4. Human Beta-Globin Amplification

3.4.1. Preparation of the Master Mix

These steps should take place in the DNA-free room; work on a small sheet of aluminum foil.

1. Thaw three vials of "Reaction Mix" and shield from light.
2. Briefly centrifuge one vial "Enzyme" and the thawed vials of "Reaction Mix."
3. Pipette 10 μl from "Enzyme" vial into each "Reaction Mix" vial, keep one in the fridge (for a maximum of 1 week), and freeze the other two.
4. Mix by pipetting up and down.
5. Label the "Reaction Mix" vial, "beta-globin Taq mix." Always keep it away from light.

3.4.2. Preparation of the PCR mix

These steps should take place in the DNA-free room; work on a small sheet of aluminum foil.

1. Thaw the primer solutions. Do not vortex or refreeze thawed primers.
2. Prepare a 10× conc. solution of pooled primers.
3. Mix the $MgCl_2$ stock solution.
4. In a 2 ml reaction tube, prepare the PCR mix by adding the following components in the order shown in the table (see Table 3).

Table 2
Program conditions for the *Pneumocystis* PCR

Program

Program name	Denaturation					
Cycles	1	Analysis mode	None			
Target (°C)	Hold (hh:mm:ss)	Slope (°C/s)	Sec target (°C)	Step size (°C)	Step delay (cycles)	Acquisition mode
95	00:10:00	20	0	0	0	None
Program name	Amplification 1					
Cycles	6	Analysis mode	None			
Target (°C)	Hold (hh:mm:ss)	Slope (°C/s)	Sec target (°C)	Step size (°C)	Step delay (cycles)	Acquisition mode
95	00:00:05	20	0	0	0	None
65	00:00:10	20	60	1	0	Single
72	00:00:20	20	0	0	0	None
Program name	Amplification 2					
Cycles	40	Analysis mode	Quantification			
Target (°C)	Hold (hh:mm:ss)	Slope (°C/s)	Sec target (°C)	Step size (°C)	Step delay (cycles)	Acquisition mode
95	00:00:05	20	0	0	0	None
60	00:00:10	20	50	2	0	Single
72	00:00:20	20	0	0	0	None
Program name	Refroidissement					
Cycles	1	Analysis mode	None			
Target (°C)	Hold (hh:mm:ss)	Slope (°C/s)	Sec target (°C)	Step size (°C)	Step delay (cycles)	Acquisition mode
40	00:01:00	20	0	0	0	None

Table 3
Preparation and final concentration of components of the beta-globin control PCR

Component	Volume (μl)	Final concentration
Water	5.1	
$MgCl_2$	1.2	4 mM[a]
B07/08 Primer mix	0.2	0.1 μM/primer
Beta-globin Taq mix	1	1×
Total volume	7.5	

[a]The $MgCl_2$ final concentration takes into account the $MgCl_2$ contained in the beta-globin Taq Mix

5. Check no residue remains in the tip after each pipetting step.
6. Do not vortex the PCR Mix.
7. Store the PCR Mix at 4°C.

3.4.3. Distribution of the PCR Mix and DNA (see Note 4)

These steps should take place in the DNA-free room; work on a small sheet of aluminum foil.

1. Place the required number of capillaries in precooled centrifuged adapters wrapped up in an aluminum foil.
2. Mix carefully the PCR mix by pipeting up and down.
3. Pipette 7.5 μl PCR mix into each precooled capillary with the same tip.
4. Take the DNA samples from the fridge and thaw the controls.
5. Mix and then briefly centrifuge the DNA samples and the controls in a standard benchtop microcentrifuge.
6. Add 2.5 μl of the DNA templates and seal with a stopper after each capillary.
7. Transfer the capillaries into the sample carousel of the LightCycler®.

3.4.4. Program the LightCycler®

Using the Conditions Shown in Table 4

3.5. Interpretation (see Note 5)

1. The sensitivity and specificity of the real-time PCR assay depend on C_t of detection. If the C_t chosen is 22, the specificity of the real-time PCR assay is 100% and if the C_t is 28, the sensitivity is 100%. Between these two cutoff points, there is a grey zone in which the PCR value and the DFA result can be discrepant.

Table 4
Program conditions for the beta-globin PCR

Program						
Program name	Denaturation					
Cycles	1	Analysis mode	None			
Target (°C)	Hold (hh:mm:ss)	Slope (°C/s)	Sec target (°C)	Step size (°C)	Step delay (cycles)	Acquisition mode
95	00:08:00	20	0	0	0	None
Program name	Amplification					
Cycles	40	Analysis mode	Quantification			
Target (°C)	Hold (hh:mm:ss)	Slope (°C/s)	Sec target (°C)	Step size (°C)	Step delay (cycles)	Acquisition mode
95	00:00:15	20	0	0	0	None
70	00:00:10	20	0	0	0	None
72	00:00:20	20	0	0	0	Single
Program name	Fusion					
Cycles	1	Analysis mode	Melting curves			
Target (°C)	Hold (hh:mm:ss)	Slope (°C/s)	Sec target (°C)	Step size (°C)	Step delay (cycles)	Acquisition mode
95	00:00:05	20	0	0	0	None
62	00:00:20	20	0	0	0	None
95	00:00:00	0,05	0	0	0	Continuous
Program name	Refroidissement					
Cycles	1	Analysis mode	None			
Target (°C)	Hold (hh:mm:ss)	Slope (°C/s)	Sec target (°C)	Step size (°C)	Step delay (cycles)	Acquisition mode
40	00:00:00	20	0	0	0	None

2. Four interpretations can be proposed to clinicians (16):
   - *Negative PCR results*: "Absence of *Pneumocystis jirovecii*. PjP diagnosis excluded."
   - *Positive PCR with Ct≥28*: "Low fungal burden: *Pneumocystis* pneumonia improbable. Prophylactic treatment could be necessary in case of an immunocompromised patient."
   - *Positive PCR with 22≤Ct<28*: "Moderate fungal burden: possible *Pneumocystis* pneumonia, interpretation depends on anamnesis, clinical and radiological signs. Specific treatment (curative or prophylactic) is necessary in case of an immunocompromised patient."
   - *Positive PCR with Ct<22*: "High fungal burden: *Pneumocystis* pneumonia. To be treated."

4. Notes

1. Collection of BAL fluid: In brief, BAL is performed by instilling three 50-ml aliquots of warm pyrogen-free normal saline (0.9%), followed by gentle suction after each aliquot is infused. The first aliquot is separated from the subsequent two, which are pooled before being dispatched between the different laboratories. This technique optimizes the separation of bronchial epithelial cells from truly alveolar material (19). The final volume recovered ranges from 75 to 100 ml.
2. In order to avoid contamination, it is of crucial importance to maintain strict physical separation (different rooms) of the various steps involved in the PCR, and to respect specific cleaning and manipulation procedures. For example, organize your bench to avoid the need for the pipette to pass above the capillaries when distributing the DNA samples.
3. In order not to waste a run, when preparing the PCR mix, do not provide for one additional reaction. This will allow you to verify that you have not made a mistake during the preparation of the PCR mix.
4. In order to improve the reproducibility, when distributing the PCR Mix into the capillaries, put the tip on the edge of the capillary without touching inside, and do not press until the second step of the pipette. Loosen the pressure only when the pipette is in the PCR mix again. When distributing the DNA sample, press until the second step of the pipette.

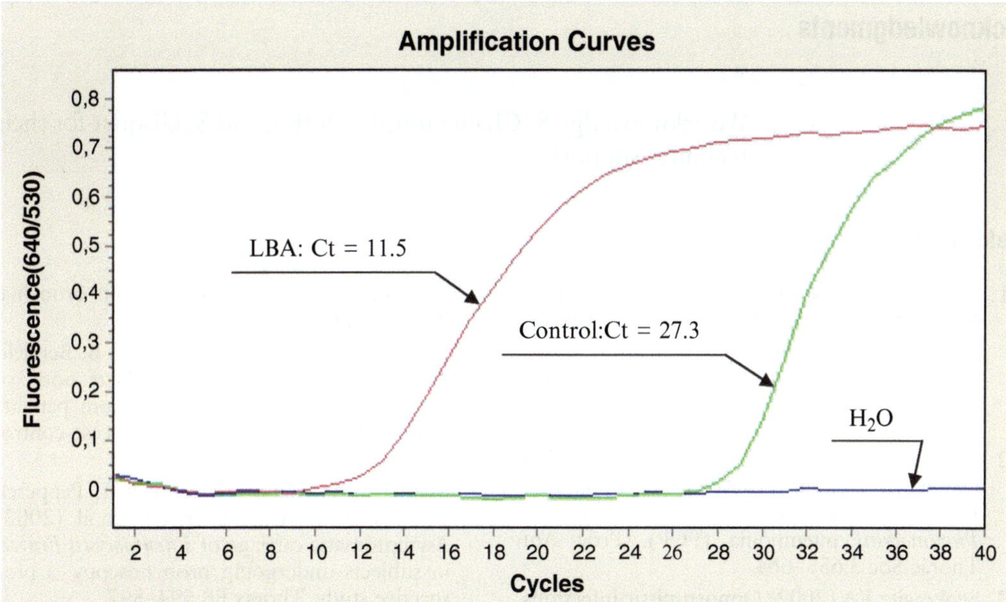

Fig. 1. Amplification curves of real-time PCR for the diagnosis of *Pneumocystis jirovecii* pneumonia.

5. All acquired fluorescence data are analyzed and the quantification of the *P. jirovecii* DNA is represented by the cycle threshold (C_t) measured by the LightCyler® software (see Fig. 1).
   - A C_t value is reported for each positive sample (20). A sample is regarded as positive if the two tubes are positive by this assay.
   - A positive result in only one tube warrants a re-amplification of the DNA previously extracted, following the same procedure. If the retest has at least one of the two tubes positive, the sample is considered positive for *P. jirovecii*.
   - A negative MSG result has to have a positive result for the human beta-globin PCR to be considered valid, to ensure the correct progress of extraction and the absence of inhibitors in the specimen. A negative result for the human beta-globin PCR requires retesting (re-extraction and re-amplification performed following the previously described procedure) of an additional aliquot of the original sample (pure and $^1/_{10}$ diluted). If the human beta-globin PCR remains negative with the pure sample but is positive with the diluted sample, the patient's specimen has to be reanalyzed for the quantification of the *P. jirovecii* DNA on the diluted sample.

Acknowledgments

We acknowledge S. Chalmeton, E. Duthu, and S. Gisquet for their technical support.

References

1. Cisneros JM, Munoz P, Torre-Cisneros J, Gurgui M, Rodriguez-Hernandez MJ, Aguado JM et al (1998) Pneumonia after heart transplantation: a multi-institutional study. Spanish Transplantation Infection Study Group. Clin Infect Dis 27:324–331
2. Huang L, Morris A, Limper AH, Beck JM (2006) An Official ATS Workshop Summary: Recent advances and future directions in *Pneumocystis* pneumonia (PCP). Proc Am Thorac Soc 3:655–664
3. Sepkowitz KA (2002) Opportunistic infections in patients with and patients without Acquired Immunodeficiency Syndrome. Clin Infect Dis 34:1098–1107
4. Wakefield AE (2002) Pneumocystis carinii. Br Med Bull 61:175–188
5. DeLorenzo LJ, Huang CT, Maguire GP, Stone DJ (1987) Roentgenographic patterns of *Pneumocystis carinii* pneumonia in 104 patients with AIDS. Chest 91:323–327
6. Huang L, Stansell J, Osmond D, Turner J, Shafer KP, Fulkerson W et al (1999) Performance of an algorithm to detect *Pneumocystis carinii* pneumonia in symptomatic HIV-infected persons. Pulmonary Complications of HIV Infection Study Group. Chest 115:1025–1032
7. Dei-Cas E (2000) *Pneumocystis* infections: the iceberg? Med Mycol 38(Suppl 1):23–32
8. Thomas CF, Limper AH (2004) Pneumocystis pneumonia. N Engl J Med 350:2487–2498
9. Broaddus C, Dake MD, Stulbarg MS, Blumenfeld W, Hadley WK, Golden JA et al (1985) Bronchoalveolar lavage and transbronchial biopsy for the diagnosis of pulmonary infections in the acquired immunodeficiency syndrome. Ann Intern Med 102:747–752
10. Huang L, Hecht FM, Stansell JD, Montanti R, Hadley WK, Hopewell PC (1995) Suspected *Pneumocystis carinii* pneumonia with a negative induced sputum examination. Is early bronchoscopy useful? Am J Respir Crit Care Med 151:1866–1871
11. Calderon EJ, Regordan C, Medrano FJ, Ollero M, Varela JM (1996) *Pneumocystis carinii* infection in patients with chronic bronchial disease. Lancet 347:977
12. Helweg-Larsen J, Jensen JS, Dohn B, Benfield TL, Lundgren B (2002) Detection of Pneumocystis DNA in samples from patients suspected of bacterial pneumonia-a case-control study. BMC Infect Dis 2:28
13. Maskell NA, Waine DJ, Lindley A, Pepperell JC, Wakefield AE, Miller RF et al (2003) Asymptomatic carriage of *Pneumocystis jiroveci* in subjects undergoing bronchoscopy: a prospective study. Thorax 58:594–597
14. Miller RF, Ambrose HE, Wakefield AE (2001) *Pneumocystis carinii* f. sp. hominis DNA in immunocompetent health care workers in contact with patients with *P. carinii* pneumonia. J Clin Microbiol 39:3877–3882
15. Nevez G, Raccurt C, Jounieaux V, Dei-Cas E, Mazars E (1999) Pneumocystosis versus pulmonary *Pneumocystis carinii* colonization in HIV-negative and HIV-positive patients. AIDS 13:535–536
16. Fillaux J, Malvy S, Alvarez M, Fabre R, Cassaing S, Marchou B et al (2008) Accuracy of a routine real-time PCR assay for the diagnosis of *Pneumocystis jirovecii* pneumonia. J Microbiol Methods 75:258–261
17. Larsen HH, Masur H, Kovacs JA, Gill VJ, Silcott VA, Kogulan P et al (2002) Development and evaluation of a quantitative, touch-down, real-time PCR assay for diagnosing *Pneumocystis carinii* pneumonia. J Clin Microbiol 40:490–494
18. Saiki RK, Scharf S, Faloona F, Mullis KB, Horn GT, Erlich HA et al (1985) (1992) Enzymatic amplification of beta-globin genomic sequences and restriction site analysis for diagnosis of sickle cell anemia. Biotechnology 24:476–480
19. Rennard SI, Ghafouri M, Thompson AB, Linder J, Vaughan W, Jones K et al (1990) Fractional processing of sequential bronchoalveolar lavage to separate bronchial and alveolar samples. Am Rev Respir Dis 141:208–217
20. Bustin SA (2000) Absolute quantification of mRNA using real-time reverse transcription polymerase chain reaction assays. J Mol Endocrinol 25:169–193

Chapter 12

Rapid Identification of Mycobacteria and Rapid Detection of Drug Resistance in *Mycobacterium tuberculosis* in Cultured Isolates and in Respiratory Specimens

Wing-Cheong Yam and Kit-Hang Gilman Siu

Abstract

Recent advances in molecular biology and better understanding of the genetic basis of drug resistance have allowed rapid identification of mycobacteria and rapid detection of drug resistance of *Mycobacterium tuberculosis* present in cultured isolates or in respiratory specimens. In this chapter, several simple nucleic acid amplification-based techniques are introduced as molecular approach for clinical diagnosis of tuberculosis. A one-tube nested IS6110-based polymerase chain reaction (PCR) is used for *M. tuberculosis* complex identification; the use of a multiplex allele-specific PCR is demonstrated to detect the isoniazid resistance; PCR-sequencing assays are applied for rifampicin and ofloxacin resistance detection and 16S rDNA sequencing is utilized for identification of mycobacterial species from cultures of acid fast bacilli (AFB). Despite the high specificity and sensitivity of the molecular techniques, mycobacterial culture remains the "Gold Standard" for tuberculosis diagnosis. Negative results of molecular tests never preclude the infection or the presence of drug resistance. These technological advancements are, therefore, not intended to replace the conventional tests, but rather have major complementary roles in tuberculosis diagnosis.

Key words: *Mycobacterium tuberculosis*, *IS6110* PCR, COBAS® TaqMan® MTB Test, PCR sequencing, Rifampicin resistance, Isoniazid resistance, Multiplex allele-specific PCR, Ofloxacin resistance, 16S rDNA sequencing

1. Introduction

Tuberculosis (TB) is a contagious disease caused by mycobacteria, mainly *Mycobacterium tuberculosis*. Overall, one-third of the world's population is currently infected with the TB bacillus. According to a WHO report, there were 9.2 million new cases and 1.7 million deaths resulting from TB in 2006 (1). Due to the risk of spread of the disease, the potential for the emergence of drug-resistant

Mark Wilks (ed.), *PCR Detection of Microbial Pathogens: Second Edition*, Methods in Molecular Biology, vol. 943, DOI 10.1007/978-1-60327-353-4_12,

strains, and the severity of the disease in patients infected with HIV-1, rapid diagnosis of *M. tuberculosis* is extremely important to allow clinical decisions on the initiation and type of antituberculosis treatment to be made, with or without isolation of the patient to prevent secondary transmission to susceptible contacts (2).

1.1. Identification of M. tuberculosis in Clinical Specimens

The isolation of an organism of the *M. tuberculosis* complex is required for the definitive diagnosis of tuberculosis. Routine cultures are time-consuming and can take up to 8 weeks. Examination of a direct smear for acid-fast bacilli (AFBs) in respiratory specimens provides a rapid diagnosis for tuberculosis, but the detection limit is about 5×10^3 bacilli per milliliter of specimen, and *M. tuberculosis* cannot be distinguished from other AFBs (3). Immunological and serological techniques are of limited use, since they are invariably confounded by previous BCG vaccination, exposure to environmental mycobacteria and also conditions which results in immunosuppression (4, 5). Nucleic acid amplification assays have been extensively evaluated and confirmed to be a reliable and rapid strategy for identification and drug susceptibility testing of *M. tuberculosis.* Recent studies have shown that turnaround times can be further shortened by allowing the detection of mycobacteria directly from clinical specimens before the culture result is available (6–9).

In this chapter, two PCR-based approaches for rapid identification of *M. tuberculosis* are described in detail. The first approach is an in-house single-tube nested PCR targeting the *M. tuberculosis* complex-specific insertion sequence *IS6110.* The assay utilizes two sets of primers with different melting temperatures (Tm), 88°C for external primers, 70°C for internal primers, and two amplification cycles with different annealing temperature, to augment sensitivity and specificity (7, 10, 11). Compared to a two tube nested protocol, this single-tube nested PCR format also reduces the risk of cross-contamination. The coupled use of dUTP and uracil-*N*-glycosylase also prevents carryover of amplicons from previous experiments (see Fig. 1a). A recent study reported that the assays were 100% specific and had overall diagnostic sensitivities

Fig. 1. (**a**) Graphical description of the principle of one-tube nested IS6110 PCR. At initial 37°C, UNG is activated and it destructs uracil-containing previous amplicon. At 94°C, 12 min, the UNG is denatured with simultaneously activation of the AmpliTaq Gold. On the first amplification cycles with annealing temperature of 72°C, external primers bind with template with high stringency and produce amplicon for next cycles. On the second amplification cycles, a much lower annealing temperature would be used for the internal primers to perform amplification with greater efficiency rather than stringency. (**b**) Gel photo showing the results of IS6110 PCR identification of *M. tuberculosis* complex. M: 100-bp DNA ladder; lane 1 and 2: respiratory specimen s positive for *M. tuberculosis* complex; lane 3, 5, 6, 7, 8, 10: respiratory specimen s negative for *M. tuberculosis* complex; lane 4: strong positive control DNA from H37Rv; lane 9: weak positive control DNA prepared by tenfold dilution of strong positive control; lane 10:mQ water as a negative control.

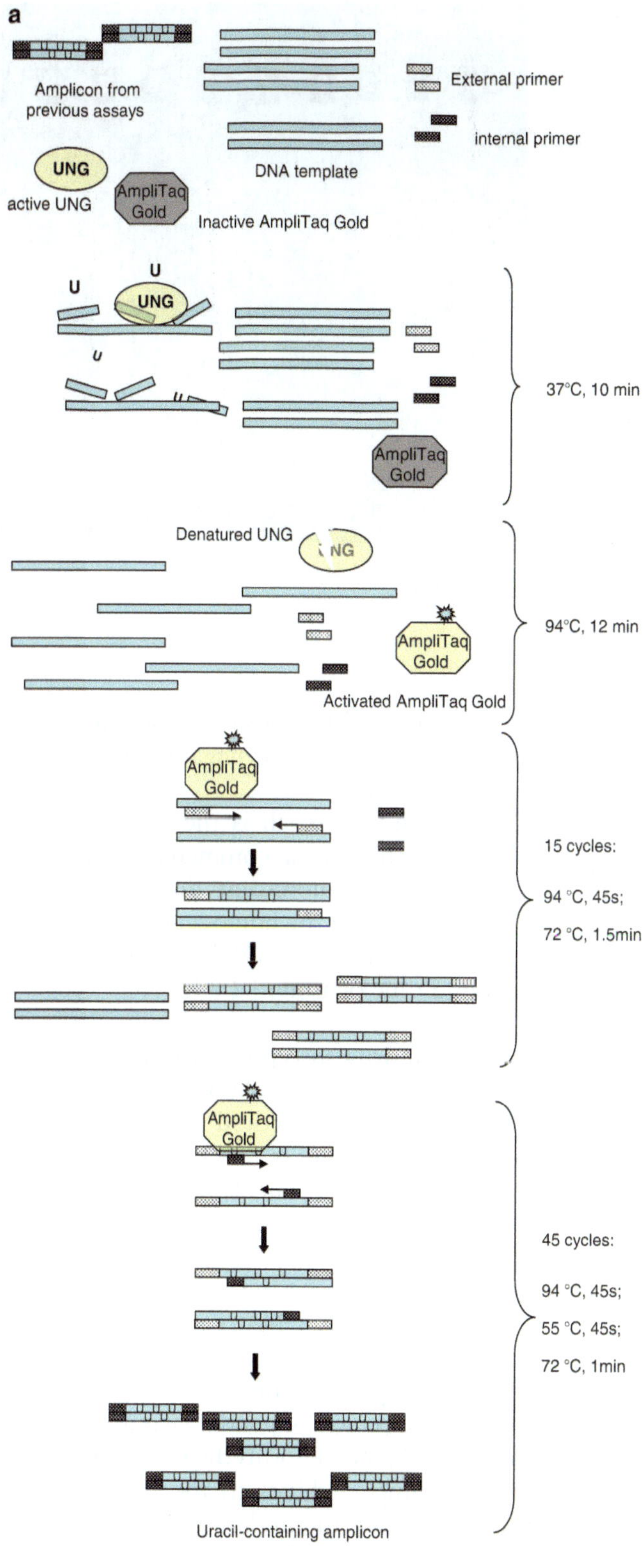
a
Amplicon from previous assays
External primer
internal primer
UNG
active UNG
AmpliTaq Gold
DNA template
Inactive AmpliTaq Gold
U
UNG
37°C, 10 min
Denatured UNG
94°C, 12 min
Activated AmpliTaq Gold
15 cycles:
94 °C, 45s;
72 °C, 1.5min
45 cycles:
94 °C, 45s;
55 °C, 45s;
72 °C, 1min
Uracil-containing amplicon

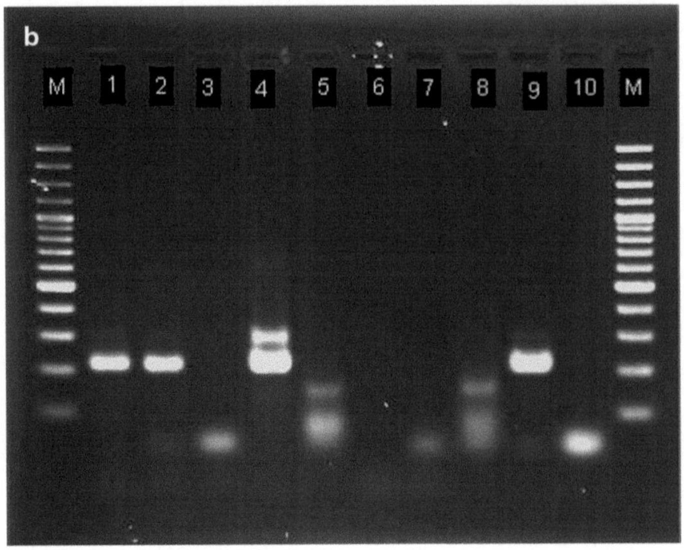

Fig. 1. (continued)

of ca. 90% for culture positive respiratory specimens, although their sensitivities for smear-negative specimens are still limited at 60–70% (7, 10, 11).

The second approach is the use of the COBAS® TaqMan® MTB Test, which is a commercial assay that integrates PCR amplification and fluorescent detection of *M. tuberculosis* 16S rRNA gene by using the homogeneous 5′ nuclease assays. The test is performed inside the COBAS® TaqMan® 48 Analyzer with the turnaround time of around 2.5 h.

1.2. Detection of Drug Resistance in M. tuberculosis

Conventional drug susceptibility test requires sufficient growth of mycobacterial colonies to allow standardization of inocula used in the agar proportional method (12), which takes another 2 weeks after a positive culture is first obtained. Recent developments in molecular biology and elucidation of the genetic basis of antituberculous drug resistance have provided new tools for the rapid detection of drug resistance in *M. tuberculosis* present in culture or even in clinical specimens.

PCR followed by DNA sequencing has been the most widely used method for detecting resistance-associated mutations. The technique is especially efficient when the mutations are confined within a small region of genes. Rifampicin (RIF), in combination with isoniazid (INH), forms the backbone of short-course chemotherapy. Rifampicin binds to the β-subunit of DNA-dependent RNA polymerase to hinder transcription and thereby kill the microorganism (13). More than 95% of rifampicin resistance was shown to be caused by missense mutations, in-frame deletions or

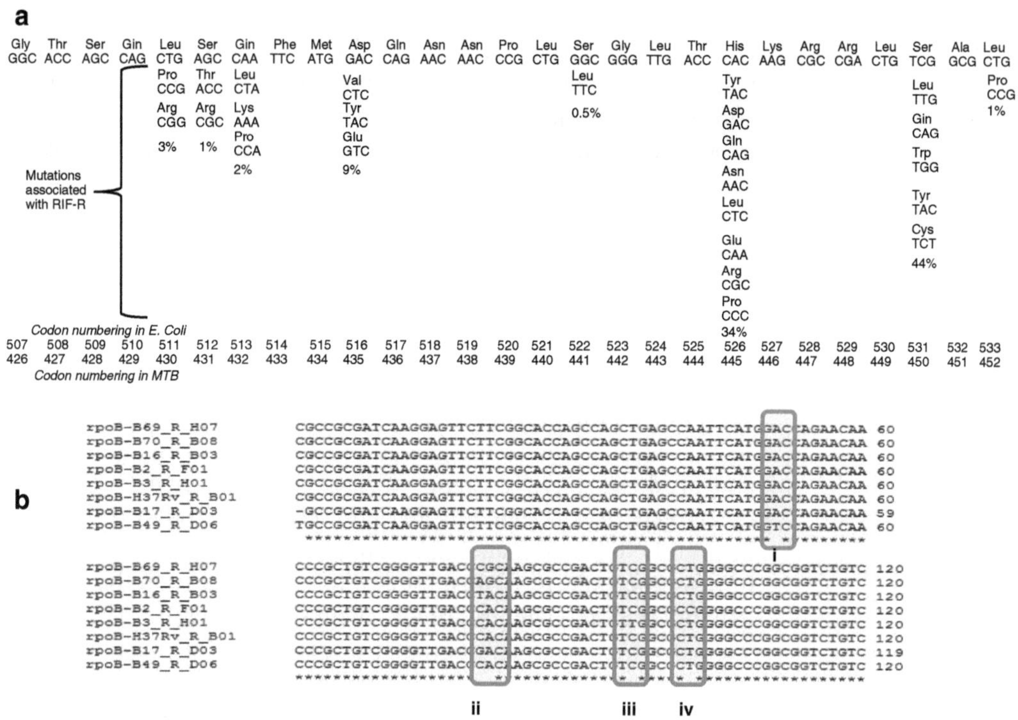

Fig. 2. (**a**) Diagram showing the regions and the frequencies of mutations associated with RIF resistance in *M. tuberculosis*. (**b**) Alignment file of ClustalW displaying the alignment results of 7 RIF-R *M. tuberculosis* clinical strains against susceptible reference strain H37Rv. Region i: GAC→GTC resulted in D516V in rpoB-B49; Region ii: CAC→CGC, CAC→AGC, CAC→TAC, and CAC→ GAC resulted in H526R in rpoB-B69, H526S in rpoB-B70, H526Y in rpoB-B16, and H526D in rpoB-B17, respectively; region iii: TCG→TTG resulted in S531L in rpoB-B3; region iv: CTG→CCG resulted in L533P in rpoB-B2.

insertions in the *rpoB* gene encoding the β-subunit. In the great majority of RIF resistant strains, mutations occurs within a 81-bp hotspot region, named as rifampicin resistance determining region (RRDR) encoding 27 amino acids and corresponding to codons 507–533 or cluster I according to *Escherichia coli* numbering. These occur in the center of the *rpoB* gene with the most common changes in codons Ser531Leu, His526Tyr, and Asp516Val (see Fig. 2a) (13). A previous study reported a molecular assay mainly based on PCR sequencing, to detect RIF-resistant *M. tuberculosis* in culture (14) by assessing the presence of resistance-associated mutations inside the RRDR. The assay was subsequently modified by Yam et al. for direct application in clinical samples, shortening the turnaround time for drug resistance detection from a month to a couple of days (15).

INH enters *M. tuberculosis* as a prodrug and requires activation via the mycobacterial catalase-peroxidase enzyme encoded by the *kat*G gene (16). The activated INH appears to disrupt the biosynthetic pathway of mycolic acid, an essential component of mycobacterial cell wall, through the inhibition of the NADH-dependent enoyl-ACP reductase enzyme (InhA) (17). Unlike rifampicin, INH resistance is likely to be mediated by several molecular mechanisms and involve numerous genes, namely *Kat*G, *mab*A-*inh*A operon, *ahp*C, *kas*A, and *ndh* (18). The predominant mechanism is associated with mutations in *kat*G, particularly in codon 315. Approximately 50% of INH-resistant *M. tuberculosis* strains were found to harbor mutations in *kat*G315 (18, 19). One particular base substitution at codon 315, Ser→Thr (AGC→ACC), was reported to be the most frequently mutated allele. The second most common resistance mechanism is that about 8–30% of isoniazid resistant *M. tuberculosis* isolates carry a mutation located at the 15th nucleotide preceding the *mabA-inhA* operon (19–21). This mutation infers the overexpression of the INH drug target, inhA, and produces INH resistance via a titration mechanism (13). To detect genetic determinants for INH resistance, the PCR sequencing can be expensive and elaborate as multiple genes are involved in resistance and resistance mutations are not clustered in a small genetic region. The multiplex allele-specific PCR (MAS-PCR) assay is a molecular strategy that utilizes three primers, with two outer primers flanking a region of interest and which invariably anneal to the conserved region of the target. The 3′ end base of the wildtype allele-specific inner primer is complementary to the wildtype targeted codon to amplify a wildtype allele-specific fragment in cases where there is no mutation present. An alteration of the targeted codon would cause a mismatch between the inner primer and the target DNA and result in no amplification of the wildtype allele-specific fragment. Because of the ease and low running cost, the assay facilitates the high-throughput analysis of multiple gene polymorphisms. The application of MAS-PCRs in detecting *katG*315 and *mabA*-15 mutations in *M. tuberculosis* clinical isolates were reported previously (21, 22). By combining the two MAS-PCR assays, a total of 77.5% of INH-resistant *M. tuberculosis* isolates were identified in a study performed in Hong Kong (21). Recently, our laboratory also extended the application of MAS-PCR assays in the detection of INH resistance in *M. tuberculosis* present in clinical specimens, providing a rapid and cost-effective measure for INH resistance detection (23).

Ofloxacin (OFX) is the common fluoroquinolone often used as a second-line drug in the treatment of multidrug resistant *M. tuberculosis* (which is defined as resistance to RIF and INH). The OFX targets and inactivate DNA gyrase, a type II DNA topoisomerase (24, 25). The DNA gyrase is encoded by two genes, *gyrA* and *gyrB*. Like RIF, resistance-associated mutations are tend to be confined within a small gene region, known as quinolone resistance-determining region (QRDR). The 120 bp-QRDR located inside the *gyrA* gene is a conserved region that corresponds to the point of interaction between the drug and the gyrase. Missense

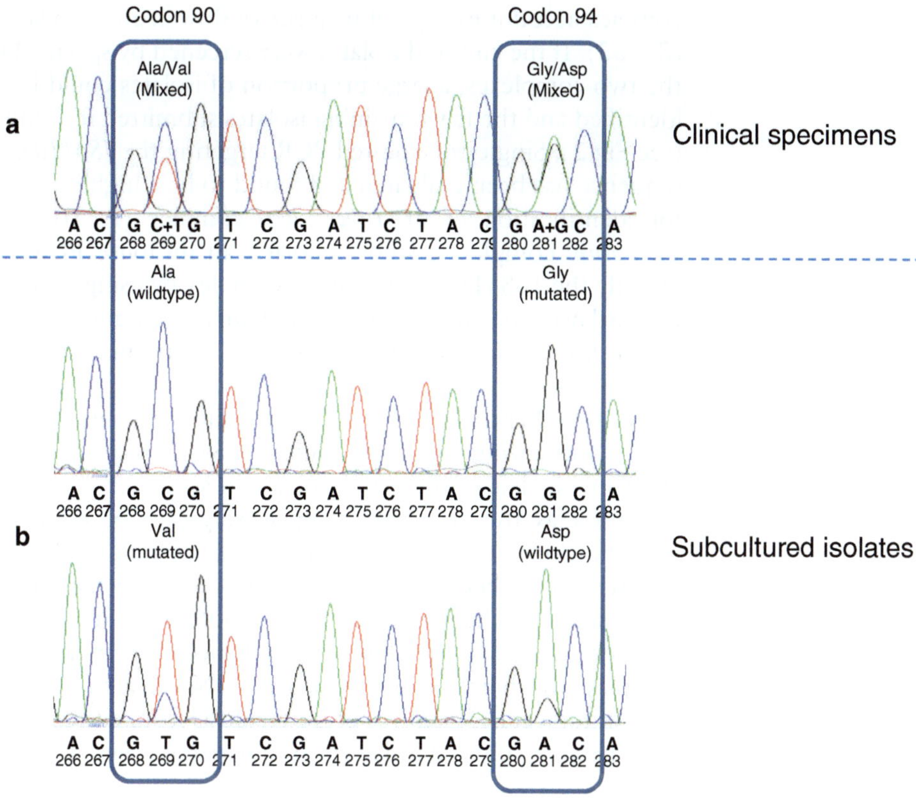

Fig. 3. (**a**) In a clinical respiratory specimen, *GyrA* PCR-sequencing assay identified a *M. tuberculosis* strain carrying mixed bases at nucleotides 269 and 281 that corresponding to codon 90 and 94 in *gyrA* gene. (**b**) In corresponding AFB culture, two subpopulation of OFX-R *M. tuberculosis* were identified. The PCR-sequencing assay identified one subpopulation harboring A90V mutation and wildtype codon 94, whereas another subpopulation harboring wildtype codon 90 and G94D mutation. Both isolates have OFX MIC of 16 μg/mL.

mutations in codon 90, 91, and 94 in the QRDR were shown to be highly associated with resistance to OFX. PCR-sequencing assays have also been developed to detect OFX-resistant *M. tuberculosis* in both culture and specimens (26) (see Fig. 3).

1.3. Genotypic Identification of Non-tuberculous Mycobacteria from Culture

Identification of mycobacteria by growth characteristics and conventional biochemical tests takes 3–5 weeks to obtain sufficient cells for analysis. Various molecular methods have been developed to facilitate the rapid and reliable identification of many mycobacterial species (27–32). Previous study characterized hypervariable regions within the 16S rDNA molecule that exhibit species-specific characteristics in mycobacteria and suggested that direct sequencing analysis of these regions could be a reliable genotypic method for identification of mycobacterial species (32). In our experience, 16S rDNA sequencing is an excellent tool for mycobacterial species identification (33). Nevertheless, the blind use of DNA sequencing would be too expensive for routine diagnosis for all samples, although this may change in the near future. Previous studies indicated that *M. tuberculosis* complex and *M. avium* complex account for a great majority of slow growers in laboratory (32, 33). If the cultured isolates were screened by specific PCR for the two complexes, a large proportion of isolates could be readily identified and the few remaining isolates submitted for sequencing (see Fig 2). Single-tube nested PCR targeting the *IS6110* insertion sequence has been evaluated and found to be a highly specific tool for identification of *M. tuberculosis* complex (7, 10, 11). For *M. avium* complex, a study reported an in-house PCR assay to identify the 16S rDNA gene of the complex by using two different forward primers, one specific for *M. avium*, while another is specific for *M. intracellulare*, and one common reverse primer (31). Slow growing isolates with negative results on both complex-specific PCRs, together with the rapid growers, should be subjected to 16S rRNA sequencing for identification (33). A sequence match of ≧99% with that of the prototype strain sequence in a repository was used as the criterion for species, group or complex level identification (34). Although sequencing has high resolving power to discriminate between mycobacterial species, certain phenotypic tests, such as chromogenicity and photoreactivity, have to be retained to resolve the closely related species or members of the complex with highly similarity of 16S rDNA sequences. The overview of molecular diagnosis of tuberculosis from either respiratory specimens or cultured isolates is summarized in Fig. 4.

2. Materials

2.1. Specimen Decontamination by Sputasol-Sodium Hydroxide

1. 4% (w/v) NaOH–sputasol mixture: Aseptically add the whole contents (7.5 mL) of a vial of Sputasol (Oxoid, Hampshire, UK) to 50 mL 8% (w/v) NaOH. Top up the volume to 100 mL using sterile distilled water. Stable for 48 h at 4°C.
2. Sodium phosphate buffer: 0.067 M sodium phosphate, pH5.3. Stored at room temperature for up to 3 months.

2.2. DNA Extraction

2.2.1. DNA Extraction by In-House Alkaline Lysis Method (for All Purposes Except Use in the COBAS® TaqMan® MTB Test)

1. Washing buffer: 0.1 M Tris/HCl, pH7.5. Stored at 4°C.
2. Lysis buffer: 1% (w/v) sodium hydroxide, 0.025% (w/v) sodium dodecyl sulfate. Stored at 4°C.
3. Neutralization buffer: 0.1 M Tris/HCl, pH7.5. Stored at 4°C.

2.2.2. DNA Extraction by AMPLICOR® Respiratory Specimen Preparation Kit (for All Purpose) (Roche Diagnostics, Basel, Schweiz)

This kit consists of a Respiratory Specimen Wash Solution (RW), Respiratory Specimen Lysis Reagent RL), and Respiratory Specimen Neutralization Reagent (RN). All reagents are stored at 4°C.

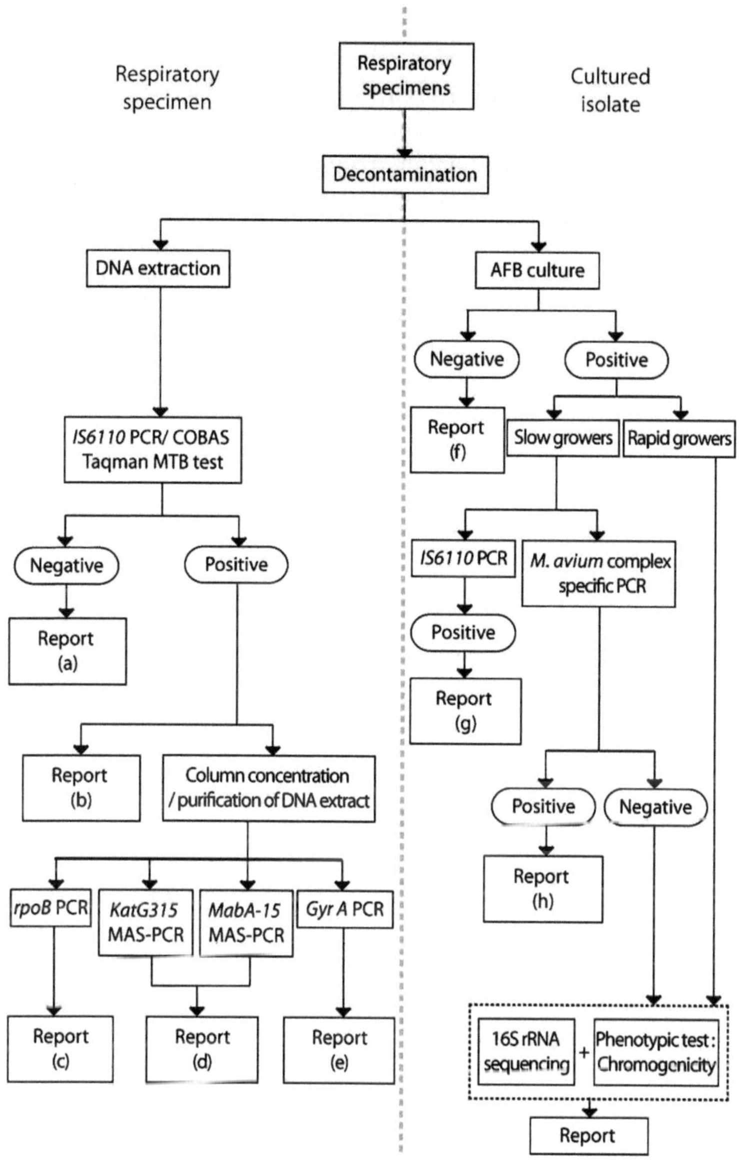

Fig. 4. Flowchart showing the strategy for molecular diagnosis of tuberculosis from respiratory specimens and identification of Mycobacteria from cultured isolates. (a) PCR for *M. tuberculosis* complex: Negative. (b) PCR for *M. tuberculosis* complex: Positive. (c) (i) Known mutation of MTB rpoB gene associated with rifampicin resistant NOT detected/(ii). Mutation, (e.g.,) S531L, of MTB rpoB gene associated with rifampicin resistance detected. (d) (i) Known mutation of MTB katG/mabA gene associated with isoniazid resistant NOT detected/(ii). Mutation, (e.g.,) S315T, of MTB katG gene associated with isonazid resistance detected. (e) (i) Known mutation of MTB gyrA gene associated with ofloxacin resistant NOT detected/(ii). Mutation, (e.g.,) G94D, of MTB gyrA gene associated with ofloxacin resistance detected. (f) No growth of AFB. (g) Culture positive for *M. tuberculosis* complex. (h) Culture positive for *M. avium* complex.

2.3. Identification of M. tuberculosis Complex

2.3.1. Identification of M. tuberculosis Complex by Single Tube Nested IS6110 Polymerase Chain Reaction

1. 10× PCR buffer (Applied Biosystems, Foster City, CA, USA): 100 mM Tris–HCl, pH 8.3, 500 mM KCl. Stored at –20°C.
2. 25 mM $MgCl_2$ solution (Applied Biosystems). Stored at –20°C.
3. 100 mM dGTP, 100 mM dATP, 100 mM dUTP, and 100 mM dCTP (Fermentas, Burlington, Ontario, Canada). Stored at –20°C.
4. Primers chosen from the insertion sequence, *IS6110.* External primers were derived from position 367 to 392 and positions 746 to 769, and internal primers were derived from positions 455 to 472 and positions 670 to 652 (see Table 1). Stored at –20°C.
5. AmpliTaq Gold DNA Polymerase (Applied Biosystems), concentration 5 U/μL. Store at –20°C.
6. Uracil-DNA-glycosylase 5 U/μL (UNG) (Roche). Keep at –20°C.
7. 1× Tris–Borate–EDTA solution: Keep at room temperature.
8. 6× Sample Loading Dye (Fermentas). Store at –20°C.
9. Molecular size marker. GeneRuler 100 bp DNA Ladder plus (Fermentas). Keep at –20°C.
10. Ethidium bromide solution, 0.5 μg/mL. Add 1 drops of 10 mg/mL ethidium bromide solution to 1 L water. Store at room temperature.

2.3.2. Identification of M. tuberculosis Complex by the COBAS® TaqMan® MTB Test (Roche Diagnostics)

This kit consists of a COBAS® TaqMan® MTB Master Mix (MTB MMX), a Mycobacterium Magnesium Reagent (MYCO Mg^{2+}), a Mycobacterium internal control (MYCO IC), a *M. tuberculosis* low positive control (MTB (+)C), and a Mycobacterium Negative Control (MYCO (–)C).

2.4. DNA Concentration and Purification by QIAquick PCR Purification Kit (Qiagen, Venlo, Netherlands)

The kit consists of a Buffer PB, a Buffer PE, and a Buffer EB. All reagents are stored at room temperature.

1. Buffer PB
2. Buffer PE
3. Buffer EB

2.5. Detection of Drug-Resistant M. tuberculosis in Respiratory Specimen

1. 10× PCR buffer (Applied Biosystems): 100 mM Tris–HCl, pH 8.3, 500 mM KCl. Stored at –20°C.
2. 25mM $MgCl_2$ solution (Applied Biosystems). Stored at –20°C.
3. 10 mM dNTP Mixes (Fermentas). Stored at –20°C.
4. PCR and sequencing primer.
   (a) PCR and sequencing primers for RIF-resistant *M. tuberculosis* detection: forward Primer TR8x, reverse primer TR9x cho-

Table 1
Primers used for molecular diagnosis of mycobacteria

Primer name	Type	Temperatures (°C)	Sequence	References
IS6110 PCR				(7, 10, 11)
External forward primer	PCR	74	5′-CCGGCCAGCACGCTAATTAACGGTTC-3′	
External reverse primer	PCR	76	5′-TGTGGCCGGATCAGCGATCGTGGT-3′	
Internal forward primer	PCR	56	5′-CTGCACACAGCTGACCGA-3′	
Internal reverse primer	PCR	58	5′-CGTTCGACGGTGCATCTG-3′	
rpoB PCR				(14, 15)
TR8x	PCR and sequencing	78	5′-TCGCCGCGATCAAGGAGTTCTTCGGC-3′	
TR9x	PCR and sequencing	87	5′-TGCACGTCGCGGACCTCCAGCCCGGCAC-3′	
katG315 MAS-PCR				(21, 22), this chapter
katG315_0Fx	PCR	71	5′-GCAGATGGGGCTGATCTACGTGAACC-3′	
katg315_5Rx	PCR	68	5′-TCCATACGACCTCGATGCCGC-3′	
katG315_4Rx	PCR	70	5′-AACGGGTCCGGGATGGTGCC-3′	
mabA MAS-PCR				(21, 22), this chapter
MabA-115Fx	PCR	70	5′-ACAAACGTCACGAGCGTAACCCCAG-3′	
MabA+4Rx	PCR	66	5′-GCAGTCACCCCGACAHCCTATCG-3″	
MabA+366Rx	PCR	70	5′-GAGGTTGGCGTTGATGACCTTCTCG-3′	

Table 1 (continued)

Primer name	Type	Temperatures (°C)	Sequence	References
GyrA PCR				(25)
GyrA1	PCR and sequencing	55	5′-CAGCTACATCGACTATGCGA-3′	
GyrA2	PCR and sequencing	55	5′-GGGCTICGGTGTACCTCAT-3′	
*M. avium*complex specific-PCR				(29, 31)
MAC1 (*M. avium*-specific forward primer)	PCR	65	5′-GGACCTCAAGACGCATGTCTTCTG-3′	
MAC2 (*M. intracellulare*-specific forward primer)	PCR	64	5′-GGACCTTTAGGCGCATGTCTTTAG-3′	
MACR (common reverse primer)	PCR	67	5′-GCTCTTTACGCCCAGTAATTCCGG-3′	
16S rRNA gene sequencing				(31)
Broad-range primer-285	PCR & sequencing	56	5′-GAGAGTTTGATCCTGGCTCAG-3′	
The mycobacteria-specific primer-264	PCR	65	5′-TGCACACAGGCCACAAGGGA-3′	
Sequencing-specific primer-259	Sequencing	61	5′-TTTCACGAACAACGCGACAA-3′	

sen from *rpoB* gene sequence, region 759807–763325 under accession no. NC_000962 (see Table 1). Stored at −20°C.

(b) *KatG* MAS-PCR primers for INH-resistant *M. tuberculosis* detection: Outer forward primer katG315_0Fx, inner reverse primer katg315_5Rx, outer reverse primer katG315_4Rx chosen from *katG* gene sequence, region 2153889–2156111 under accession no. NC_000962 (see Table 1). Stored at −20°C.

(c) MabA MAS-PCR primers for INH-resistant *M. tuberculosis* detection: Outer forward primer MabA-115Fx, inner reverse primer MabA+4Rx, outer reverse primer MabA+366Rx chosen from region 1673325 to 1673778 under accession no. NC_000962 (see Table 1). Stored at −20°C.

(d) PCR and sequencing primers for OFX-resistant *M. tuberculosis* detection: Forward primer GyrA1 and reverse primer GyrA2 chosen from the reference *gyrA* sequence with accession no. L27512 (see Table 1). Stored at −20°C.

(e) AmpliTaq Gold DNA Polymerase (5 U/μL) (Applied Biosystems). Stored at −20°C.

(f) 1× Tris–Borate–EDTA solution: Dilute 10× Tris–Borate–EDTA solution (Bio-rad, Hercules, CA, USA) tenfold in mQH_2O for working solution 1× TBE. Keep at room temperature.

(g) 6× Sample Loading Dye (Fermentas). Keep at −20°C.

(h) Molecular size marker. GeneRuler 100 bp DNA Ladder plus (Fermentas). Keep at −20°C.

(i) Ethidium bromide solution, 0.5 μg/mL. Store at room temperature: Add 1 drop of 10 mg/mL Ethidium bromide solution to 1 L water.

(j) BigDye® Terminator v 1.1 Mix (Applied Biosystems). Stored at −20°C.

(k) BigDye® Terminator v3.1, v1.1 Sequencing Buffer (5×). (Applied Biosystems). Stored at 4°C.

(l) DyeEx™ 2.0 Spin Kit (Qiagen, Venlo, Netherlands). Stored at room temperature.

(m) Hi-Di™ Formamide. (Applied Biosystems). Stored at −20°C.

2.6. Genotypic Identification of Non-tuberculous Mycobacteria from Culture

2.6.1. In-House PCR for M. avium Complex Identification

1. 10× PCR buffer (Applied Biosystems): 100 mM Tris–HCl, pH 8.3, 500 mM KCl. Stored at −20°C.
2. 25 mM $MgCl_2$ solution (Applied Biosystems). Stored at −20°C.
3. 10 mM dNTP mix (Fermentas). Stored at −20°C.
4. PCR primers for *M. avium* complex identification: *M. avium*-specific forward primer MAC1 chosen from *M. avium* 16S rDNA, region 1487542–1489058 under accession no. NC_008595, *M. intracellulare*-specific forward primer MAC2

chosen from *M. intracellulare* 16S rRNA, region 53942–55231 under accession no. ABIN01000139, and common reverse primer MACR chosen from the conserved region of both 16S rDNA gene (see Table 1).

5. AmpliTaq Gold DNA Polymerase 5 U/μL (Applied Biosystems. GeneRuler 100 bp DNA Ladder plus (keep at −20°C.
6. Ethidium bromide solution (0.5 μg/m). Store at room temperature: add ~100 μL of 10 mg/mL ethidium bromide solution to 1 L water.

2.6.2. 16S rDNA PCR Sequencing for Identification of Mycobacteria

1. 10× PCR buffer (Applied Biosystem): 100 mM Tris–HCl, pH 8.3, 500 mM KCl. Stored at −20°C.
2. 25mM $MgCl_2$ solution (Applied Biosystem). Stored at −20°C.
3. 10mM dNTP mix (Fermentas). Stored at −20°C.
4. The broad-range primer 285 and the mycobacteria-specific primer 264 were used for amplification, corresponding to *E. coli* 16S rDNA positions 9–30 and 1,027–1,046, respectively. Primer 285 and reverse primer 259 corresponding to *E. coli* 16S rDNA position 590–609 were used as the sequencing primers (see Table 1).
5. AmpliTaq Gold DNA Polymerase 5 U/μL (Applied Biosystems) Stored at −20°C.
6. 1× Tris–Borate–EDTA solution, store at room temperature.
7. 6× Sample Loading Dye (Fermentas). Keep at −20°C.
8. Molecular size marker. GeneRuler 100 bp DNA Ladder plus (Fermentas). Keep at −20°C.
9. Ethidium bromide solution, 0.5 μg/mL. Store at room temperature. Add 1 drop of 10 mg/mL ethidium bromide solution to 1 L water.
10. BigDye® Terminator v 1.1 Mix (Applied Biosystems). Stored at −20°C.
11. BigDye® Terminator v3.1, v1.1 Sequencing Buffer (5×). (Applied Biosystems). Stored at 4°C.
12. DyeEx™ 2.0 Spin Kit (Qiagen, Venlo, Netherlands). Stored at room temperature.
13. Hi-Di™ Formamide. (Applied Biosystem). Stored at −20°C.

3. Methods

3.1. Specimen Decontamination by Sputasol-Sodium Hydroxide

1. Equal parts (usually 3–5 mL) of a 4% NaOH–sputasol mixture are added to a respiratory specimen. The mixture is then mixed well by shaking.

2. The mixture is kept for 30 min at room temperature during which time the bottle is agitated every 10 min on a mechanical mixer (see Note 1).
3. The mixture is then topped up with 0.067 M sodium phosphate buffer, pH5.3, to a final volume around 20 mL, and then it is centrifuged at 3,000×*g* for 30 min.
4. The supernatant is decanted and the sediment is resuspended in 0.5 mL 0.067 M sodium phosphate buffer, pH5.3. 200 μL of the sediment is transferred to a 1.5 mL microtube and stored at 20°C before DNA extraction.

3.2. DNA Extraction by In-House Alkaline Lysis Method (for All Purpose Except COBAS® TaqMan® MTB Test) or by Roche AMPLICOR® Respiratory Specimen Preparation Kit (for all Purpose) (see Note 2)

1. The digested specimen (approx. 200 μL) is thawed at room temperature and is spun at 18,600×*g* in a microcentrifuge for 10 min. The supernatant is then decanted and discarded.
2. The sediment is washed with 500 μL in-house wash buffer or RW buffer and is vortexed to mix.
3. The mixture is then spun at 18,600×*g* in a microcentrifuge for 10 min. The supernatant is then decanted.
4. 100 μL in-house lysis solution or RL buffer is added to the sediment and the mixture is vortexed thoroughly.
5. The tube is incubated at 60°C for 45 min in dry heat block and is then briefly centrifuged to remove the droplets on the lid.
6. 100 μL in-house neutralization solution or RN buffer is added and vortexed briefly.
7. The mixture is centrifuged at 18,600×*g* in a microcentrifuge for 10 min. The supernatant containing mycobacterial DNA is transferred to a new microtube (see Note 3).

3.3. Identification of M. tuberculosis Complex

3.3.1. Identification of M. tuberculosis Complex by Single Tube Nested IS6110 Polymerase Chain Reaction (see Note 4)

1. For each PCR reaction, a volume of 90 μL reaction mixture is firstly prepared. It consists of 1× PCR buffer (10 mM Tris–HCl (pH 8.3), 50 mM KCl), 2 mM $MgCl_2$, 0.15 mM dATP, dGTP, dCTP, and 0.45 mM dUTP (see Note 5), 0.02 μM external primers, 0.75 μM internal primers (see Table 1), 2 U AmpliTaq Gold, and 0.5 U Uracil-DNA-glycosylase.
2. A volume of 10 μL extracted DNA specimens weak and strong positive controls or negative control rewarmed to room temperature are added to the 90 μL of reaction mixture.
3. The PCR reaction mixture is transferred to the PCR machine (GeneAmp® PCR system 9700) and the reaction cycle is set as follow: a single cycle of 37°C for 10 min, followed by 94°C for 12 min. The temp-cycling condition consists of 94°C for 45 s and 72°C for 1.5 min for the first 15 cycles, then 94°C for 45 s, 55°C for 45 s, and 72°C for 1 min for 45 cycles. This is followed by 72°C for 10 min. and the reaction products stored at 4°C (see Note 6).

4. After completion of the PCR reaction, the PCR product is pulse spun. A 10 μL portion of the PCR product is electrophoresed through a 2% agarose gel in TBE buffer.
5. The gel is stained with ethidium bromide for 15 min with mild shaking and the size of any bands determined by comparison with to the standard molecular size marker A positive result is indicated by the presence of target bands of 216 bp and/or 315/304 bp under UV illumination. For negative result, no target bands are observed (see Fig. 1b and Note 7).

3.3.2. Identification of M. tuberculosis Complex by the COBAS® TaqMan® MTB Test

The test is performed according to manufacturer's instruction. In brief:

1. The working controls, MTB (+) C and MYCO (−) C, are prepared as follow: 50 μL of each of the controls are transferred into a 1.5 mL microtube, respectively. 200 μL of RL are added to each controls and is vortexed to mix. 100 μL of each mixture is transferred to new 1.5 mL microtubes, respectively, followed by incubation at 60°C in a dry heat block for 45 min. The tubes are pulse spun to remove condensate from cap. 100 μL of RN is added and vortexed briefly to mix. Once opened, the working MTB (+) C and MYCO (−) C solutions should be used within 4 weeks or until expiration date, whichever comes first.
2. The working reaction mixture is prepared as follow: the MTB MMX, MYCO Mg^{2+}, and MYCO IC (see Note 8) are equilibrated at ambient temperature for at least 30 min. K-carrier is placed in a K-carrier holder and new K-tubes are placed in the K-carrier without touching the sides of the K-tubes. For 12 tests, 200 μL of MYCO Mg^{2+} and 50 μL MYCO IC are added into *one vial* MTB MMX. (By proportion, for 1 test added, add 16.7 μL of MYCO Mg^{2+}, 4.2 μL MYCO IC, and 38.3 μL MTB MMX) (see Note 9). The vial is capped and mixed well by inverting ten times. 50 μL of working MMX is transferred into each K-tube.
3. 50 μL of extracted specimen or controls are added to the appropriate K-tube containing working MMX using a micropipettor with an aerosol barrier. Each specimen or controls are gently mixed up and down three times with the micropipettor without generating bubbles. The reaction mixtures are now ready to load.
4. The COBAS® TaqMan® 48 Analyzer is loaded and operated according to the manufacturer's guidance: The workstation computer and the COBAS® TaqMan® 48 Analyzer should be turned on at least 30 min prior to starting the run. Quality Control information, K-carrier ID, Batch, and sample ID are entered into the analyzer prior to sample loading. The cover

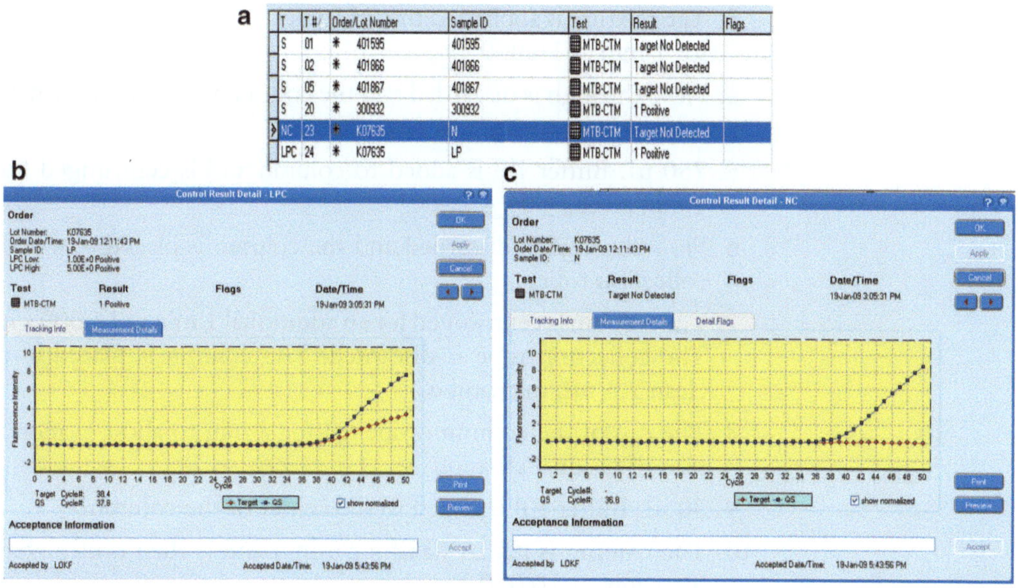

Fig. 5. (**a**) COBAS® TaqMan® MTB test result table. Sample positive with *M. tuberculosis* DNA was showed as "1 positive"; whereas sample negative for *M. tuberculosis* DNA was showed as "Target Not Detected." Samples were valid if no flags are present. (**b**) Amplification curve showing the fluorescent intensity against PCR cycles for the Internal control (*Square*) and a positive target (*rhombus*). (**c**) Amplification curve showing the fluorescent intensity against PCR cycles for the Internal control (*Square*) and a negative target (*rhombus*).

of the thermal cycler is opened when all the information is entered. The loaded K-carrier containing the capped K-tubes with working Master Mix and specimens and controls is transferred into the Thermal Cycler. The reaction is then started. Amplification and detection are automatically performed by the COBAS® Taq Man® 48 Analyzer and are complete in about 2.5 h.

5. After the run, the results window or printout must be checked for "flags" and comments to ensure that the run is valid. The run is valid if no flags appear for the COBAS® TaqMan® MTB controls. The run is not valid if flags appear for the COBAS® Taq Man® MTB Controls. When the run is invalid, the entire run including specimen and control preparation must be repeated for amplification and detection. For a valid run, check each individual specimen for flags or comments on the result printout. A valid run may include both valid and invalid specimen results depending on whether flags and/or comments are obtained for the individual specimens (see Fig. 5 and Note 10).

3.4. DNA Concentration and Purification by QIAquick PCR Purification Kit (see Note 11)

1. 100 μL of DNA sample extracted by in-house alkaline lysis method or by Roche AMPLICOR® Respiratory Specimen Preparation Kit is transferred to a 1.5 mL microtube.
2. 5 volumes (500 μL) of PB solution are added to 1 volume (100 μL) of DNA extract and mixed well.

3. The mixture is applied to the QIAquick column and centrifuged for 1 min at 16,000 × *g*.
4. Flow-through is discarded and the column is placed to the new tube.
5. 750 μL Buffer PE is added to column and is centrifuged for 1 min at 16,000 × *g*.
6. Flow-through is discarded and the column is placed to a new collection tube.
7. The column is centrifuged for an additional 1 min at 16,000 × *g*. The collection tube is discarded. The column is placed in a clean 1.5 mL microtube.
8. The cap of the column is opened for 2 min to allow evaporation of residual ethanol.
9. 30 μL Buffer EB is added to the center of the column.
10. The column is left to stand for 5 min and is then centrifuged for 1 min at 16,000 × *g*.
11. The flow-through is the purified DNA extract which is used in the drug-resistance detection assay.

3.5. Detection of drug resistance in Mycobacterium tuberculosis by molecular methods

3.5.1. Detection of RIF-Resistant M. tuberculosis by PCR Sequencing (see Note 12)

1. For each PCR reaction, a volume of 40 μL reaction mixture is firstly prepared. It consists of 1× PCR buffer (10 mM Tris–HCl (pH 8.3), 50 mM KCl), 1.25 mM $MgCl_2$, 0.2 mM (each) dNTP, 0.4 μM (each) primers, TR8x and TR9x (see Table 1 and Note 13), and 2 U AmpliTaq Gold.
2. A volume of 10 μL purified DNA extract is add to the 40 μL of reaction mixture.
3. PCR reaction mixture is transferred to the PCR machine (GeneAmp® PCR system 9700) and the reaction cycle is set as follow: a single cycle of 94°C for 12 min. The reaction mixture is then subjected to 40 cycles of amplification (denaturation at 94°C for 1 min, and annealing and extension at 72°C for 2 min), followed by a final 7-min extension at 72°C and is finally stored at 4°C.
4. After completion of the PCR reaction, the PCR product is pulse spun. A 10 μL of PCR product is electrophoresed through a 2% agarose gel in TBE buffer.
5. The gel is stained with ethidium bromide for 15 min with mild shaking, a positive result is indicated by the presence of a target band of 157-bp under UV illumination. For negative result, no target bands are observed.
6. For positive sample, the remaining 40 μL PCR product is subjected to purification by QIAquick PCR Purification kit prior to sequencing. Five times volume (200 μL) of PB solution is added to 1 volume (40 μL) of PCR product and mixed well.

The product is then purified by following steps 3–11 in Subheading 3.4.

7. Two cycle sequencing reactions, one forward and one reverse reaction, are set up for each sample. A 20 μL sequencing mixture should be freshly prepared. It consists of 1× sequencing buffer, 1/8× BigDye® Terminator v 1.1 Mix, (see Note 14), 0.16 μM of TR8x or TR9x, and 3 μL of PCR product (see Note 15).
8. The PCR reaction mixture is transferred to the PCR machine (GeneAmp® PCR system 9700) and the reaction setup as follows. After a single cycle of heating to 96°C for 1 min, the reaction mixture is then subjected to 25 cycles of amplification (denaturation at 96°C for 10 s, and annealing at 50°C for 5 s and extension at 60°C for 4 min) and is then stored at 4°C.
9. The sequencing product is then purified using Dye-Ex spin column (referring to manufacturer protocol).
   (a) The spin column is gently vortexed to resuspend the resin and then the cap of the column is loosened by a quarter turn.
   (b) The bottom closure of the spin column is snapped off and the column is placed in the collection tube.
   (c) The column is centrifuged at 750×*g* for 3 min. The column is then transferred to a clean microtube.
   (d) A total of 20 μL of the sequencing product is transferred to the gel bed of the column.
   (e) The purified product is collected into the microtubes by centrifuging at 750×*g* for 3 min.
   (f) All of purified product is added to 0.2-mL wells of a 96-well plate containing 15 μL Hi-Di™ formamide.
   (g) The plate is heated at 95°C for 2–3 min to denature (do not exceed 5 min). Hereafter, it is immediately placed in ice block for 3 min. The product is now ready for sequencing.
10. The purified sequencing product is loaded to the Applied Biosystems 3130 Genetic Analyzer according to manufacturer's instruction.
11. The generated sequences are assembled using Pre Gap 4 software, followed by editing using Gap v 4.10 software of the Staden package, which is a free software downloaded from http://staden.sourceforge.net/
12. After editing, the sequence is saved as a *fasta* file and is compared with the reference *rpoB* sequence, region 76786 to 76942 under accession no. NC_000962 using an online program clustalW (http://www.ebi.ac.uk/Tools/clustalw2/index.html) (see Fig. 2b).

13. For samples harboring mutations, the regions of mutation and the consequent amino acid substitutions should be reported clearly, e.g., resistance-associated mutation Ser531Leu present. For a sample without a known mutation, it should be reported as "no known mutation found" (see Note 16).

3.5.2. Detection of INH-Resistant M. tuberculosis by MAS-PCR (see Note 12)

1. For each *KatG* MAS-PCR reaction, a volume of 22.5 μL reaction mixture is firstly prepared. It consists of 1× PCR buffer (10 mM Tris–HCl (pH 8.3), 50 mM KCl), 2 mM $MgCl_2$, 0.2 mM (each) dNTP, 0.6 μM outer forward primer katG0Fx, 0.8 μM outer reverse primer katG4Rx, 0.6 μM inner reverse primer (see Table 1), (see Note 13), and 1.25 U of Ampli*Taq* Gold Polymerase.
2. For each *MabA* MAS-PCR reaction, a volume of 22.5 μL reaction mixture is firstly prepared. It consists of 1× PCR buffer (10 mM Tris–HCl (pH 8.3), 50 mM KCl), 2 mM $MgCl_2$, 0.2 mM (each) dNTP, 0.85 μM outer forward primer katG0Fx, 0.17 μM outer reverse primer katG4Rx, 0.85 μM inner reverse primer (see Table 1), and 1.25 U of Ampli*Taq* Gold Polymerase.
3. A volume of 2.5 μL purified DNA extract is add to the 22.5 μL of reaction mixtures of KatG MAS-PCR and MabA MAS PCR, respectively.
4. PCR reaction mixtures are transferred to the PCR machine (GeneAmp® PCR system 9700) and the reaction cycle is set as follow: initial denaturation at 96°C for 8 min; 6 cycles at 95°C for 1 min, 72°C for 1 min, 72°C for 30 s; 6 cycles at 95°C for 1 min, 70°C for 40 s, 72°C for 30 s; 25 cycles at 94°C for 1 min, 68°C for 30 s, 72°C for 30 s; followed by final elongation at 72°C for 7 min and is finally stored at 4°C.
5. After completion of the PCR reaction, the PCR product is pulse spun and 10 μL of each PCR product is electrophoresed through a 2% agarose gel in TBE buffer.
6. The gel is stained with ethidium bromide for 30 min with mild shaking, the size of band measured and compared to the standard molecular size marker. For *KatG* MAS-PCR, wildtype *katG*315 is indicated by the presence of target bands of 296-bp with or without the band of 435-bp under UV illumination; the mutated katG315 is indicated by a single band of 435-bp. For *MabA* MAS-PCR, wildtype *mabA*-15 is indicated by the target bands of 123-bp with or without the band of 454-bp; the mutated mabA-15 is indicated by a single band of 454-bp (see Note 16 and Fig. 6a, b).

3.5.3. Detection of OFX-Resistant M. tuberculosis by PCR Sequencing (see Note 12)

1. For each PCR reaction, a volume of 45 μL reaction mixture is firstly prepared. It consists of 1× PCR buffer (10 mM Tris–HCl (pH 8.3), 50 mM KCl), 1.5 mM $MgCl_2$, 0.2 mM (each) dNTP,

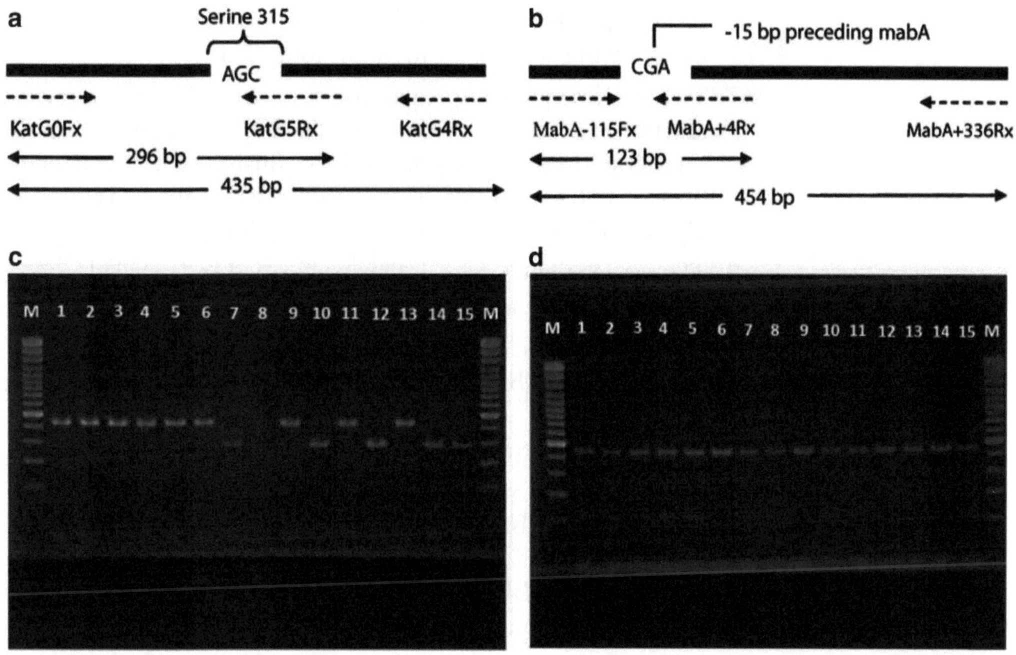

Fig. 6. (**a**) Schematic diagrams of KatG315 MAS-PCR. (**b**) Schematic diagrams of MabA-15 MAS-PCR. (**c**) Results of KatG315 MAS-PCR typing. M: 100-bp DNA ladder; lane 1–6, 9, 11, 13: sample with mutated katG315 allele; lane 7, 10, 12, 14, 15: samples with wildtype katG315 allele; lane 8: negative result on MAS-PCR. (**d**) Results of MabA-15 MAS-PCR typing. M: 100-bp DNA ladder; lane 6, 9, 11–14: sample with mutated MabA-15 allele; lane 1–5, 7, 8, 10, 15:: samples with wildtype MabA-15 allele.

0.3 μM (each) primers, GyrA1 and GyrA2 (Table 1), and 1 U AmpliTaq Gold.

2. A volume of 5 μL purified DNA extract is add to the 45 μL of reaction mixture.
3. PCR reaction mixture is transferred to the PCR machine (GeneAmp® PCR system 9700) and the reaction setup as follows: after a single cycle of heating to 94°C for 12 min, the reaction mixture is then subjected to 45 cycles of amplification (denaturation at 94°C for 1 min, and annealing at 55°C for 1 min and extension at 72°C for 2 min), followed by a final 10-min extension at 72°C and is then stored at 4°C (see Note 17).
4. After completion of the PCR reaction, the PCR product is pulse spun and a 5 μL of PCR product electrophoresed through a 2% agarose gel in TBE buffer.
5. The gel is stained with ethidium bromide for 15 min with mild shaking. Compared to the standard molecular size marker (GeneRuler 100 bp DNA Ladder Plus), a positive result is indicated by the target bands of 320-bp under UV illumination.
6. For positive sample, the remaining 45 μL PCR product is subjected to purification by QIAquick PCR Purification kit. 5 volumes (225 μL) of PB solution are added to 1 volume (45 μL)

of PCR product and mixed well. The product is then purified by following steps 3–11 in Subheading 3.4.

7. Two cycle sequencing reactions, one forward and one reverse reaction, are set up for each sample. A 20 μL sequencing mixture should be freshly prepared. It consists of 1× sequencing buffer, 1/8× BigDye® Terminator v 1.1 Mix, 0.16 μM of GyrAl or GyrA2 and 3 μL of PCR product.
8. The sequencing reaction is then performed and purified according to Subheading 3.5.1 steps 8–10.
9. The generated sequences are assembled using Pre Gap 4 software, followed by editing using Gap v 4.10 software of the staden package.
10. After editing, the sequence is saved as a *fasta* file and is compared with the reference *gyrA* sequence with accession no. L27512 in GenBank using an online program clustalW.

3.6. Genotypic Identification of Non-tuberculous Mycobacteria from Culture

3.6.1. DNA Extraction by In-House Alkaline Lysis Method for Cultured Isolate

1. The colony on LJ medium is lightly touched and is suspended in 500 μL in-house wash buffer.
2. The isolate DNA is then extracted by following steps 3–7 in Subheading 3.2.

3.6.2. In-House PCR for M. avium Complex Identification

1. For each PCR reaction, a volume of 100 μL reaction mixture is firstly prepared. It consists of 1× PCR buffer (10 mM Tris–HCl (pH 8.3), 50 mM KCl), 2 mM $MgCl_2$, 0.2 mM (each) dNTP, 0.4 μM (each) primers, MAC1, MAC2, and MACR (see Table 1), and 2 U AmpliTaq Gold.
2. A volume of 10 μL purified DNA extract is add to the 90 μL of reaction mixture.
3. PCR reaction mixture is transferred to the PCR machine (GeneAmp® PCR system 9700) and after heating at 94°C for 10 min, the reaction mixture is then subjected to 50 cycles of amplification (denaturation at 94°C for 1 min, annealing at 68°C for 1 min, and extension at 72°C for 1 min), followed by a final 10-min extension at 72°C and is finally stored at 4°C.
4. After completion of the PCR reaction, the PCR product is pulse spun and 10 μL of PCR product is electrophoresed through a 2% agarose gel in TBE buffer.
5. The gel is stained with ethidium bromide for 15 min with mild shaking. Compared to the standard molecular size marker (GeneRuler 100 bp DNA Ladder Plus), a positive result is indicated by the target bands of 390-bp under UV illumination. For negative result, no target bands are observed (see Note 18).

3.6.3. 16S rDNA PCR Sequencing for Identification of Mycobacteria (see Note 19)

1. For each PCR reaction, a volume of 45 μL reaction mixture is firstly prepared. It consists of 1× PCR buffer (10 mM Tris–HCl (pH 8.3), 50 mM KCl), 2 mM $MgCl_2$, 0.2 mM (each) dNTP, 0.4 μM (each) primers, primer-285 and primer-264 (see Table 1), and 2 U AmpliTaq Gold.
2. A volume of 5 μL purified DNA extract is added to the 45 μL of reaction mixture.
3. The PCR reaction mixture is transferred to the thermal cycler (GeneAmp® PCR system 9700) and the reaction setup as follows: After a single cycle of heating to 94°C for 10 min, the reaction mixture is then subjected to 35 cycles of amplification (denaturation at 94°C for 1 min, annealing at 65°C for 1 min, and extension at 72°C for 2 min), followed by a final 10-min extension at 72°C and is then stored at 4°C.
4. After completion of the PCR reaction, the PCR product is pulse spun. A 5 μL of PCR product is electrophoresed through a 2% agarose gel in TBE buffer.
5. The gel is stained with ethidium bromide for 15 min with mild shaking. Compared to the standard molecular size marker (GeneRuler 100 bp DNA Ladder Plus), a positive result is indicated by the presence of a target band under UV illumination of about 1040-bp, depending on species.
6. If a positive result is obtained, the remaining 40 μL PCR product is subjected to purification by QIAquick PCR Purification kit. 5 volumes (200 μL) of PB solution are added to 1 volume (40 μL) of PCR product and mixed well. The product is then purified by following steps 3–11 in Subheading 3.4.
7. Two cycle sequencing reactions, one forward and one reverse reaction, are set up for each sample. A 20 μL sequencing mixture should be freshly prepared. It consists of 1× sequencing buffer, 1/8× BigDye® Terminator v 1.1 Mix, 0.16 μM of primer-285 or 16SrDNA sequencing primer-259 (see Table 1), and 3 μL of PCR product.
8. The sequencing reaction is then performed and purified according to Subheading 3.5.1 steps 8–10.
9. The generated sequences are assembled using PreGap 4 software, followed by editing using Gap v 4.10 software of the Staden package.
10. After editing, the sequence is saved as a *fasta* file and is searched for sequence match in Ribosomal Differentiation of Medical Microorganisms (RIDOM) (http://www.ridom-rdna.de/).
11. A sequence match of ≧99% with that of the prototype strain sequence in a repository should be used as the criterion for species, group or complex level identification (see Note 20).

4. Notes

1. Sodium hydroxide at a concentration of 4% is toxic not only for contaminants but also for some mycobacteria. Timing of the digestion–decontamination procedure, therefore, is crucial for the maximum recovery of AFB in culture. The sensitivity of PCR-based technique, however, is not decreased by the contamination step as nucleic acid still remains intact even when the bacteria are dead. Prolonged decontamination would therefore result in inconsistent results between AFB culture and PCR-based detection.
2. In our experience, both control and sample DNA extracted by in-house alkaline lysis method give invalid or negative result on COBAS® TaqMan® MTB Test, although the same extraction method gave successful amplification in other PCR-based assays. The same samples when extracted using the Roche AMPLICOR® Respiratory Specimen Preparation Kit restores the positive signal in COBAS® TaqMan® MTB Test as well as other PCR assays. The commercial extraction kit may contain internal control or signaling material that is essential for the COBAS® TaqMan® 48 Analyzer to recognize the DNA extract as a valid sample for test run. On the basis of this observation, the in-house alkaline lysis method is used to extract DNA for all PCR-based assays except COBAS® TaqMan® MTB Test, while the Roche AMPLICOR® Respiratory Specimen Preparation Kit can be used for DNA extraction for all subsequent assays.
3. DNA extracts can be stored at 4°C for in-house or COBAS® TaqMan® MTB Test done on same day. If PCR is not processed on the same day or for long-term storage, the DNA extract should be kept at −20°C.
4. As good laboratory practice, *M. tuberculosis* DNA positive and negative controls should be included in each PCR run. To prepare a strong positive control, a uniform suspension (McFarland standard 2) of *M. tuberculosis* H37Rv is made in 10% (w/v) TE buffer, pH 8.0. After heating at 121°C for 15 min and centrifugation at ≥12,500 × *g* for 10 min, the supernatant is diluted by 10^{-5}. The weak positive control is prepared by making a further 1/10 dilution of the strong positive control in 10% TE buffer. Both positive controls are aliquoted in microtubes and kept frozen at −20°C before use. Do not use for more than 6 months. Sterile mQ water is used as negative control. Water is aliquoted in microtubes and stored at 4°C before use.
5. The substitution of dUTP for dTTP in PCR results in uracil-containing PCR products. The contaminating PCR products from previous reaction would be eliminated by UNG through

the excision of uracil from the PCR products before the start of PCR cycle, thereby preventing false positives. The integrity of DNA template for current reaction would not be affected as it does not carry dUTP. However, the efficiency for dUTP incorporation is much less than that of dATP, dTTP, dCTP, dGTP, a 3× concentration of dUTP is therefore essential to maintain the amplification efficiency when it is used in place of dTTP in PCR.

6. The application of "Hotstart" polymerase is essential for this PCR protocol. According to the manufacturer's description, AmpliTaq Gold, as a hotstart enzyme, is inactive in its native form. During the first stage of incubation at 37°C, only UNG is activated to destroy amplicons containing uracil. Subsequent heating at 95°C for 12 min destroys the UNG and simultaneously activates AmpliTaq Gold. During the first 15 cycles, the high annealing temperature (72°C) favors the external primers with higher Tm, to bind with the DNA template with high stringency and to produce small quantity of specific amplicons for the subsequent cycles. Furthermore, the high temperature also prevents any residual activity of UNG which might "nick" the amplifying products. In latter cycles, a much lower annealing temperature is used for the internal primers to perform amplification with greater efficiency rather than stringency. By this algorithm, higher yield of specific PCR product with low background can be achieved (see Fig. 1).
7. Specimens are presumptive negative for *M. tuberculosis.* This negative result does not preclude *M. tuberculosis* infection since insufficient nucleic acids and the presence of PCR inhibitors can also give rise to a negative result.
8. Mycobacterium Internal Control is a proprietary noninfectious, recombinant linearized plasmid DNA with primer binding regions identical to those of the *M. tuberculosis* target sequence, a randomized internal sequence of similar length and base composition as the *M. tuberculosis* target sequence, and a unique probe binding region that differentiates the Mycobacterium Internal Control amplicon from target amplicon. The Mycobacterium Internal Control Reagent is included in the COBAS® TaqMan® MTB Test and is introduced into each amplification reaction to be co-amplified with MTB DNA from the clinical specimen. The Mycobacterium Internal Control is designed to ensure identification of specimens that contain inhibitors that would interfere with the amplification and detection of the MTB target sequence.
9. The Amplification and Detection Reagents are packaged in 12-test, single use vials. For the most efficient use of reagents, specimens and controls should be processed in batches that are multiples of 12.

10. The run is not valid if the following flags appear for the controls sample: Flag (_N_NC_INVALID) for negative control, which interpreted as contamination or no internal control signal; flag (_L_LPC_INVALID) for low positive control, which interpreted as control not within range, no target signal, or no IC signal. For a valid run, check each individual specimen for flags. For valid specimen with result as "Target Not Detected," it should be noted that the specimen is only presumptive negative for *M. tuberculosis.* This negative result does not preclude *M. tuberculosis* infection because results depend on adequate specimen collection, absence of inhibitors, and sufficient DNA to be detected.
11. Our experience indicates that purification of DNA extract by the QIAquick PCR Purification kit significantly increases the sensitivities of the subsequent PCR-based assays for drug resistance detection. A previous study showed that the sensitivities of the rpoB PCR in detecting RIF-resistant *M. tuberculosis* from smear-positive and smear-negative specimens are of 100% and 61%, respectively, which is significantly higher than those of 82% and 46% prior to purification. The increase in sensitivity may be due to the removal of inhibitor or impurity present in the DNA extracts and also the increase of concentration of the nucleic acid in the specimen as the original volume, 100 μL, is reduced to a final volume of 30 μL.
12. Like other resistance genes, *rpoB, katG, MabA, and gyrA* share a highly similarity of sequence homology between *M. tuberculosis* and other non-tuberculous mycobacteria (NTM). Our previous study of *rpoB* PCR assay showed that several false amplification of *rpoB* gene occurred on samples containing NTM but not *M. tuberculosis.* The result reflected that the molecular assay detecting drug resistance, like *rpoB, katG, mabA*, and *gyrA* assays, should be performed in combination with PCR assays specific for *M. tuberculosis* identification to achieve 100% specificity(15) (see Fig. 4).
13. The PCR-sequencing assay for rifampicin resistance detection and the MAS-PCR assay for INH resistance detection for *M. tuberculosis* present in respiratory specimen are modified from Telenti et al. and Mokrousov et al., respectively (14, 22). Both original assays exhibited specific results for cultured isolates. The application of the original assays on direct specimens, however, exhibited nonspecific amplification despite several attempts at optimization. In the modified assays, the primers are extended for several nucleotides to achieve a higher Tm. The annealing temperatures of both assays are also increased up to 72°C. These modifications would increase the stringency of the assay so as to minimize the nonspecific binding between primers and other bacterial DNA that may present in the specimens.

14. As a cost-reduction measure, our experience demonstrates that an adequate result could be obtained even by using 1/8 of recommended reaction volume of BigDye® Terminator, i.e., add only 1 μL of BigDye instead of 8 μL in a 20 μL reaction mix, for DNA sequencing.
15. According to the manufacturer's protocol for BigDye® Terminator v 1.1 Mix, it is recommended to quantitate the amount of purified DNA by measuring the absorbance at 260 nm. A DNA amount of 1–3 ng is sufficient for cycle sequencing reaction. In general, higher DNA quantities give higher signal intensities. According to our experience, addition of 3 μL PCR products to sequencing mixture usually gives optimal sequencing signal intensities.
16. Most of the molecular assays for the detection of drug resistance are based on the presence of hot-spot mutations, which account for a great majority, but not all, resistant isolates. The presence of hot-spot mutation indicates true resistance. However, the absence of these mutations *does not* preclude drug resistance. The isolates can still be resistant to drugs because of some novel mutations.
17. The *GyrA* PCR described by Takiff et al. (26) is modified for direct detection of OFX-resistant *M. tuberculosis* in respiratory specimens by extending the amplification cycles from 35 to 45.
18. This negative result does not preclude *M. avium* infection since insufficient nucleic acids and the presence of inhibitors can also give rise to a negative result.
19. A standard reference strain of known species of Mycobacteria should be run in parallel as a control. *Mycobacterium smegmatis* (ATCC 700084) is used for control in our laboratory.
20. There are no standardized guidelines available to establish the cutoffs for computer-assisted analysis of sequence similarity for 16S rRNA-based bacterial identification. Several difference rates, such as ≦0.5, ≦1, and ≦3% have been suggested for discrimination at species level. Since bacterial genera do not evolve at the same speed, different cutoff values may be necessary for different bacterial genera. Our laboratory uses a ≦1% difference as a suitable cutoff as suggested by Drancourt et al. (34).

References

1. World_Health_Organization (2008), Global tuberculosis control—surveillance, planning, financing. WHO/HTM/TB/2008.393, World Health Organization, Geneva, Switzerland.
2. Cheng VC, Yam WC, Hung IF, Woo PC, Lau SK, Tang BS, Yuen KY (2004) Clinical evaluation of the polymerase chain reaction for the rapid diagnosis of tuberculosis. J Clin Pathol 57:281–285

3. Shinnick TM, Good RC (1995) Diagnostic mycobacteriology laboratory practices. Clin Infect Dis 21:291–299
4. Grange JM (1984) *Mycobacterium avium*. Adv Tuberc Res 21:1–78
5. Lu CZ, Qiao J, Shen T, Link H (1990) Early diagnosis of tuberculous meningitis by detection of anti-BCG secreting cells in cerebrospinal fluid. Lancet 336:10–13
6. Somoskovi A, Dormandy J, Mitsani D, Rivenburg J, Salfinger M (2006) Use of smear-positive samples to assess the PCR-based genotype MTBDR assay for rapid, direct detection of the *Mycobacterium tuberculosis* complex as well as its resistance to isoniazid and rifampin. J Clin Microbiol 44:4459–4463
7. Yuen KY, Yam WC, Wong LP, Seto WH (1997) Comparison of two automated DNA amplification systems with a manual one-tube nested PCR assay for diagnosis of pulmonary tuberculosis. J Clin Microbiol 35:1385–1389
8. Hillemann D, Rusch-Gerdes S, Richter E (2007) Evaluation of the GenoType MTBDRplus assay for rifampin and isoniazid susceptibility testing of *Mycobacterium tuberculosis strains* and clinical specimen. J Clin Microbiol 45: 2635–2640
9. Barber R (2008) Evaluation of the BD ProbeTec ET system for the direct detection of *Mycobacterium tuberculosis* from clinical samples. Br J Biomed Sci 65:7–12
10. Chan CM, Yuen KY, Chan KS, Yam WC, Yim KH, Ng WF, Ng MH (1996) Single-tube nested PCR in the diagnosis of tuberculosis. J Clin Pathol 49:290–294
11. Yam WC, Yuen KY, Seto WH (1998) Direct detection of *Mycobacterium tuberculosis* in respiratory specimens using an automated DNA amplification assay and a single tube nested polymerase chain reaction (PCR). Clin Chem Lab Med 36:597–599
12. Pfyffer GE, Brown-Elliot BA (2003) Jost KC and Wallace RJ Jr (2003) *Mycobacterium*: general characteristics, isolation, and staining procedures. In: Murray PR, Baron EJ, Jorgensen JH, Pfaller MA, Yolken RH (eds) Manual of Clinical Microbiology, 8th edn. ASM Press, Washington, DC, pp 560–584
13. Ramaswamy S, Musser JM (1998) Molecular genetic basis of antimicrobial agent resistance in *Mycobacterium tuberculosis*: 1998 update. Tuber Lung Dis 79:3–29
14. Telenti A, Honore N, Bernasconi C, March J, Ortega A, Heym B, Takiff HE, Cole ST (1997) T. Genotypic assessment of isoniazid and rifampin resistance in *Mycobacterium tuberculosis*: a blind study at reference laboratory level. J Clin Microbiol 35:719–723
15. Yam WC, Tam CM, Leung CC, Tong HL, Chan KH, Leung ET, Wong KC, Yew WW, Seto WH, Yuen KY, Ho PL (2004) Direct Detection of Rifampin-Resistant *Mycobacterium tuberculosis* in Respiratory Specimens by PCR-DNA sequencing. J Clin Microbiol 42: 4438–4443
16. Zhang Y, Heym B, Allen B, Young D, Cole S (1992) The catalase- peroxidase gene and isoniazid resistance to *Mycobacterium tuberculosis*. Nature 358:591–593
17. Banerjee A, Dubnau E, Quemard A, Balasubramanian V, Um KS, Wilson T, Collins D, de Lisle G, Jacobs WR Jr (1994) inhA, a gene encoding a target for isoniazid and ethionamide in *Mycobacterium tuberculosis*. Science 263:227–230
18. Ramaswamy SV, Reich R, Dou SJ, Jasperse L, Pan X, Wanger A, Quitugua T, Graviss EA (2003) Single Nucleotide Polymorphisms in Genes Associated With Isoniazid Resistance in *Mycobacterium Tuberculosis*. Antimicrob Agents Chemother 47:1241–1250
19. Hazbon MH, Brimacombe M, Bobadilla del Valle M, Cavatore M, Guerrero MI, Varma-Basil M, Billman-Jacobe H, Lavender C, Fyfe J, Garcia-Garcia L, Leon CI, Bose M, Chaves F, Murray M, Eisenach KD, Sifuentes-Osornio J, Cave MD, Ponce de Leon A, Alland D (2006) Population genetics study of isoniazid resistance mutations and evolution of multidrug-resistant *Mycobacterium* tuberculosis. Antimicrob Agents Chemother 50: 2640–2649
20. Musser JM, Kapur V, Williams DL, Kreiswirth BN, van Soolingen D, van Embden JD (1996) Characterization of the catalase-peroxidase gene (katG) and inhA locus in isoniazid-resistant and susceptible strains of *Mycobacterium tuberculosis* by automated DNA sequencing: restricted array of mutations associated with drug resistance. J Infect Dis 173:196–202
21. Leung ET, Ho PL, Yuen KY, Woo WL, Lam TH, Kao RY, Seto WH, Yam WC (2006) Molecular characterization of isoniazid resistance in *Mycobacterium tuberculosis*: identification of a novel mutation in inhA. Antimicrob Agents Chemother 50:1075–1078
22. Mokrousov I, Otten T, Filipenko M, Vyazovaya A, Chrapov E, Limeschenko E, Steklova L, Vyshnevskiy B, Narvskaya O (2002) J Clin Microbiol 40:2509–2512, Detection of isoniazid-resistant *Mycobacterium tuberculosis* strains by a multiplex allele-specific PCR assay targeting katG codon 315 variation
23. Siu KHG, Tam YH, Ho PL, Lee SGA, Que TL, Tse WSC, Yip KT, Lam TH, Cheng CCV, Yuen KY, Yam WC (2011) Direct detection of isoniazid-resistant *Mycobacterium tuberculosis* in respiratory specimens by multiplex allele-

specific polymerase chain reaction. Diagn Microbiol Infect Dis 69:51–58

24. Rattan A, Kumari S, Khanna N, Pandhi RK, Ray K, Bala M, Misra RS (1998) Emergence of fluoroquinolone resistant *Neisseria gonorrhoeae* in New Delhi, India. Sex Transm Infect 74:229
25. Cynamon MH, Sklaney M (2003) Gatifloxacin and ethionamide as the foundation for therapy of tuberculosis. Antimicrob Agents Chemother 47:2442–2444
26. Takiff HE, Salazar L, Guerrero C, Philipp W, Huang WM, Kreiswirth B, Cole ST, Jacobs WR Jr, Telenti A (1994) Cloning and nucleotide sequence of *Mycobacterium tuberculosis* gyrA and gyrB genes and detection of quinolone resistance mutations. Antimicrob Agents Chemother 38:773–780
27. Goto M, Oka S, Okuzumi K, Kimura S, Shimada K (1991) Evaluation of Acridinium-Ester-Labeled DNA Probes for Identification of *Mycobacterium tuberculosis* and *Mycobacterium avium-Mycobacterium intracellulare* Complex in Culture. J Clin Microbiol 29:2473–2476
28. Cloud JL, Neal H, Rosenberry R, Turenne CY, Jama M, Hillyard DR, Carroll KC (2002) Identification of *Mycobacterium* spp. by using a commercial 16S ribosomal DNA sequencing kit and additional sequencing libraries. J Clin Microbiol 40:400–406
29. Sarkola A, Makinen J, Marjamaki M, Marttila HJ, Viljanen MK, Soini H (2004) Prospective evaluation of the GenoType assay for routine identification of mycobacteria. Eur J Clin Microbiol Infect Dis 23:642–645
30. Patel JB, Leonard DG, Pan X, Musser JM, Berman RE, Nachamkin I (2000) Sequence-based identification of *Mycobacterium* species using the MicroSeq 500 16 S rDNA bacterial identification system. J Clin Microbiol 38:246–251
31. Li Z, Bai GH, von Reyn CF, Marino P, Brennan MJ, Gine N, Morris SL (1996) Rapid detection of *Mycobacterium avium* in stool samples from AIDS patients by immunomagnetic PCRJ. Clin Microbiol 34:1903–1907
32. Kirschner P, Springer B, Vogel U, Meier A, Wrede A, Kiekenbeck M, Bange FC, Bottger EC (1993) Genotypic identification of mycobacteria by nucleic acid sequence determination: report of a 2-year experience in a clinical laboratory. J Clin Microbiol 31:2882–2889
33. Yam WC, Yuen KY, Kam SY, Yiu LS, Chan KS, Leung CC, Tam CM, Ho PO, Yew WW, Seto WH, Ho PL (2006) Diagnostic application of genotypic identification of mycobacteria. J Med Microbiol 55:529–536
34. Drancourt M, Bollet C, Carlioz A, Martelin R, Gayral JP, Raoult D (2000) 16S ribosomal DNA sequence analysis of a large collection of environmental and clinical unidentifiable bacterial isolates. J Clin Microbiol 38:3623–3630

Chapter 13

Direct Detection of *Mycobacterium ulcerans* in Clinical Specimens and Environmental Samples

Caroline J. Lavender and Janet A.M. Fyfe

Abstract

Mycobacterium ulcerans is a slow-growing environmental bacterium that causes a severe skin disease known as Buruli ulcer. Rapid detection of *M. ulcerans* in clinical specimens is essential to ensure early diagnosis and prevention of disability. This chapter describes a real-time PCR method for the direct detection of *M. ulcerans* from swabs, fresh tissue biopsies, and fixed tissue sections, which are the most common types of specimens used in the diagnosis of Buruli ulcer. The chapter also briefly describes methods for PCR detection of *M. ulcerans* in environmental samples, as reliable detection of *M. ulcerans* in the environment is becoming increasingly important for understanding the ecology and transmission of this important pathogen.

Key words: *Mycobacterium ulcerans*, Mycobacteria, Direct detection, Buruli ulcer, Clinical, Environmental, Real-time PCR

1. Introduction

Buruli ulcer (BU) is a skin disease caused by infection with *Mycobacterium ulcerans*, a slow-growing environmental mycobacterium (1). Although not generally a fatal condition, BU lesions can become extensive and heal by scarring. Therefore, rapid diagnosis is essential to ensuring early treatment to minimize disability. The most commonly used methods for the detection of *M. ulcerans* in clinical samples are culture, smear for acid-fast bacilli (AFB), and PCR. Culture remains the gold standard for the laboratory confirmation of BU, but due to the very slow growth of the organism it may take 8–12 weeks. Treatment needs to be initiated much sooner than this to ensure optimal outcome for the patient. Thus, in those regions with access to

Mark Wilks (ed.), *PCR Detection of Microbial Pathogens: Second Edition*, Methods in Molecular Biology, vol. 943, DOI 10.1007/978-1-60327-353-4_13,

molecular diagnostics, the use of PCR for diagnosis of BU has been a major step forward. Provided it is performed in a laboratory with high standards to avoid false positives, PCR is a reliable and rapid method for laboratory confirmation of *M. ulcerans* disease (2, 3). The most commonly used target sequence for *M. ulcerans* diagnostic PCR is IS*2404*, a multicopy insertion sequence (approximately 200 copies per genome) that encodes a 328-amino acid transposase (4), and a variety of PCR methods have been developed for this target (2, 3, 5–8).

This chapter describes the real-time PCR method for the direct detection of *M. ulcerans* in clinical specimens (swabs, fresh tissue biopsies, and paraffin-embedded formalin-fixed tissue sections), which was developed by our laboratory and has been used to diagnose cases of *M. ulcerans* disease in the Australian State of Victoria since 2002 (2). This real-time PCR assay, which targets IS*2404* and is multiplexed with an internal positive control (IPC) to monitor inhibition, has facilitated rapid diagnosis of *M. ulcerans* disease with greater confidence than conventional PCR. The chapter does not describe the gel-based PCR method recommended by the World Health Organization (WHO), as this method is described in detail in the WHO manual on the diagnosis of *M. ulcerans* disease (5). The DNA extraction protocol described in this chapter was also developed in our laboratory and differs slightly from the DNA extraction protocol in the WHO manual. In our experience, this DNA extraction method, which consists of an initial alkaline lysis step followed by a column DNA clean-up procedure, minimizes PCR inhibition, particularly for swabs in transport medium.

Culturing *M. ulcerans* directly from the environment is extremely difficult (6), so PCR performed on DNA extracted from environmental samples has proved a major advance since its first use in the mid-1990s (2, 9, 10). However, although IS*2404* PCR is highly specific for testing clinical specimens, its application to the analysis of environmental samples is less straightforward due to PCR inhibitors and the existence of other environmental mycobacteria that may carry IS*2404* (2). For this reason, we also describe here a confirmatory PCR targeting two additional sequences in the *M. ulcerans* genome, another insertion sequence, IS*2606* (90 copies per genome), and a sequence encoding the ketoreductase B domain (KR), located on the plasmid *p*MUM001 within the mycolactone polyketide synthase genes. This assay, when used in conjunction with the IS*2404*/IPC real-time PCR, provides high confidence that *M. ulcerans* is being detected (2, 11). The direct detection of *M. ulcerans* in environmental samples has become increasingly important as the research community seeks to elucidate the mode(s) of transmission and environmental reservoir(s) of this neglected tropical disease.

2. Materials

2.1. Extraction of DNA from Swabs and Fresh Tissue

The equipment and reagents listed below should be dedicated to DNA extraction (see Note 1).

1. Biosafety cabinet.
2. Laboratory bench paper.
3. Lab coat.
4. Gloves.
5. Permanent marker.
6. Small sterile Petri dishes.
7. Sterile scalpels and forceps.
8. Sterile graduated plastic transfer pipettes.
9. Sterile fine tip plastic transfer pipettes.
10. Sterile 15 ml glass bottle containing 50–100 3 mm glass beads (Ajax Finechem, Seven Hills, Australia) ("bead bottle").
11. Sterile phosphate-buffered saline (PBS). Store in single use aliquots at room temperature (see Note 2).
12. Vortex mixer.
13. Respiratory Specimen Preparation Kit (Roche, Meylan, France) or equivalent alkaline lysis kit/method. Store Roche Respiratory Specimen Preparation Kit at 4°C or according to the manufacturer's instructions if using another method (see Note 3).
14. QIAamp® DNA Mini Kit (Qiagen, Inc., Valencia, CA) or equivalent DNA clean-up kit (see Note 3). Before using buffers AW1 and AW2 for the first time, add the appropriate amount of ethanol as indicated on the bottle. Store reagents in small volume aliquots at room temperature (see Note 2).
15. Analytical grade absolute ethanol (96–100%).
16. Pipettes (capable of measuring 20–1,000 μl).
17. 20, 200, and 1,000 μl sterile filter pipette tips.
18. Sterile Sarstedt or equivalent screw capped microcentrifuge tubes (see Note 4).
19. Sterile 1.5 ml Eppendorf tubes or equivalent.
20. Microcentrifuge tube rack(s).
21. Microcentrifuge.
22. Heat block(s) or water bath(s) set to 60 and 56°C.
23. 10% bleach solution (see Note 5).
24. Fridge and freezer set to 4°C and –20°C, respectively.

2.2. Extraction of DNA from Paraffin-Embedded Formalin-Fixed Tissue Sections

The equipment and reagents listed below should be dedicated to DNA extraction (see Note 1).

1. Biosafety cabinet.
2. Laboratory bench paper.
3. Lab coat.
4. Gloves.
5. Permanent marker.
6. Sterile fine tip plastic transfer pipettes.
7. Pipettes (capable of measuring 20–1,000 μl).
8. 20, 200, and 1,000 μl sterile filter pipette tips.
9. Histolene (Fronine Pty Ltd). Store in single use aliquots at room temperature (see Note 2).
10. Absolute ethanol.
11. Sterile 1.5 ml Sarstedt tubes or equivalent (see Note 4).
12. Sterile microfuge tubes or equivalent.
13. Microcentrifuge tube rack(s).
14. QIAamp® DNA Mini Kit (Qiagen) or equivalent DNA clean-up kit (see Note 3). Store reagents in small volume aliquots at room temperature.
15. Digestion buffer (50 mM Tris–HCl pH 7.5, 10 mM EDTA, 0.5% SDS, 50 mM NaCl, 300 μg/ml Proteinase K). Store in single use aliquots (see Note 2).
16. Microcentrifuge.
17. Heating block(s) set at 56 and 70°C.
18. 10% Bleach solution (see Note 5).
19. Fridge and freezer set to 4°C and −20°C, respectively.

2.3. Extraction of DNA from Environmental Samples

The equipment and reagents listed below should be dedicated to DNA extraction (see Note 1).

1. Biosafety cabinet.
2. Laboratory bench paper.
3. Lab coat.
4. Gloves.
5. Permanent marker.
6. Pipettes (capable of measuring 20–1,000 μl).
7. 20, 200, and 1,000 μl sterile filter pipette tips.
8. Sterile fine tip transfer pipettes.
9. Sterile microfuge tubes or equivalent.
10. Microcentrifuge tube rack(s).

11. Absolute ethanol (96–100%).
12. FastDNA® Spin Kit for Soil (Qbiogene, Inc., Carlsbad, CA) (see Note 6). Before using SEWS-M solution for the first time, add 100 ml of absolute ethanol as indicated on the bottle. Store reagents in small volume aliquots at room temperature (see Note 2).
13. FastPrep® Instrument.
14. Microcentrifuge.
15. 10% Bleach solution (see Note 5).
16. Fridge and freezer set to 4°C and −20°C, respectively.

2.4. Detection of M. ulcerans DNA by Real-Time PCR

2.4.1. Mastermix Preparation (Perform in Dedicated PCR Mastermix Room) (see Note 1)

1. Biosafety cabinet.
2. Laboratory bench paper.
3. Lab coat.
4. Gloves (powder free).
5. TaqMan® Universal PCR Mastermix (Applied Biosystems, Foster City, CA), ABsolute QPCR ROX Mix (ABgene Ltd, Epsom, Victoria) or equivalent (stored at 4°C).
6. Primers IS*2404*TF/R and/or IS*2606*TF/R and KRTF/R (Table 1). Prepared as 18 μM working solutions and stored in ~100 μl aliquots at −20°C (see Note 2).
7. TaqMan® probe IS*2404*TP and/or IS*2606*TP and KRTP (Table 1). Prepared as 5 μM working solutions and stored in ~100 μl aliquots at −20°C (see Note 2).
8. Exogenous Internal Positive Control (IPC) reagents (Applied Biosystems) stored at −20°C. Store 10XExo IPC Mix in ~200 μl aliquots (see Note 2).
9. Nuclease-free water (NFW). Store at room temperature.
10. 96-Well optical reaction plate (ABI PRISM, Axygen, ABgene or equivalent).
11. Support base for 96-well place.
12. Pipettes (capable of measuring 20–1,000 μl).
13. 20, 200, and 1,000 μl sterile plugged pipette tips.
14. Sterile 1.5 or 5 ml tubes.
15. Microcentrifuge tube rack(s).

2.4.2. Addition of DNA Template (Perform in Dedicated PCR Template Room) (see Note 1)

1. Biosafety cabinet.
2. Laboratory bench paper.
3. Lab coat.
4. Gloves (powder free).

Table 1
Primers and probes designed for real-time PCR assays targeting IS*2404*, IS*2606*, and KR

Primer or probe[a]	Sequence (5′–3′)
IS*2404*TF	AAAGCACCACGCAGCATCT
IS*2404*TR	AGCGACCCCAGTGGATTG
IS*2404*TP	6FAM-CGTCCAACGCGATC-MGBNFQ
IS*2606*TF	CCGTCACAGACCAGGAAGAAG
IS*2606*TR	TGCTGACGGAGTTGAAAAACC
IS*2606*TP	VIC-TGTCGGCCACGCCG-MGBNFQ
KRTF	TCACGGCCTGCGATATCA
KRTR	TTGTGTGGGCACTGAATTGAC
KRTP	6FAM-ACCCCGAAGCACTG-MGBNFQ

[a] *TF* TaqMan forward primer, *TR* TaqMan reverse primer, *TP* TaqMan probe

5. 96-Well plate containing dispensed mastermix.
6. Support base for 96-well plate.
7. DNA extracts following specimen preparation.
8. Positive control DNA (*M. ulcerans* genomic DNA, 10 pg/μl).
9. Micropipette 0.5–10 μl.
10. 10 μl Sterile filter pipette tips.
11. Nuclease-free water.
12. Optical adhesive covers (ABI or equivalent).
13. Adhesive cover applicator.

2.4.3. Amplification and Detection of PCR Products

1. Real-time PCR instrument (e.g., ABI Prism 7000 Sequence Detector, Eppendorf Mastercycler Realplex or equivalent).
2. Optical adhesive cover compression pad (if required).

3. Methods

3.1. Extraction of DNA from Swabs and Fresh Tissue

1. Turn on biosafety cabinet and cover work surface with laboratory bench paper.
2. Set heating block to 60°C.
3. Label a sterile bead bottle (both bottle and lid) for each specimen and a reagent control (see Note 7) using a permanent marker.

4. For fresh tissue biopsies: cut tissue into small pieces using a sterile scalpel in a Petri dish and transfer to the bead bottle.
5. For swabs: place swab in bead bottle and break off the shaft.
6. Add 2 ml sterile PBS to each bead bottle and vortex vigorously for 2–3 min.
7. Transfer 1 ml of suspension to a labeled Sarstedt tube using either a sterile graduated plastic transfer pipette or 1 ml pipette (the remaining suspension can be stored at 4°C and used for culture and/or staining if required).
8. Centrifuge tubes at maximum speed in a microcentrifuge for 10 min (noting the orientation of the tubes) (see Note 8).
9. Remove the supernatant using a sterile fine tip transfer pipette, being careful to avoid the pellet.
10. Resuspend the pellet with 500 μl RW (from Roche Respiratory Specimen Preparation Kit) and vortex briefly.
11. Centrifuge tubes at maximum speed for 10 min.
12. Remove the supernatant using a fine tip transfer pipette, being careful to avoid the pellet.
13. Resuspend the pellet in 100 μl RL (from Roche Respiratory Specimen Preparation Kit) and vortex briefly.
14. Incubate the tubes at 60°C for 45 min (see Note 9).
15. Remove tubes from heating block and allow to cool.
16. Reset heating block to 56°C.
17. Centrifuge briefly (10 s) to remove liquid from the lids.
18. Neutralize the lysate with the addition of 100 μl RN (from Roche Respiratory Specimen Preparation Kit).

    Note: The following steps are performed according to the protocol for crude cell lysates supplied with the QIAamp DNA Mini Kit (see Note 10).
19. Add 20 μl Proteinase K (from QIAamp DNA Mini Kit) per 200 μl lysate.
20. Add 200 μl Buffer AL (from QIAamp DNA Mini Kit) per 200 μl lysate.
21. Mix immediately by pulse-vortexing for 15 s.
22. Incubate at 56°C for 10 min (see Note 11).
23. Remove tubes from heating block and allow to cool.
24. Centrifuge briefly (10 s) to remove liquid from the lids.
25. Add 200 μl absolute ethanol per 200 μl lysate.
26. Mix again by pulse-vortexing and spin briefly to remove liquid from the lids.

27. Add the lysate (maximum volume 620 μl) to labeled QIAamp spin columns (in a 2 ml collection tube) without wetting the rim.
28. Centrifuge at maximum speed for 1 min (see Note 8).
29. Place the spin column in a clean 2 ml collection tube and discard the tube containing the filtrate.
30. Carefully open the spin column and add 500 μl Buffer AW1 (from QIAamp DNA Mini Kit) without wetting the rim.
31. Centrifuge at maximum speed for 1 min.
32. Place the spin column in a clean 2 ml collection tube and discard the tube containing the filtrate.
33. Carefully open the spin column and add 500 μl Buffer AW2 (from QIAamp DNA Mini Kit) without wetting the rim.
34. Centrifuge at maximum speed for 3 min.
35. Place spin column in a labeled 1.5 ml Eppendorf tube and discard the collection tube containing the filtrate.
36. Carefully open the spin column and add 50 μl Buffer AE (from QIAamp DNA Mini Kit).
37. Incubate for 1 min at room temperature and centrifuge at 6,000 × *g* for 1 min (see Note 12).
38. Discard spin column.
39. Store Eppendorf tubes containing DNA extracts at 4°C if PCR is to be performed on the same day or at −20°C if PCR is to be performed at a later date.
40. Clean pipettes, centrifuge, and biosafety cabinet with 10% bleach solution (or equivalent) and UV irradiate biosafety cabinet prior to next extraction.

3.2. Extraction of DNA from Paraffin-Embedded Formalin-Fixed Tissue Sections (see Note 13)

1. Turn on biosafety cabinet and cover work surface with laboratory bench paper.
2. Label a sterile Sarstedt tube (both tube and lid) for each specimen and a reagent control (see Note 7) using a permanent marker.
3. Transfer at least 6 × 20 μm-thick paraffin-embedded tissue sections to the corresponding Sarstedt tube (see Note 14).
4. Add 1 ml Histolene and incubate at room temperature until sections have dissolved (1–5 min).
5. Centrifuge tubes at maximum speed for 5 min to pellet tissue.
6. Remove supernatant using a fine tip transfer pipette, being careful to avoid the pellet. Discard supernatant into a container suitable for volatile solvents.

7. Resuspend pellet in 1 ml absolute ethanol and incubate for 5 min at room temperature.
8. Centrifuge at maximum speed for 5 min.
9. Remove supernatant using a fine tip transfer pipette, being careful to avoid the pellet, and briefly air dry.
10. Resuspend the sample in 180 μl digestion buffer (50 mM Tris–HCl pH 7.5, 10 mM EDTA, 0.5% SDS, 50 mM NaCl, 300 μg/ml Proteinase K).
11. Add 20 μl Proteinase K (QIAamp DNA Mini Kit) and incubate at 56°C for 24 h (see Note 15).
12. Remove tubes from heat block, allow to cool and centrifuge briefly to remove liquid from the lids.
13. Add 200 μl Buffer AL (QIAamp DNA Mini Kit) and mix by pulse-vortexing.
14. Incubate at 70°C for 10 min.
15. Complete the extraction by following steps 23–40 described in Subheading 3.1.

3.3. Extraction of DNA from Environmental Samples

The following DNA extraction procedure is a slight modification of the protocol provided with the FastDNA Spin Kit for Soil® (Qbiogene, Inc., Carlsbad, USA) (see Note 6).

1. Turn on biosafety cabinet and cover work surface with laboratory bench paper.
2. Label a kit-supplied Lysing matrix E tube for each sample plus a reagent control (see Note 7) using a permanent marker.
3. For liquid samples (e.g., water or biofilm): add up to 1 ml of sample to lysing tube, centrifuge at maximum speed for 10 min then remove supernatant using a fine tipped transfer pipette.
4. For solid samples (e.g., soil or feces): add up to 500 mg of sample to lysing tube.
5. Add 978 μl sodium phosphate Buffer and 122 μl MT Buffer to sample in Lysing Matrix E tube. To avoid leakage, ensure that the tube is not overfilled and that the lid is screwed on tightly.
6. Homogenize in the FastPrep® Instrument for 40 s at a speed setting of 6.0.
7. Centrifuge at maximum speed for 15 min to pellet debris.
8. Transfer supernatant to a clean microcentrifuge tube.
9. Add 250 μl PPS (protein precipitation solution) and mix by shaking the tube by hand 10 times.
10. Centrifuge at maximum speed for 5 min to pellet precipitate.
11. Transfer 200 μl supernatant to a clean microcentrifuge tube (see Note 10).

12. Resuspend Binding Matrix suspension and add 500 μl to supernatant.
13. Place on rotator or invert by hand for 2–5 min to allow binding of DNA.
14. Transfer approximately 700 μl of the mixture to a SPIN™ Filter and centrifuge at maximum speed for 1 min (see Notes 8 and 16).
15. Empty the catch tube, add 500 μl prepared SEWS-M to the SPIN Filter and gently resuspend the pellet using the force of the liquid from the pipette tip.
16. Centrifuge at maximum speed for 1 min. Empty the catch tube and replace SPIN Filter.
17. Without any addition of liquid, centrifuge a second time at maximum speed for 2 min to dry the matrix of residual wash solution.
18. Discard the catch tube and place SPIN Filter in a clean catch tube.
19. Air dry the SPIN™ Filter for 5 min at room temperature.
20. Gently resuspend Binding Matrix (above the SPIN filter) in 50–100 μl of DES (DNase/Pyrogen-Free Water) (see Note 17). Centrifuge at maximum speed for 1 min.
21. Discard the SPIN filter.
22. Store catch tubes containing DNA extracts at 4°C if PCR is to be performed on the same day or at −20°C if PCR is to be performed at a later date.
23. Clean pipettes, centrifuge, FastPrep Instrument and biosafety cabinet with 10% bleach solution (or equivalent) and UV irradiate biosafety cabinet prior to next extraction.

3.4. Detection of *M. ulcerans* DNA by Real-Time PCR

3.4.1. Preparation of TaqMan Mastermix (Perform in Dedicated PCR Mastermix Room)

1. Remove reagents (mastermix, primers, probe, IPC) from freezer and allow to thaw.
2. Estimate the volume of mastermix required based on the number of reactions to be performed (i.e., four non-template controls [NTC], two positive controls, and duplicate reactions for each specimen and reagent control) (see Note 7).
3. Prepare mastermix required according to Table 2.
4. Dispense 24 μl volumes of the prepared mastermix into the appropriate wells of the 96-well plate (which is sitting in the support base).
5. Remove the plate from the support base and take the plate to the PCR template room (the support base should remain in the mastermix room), being careful not to touch the top or bottom of the plate.

Table 2
Preparation of mastermix for *M. ulcerans* real-time PCR assays

Reagent	Volume per reaction IS2404/IPC[a] assay	IS2606/KR assay
2× TaqMan Universal PCR mastermix	12.5 μl	12.5 μl
Primer IS*2404*TF (18 μM)	1.25 μl	–
Primer IS*2404*TR (18 μM)	1.25 μl	–
Probe IS*2404*TP (5 μM)	1.25 μl	–
Primer IS*2606*TF (18 μM)	–	1.25 μl
Primer IS*2606*TR (18 μM)	–	1.25 μl
Probe IS*2606*TP (5 μM)	–	1.25 μl
Primer KRTF (18 μM)	–	1.25 μl
Primer KRTR (18 μM)	–	1.25 μl
Primer KRTP (5 μM)	–	1.25 μl
50× ExoIPC DNA	0.5 μl	–
10× ExoIPC Mix	2.5 μl	–
Nuclease-free water (NFW)	4.75 μl	4 μl
Total	24 μl	24 μl

[a] *IPC* internal positive control

3.4.2. Adding DNA Extracts and Controls to the Mastermix (Perform in Dedicated PCR Template Room)

1. In the template room, place the 96-well plate containing the dispensed mastermix into the second support base.
2. Add 1 μl NFW to NTC wells (e.g., A1, A2, A3, A4), 1 μl *M. ulcerans* DNA to positive control wells (e.g., A5, A6), and 1 μl DNA extracts from the test samples and reagent/extraction control (tested in duplicate).
3. Check that there are no bubbles at the base of each well and remove any bubbles with a sterile pipette tip.
4. Seal the plate with an optical adhesive cover using the adhesive cover applicator, being careful not to touch the top or bottom of the plate.

3.4.3. Amplification and Detection of PCR Products

1. Transfer the plate to the real-time PCR instrument.
2. If required, place the optical cover compression pad on top of the plate.

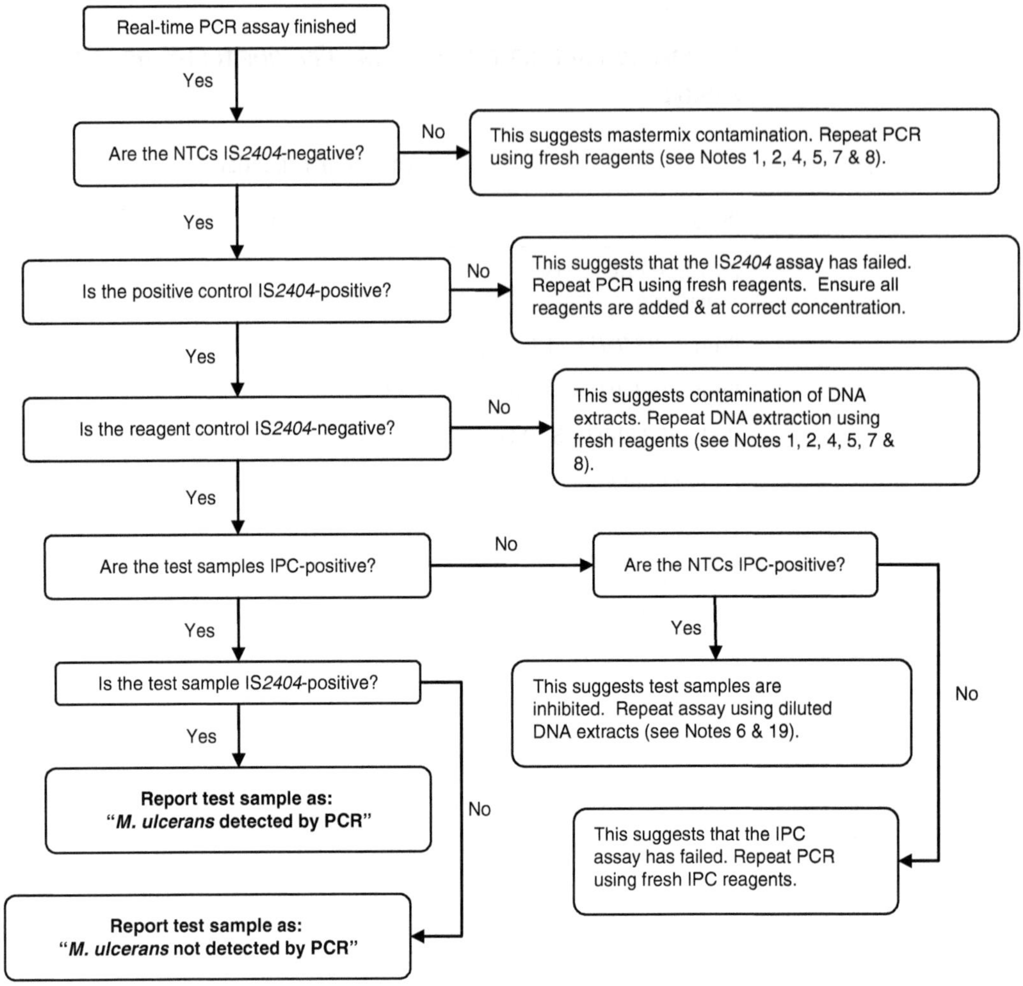

Fig. 1. Steps for interpreting, validating, and troubleshooting the real-time PCR targeting IS*2404* for the direct detection of *M. ulcerans*.

3. Set up and launch the run according to the software program and instructions supplied with the real-time PCR instrument. The parameters should be set to include detection of both VIC and 6-FAM fluorescent dyes. The amplification program employed in our laboratory using the ABI Prism 7000 Sequence Detector Instrument is 1 cycle of 50°C for 2 min, 1 cycle of 95°C for 15 min, and 40 cycles of 95°C for 15 s and 60°C for 1 min. Note that the volume of the reactions is 25 μl.

3.4.4. Interpretation of Results

1. Follow the steps outlined in Fig. 1 for interpreting, validating, and troubleshooting the results of the PCR assay (see Note 18).
2. Test samples that generate amplification curves in both replicates for the IS*2404* assay are positive for *M. ulcerans* DNA. Because culture is the gold standard for diagnosis of *M. ulcerans*

disease, a positive PCR result (for clinical specimens) should be accompanied with the comment: "This result does not necessarily indicate the presence of viable organisms. Routine culture is recommended in addition to this test."

3. Test samples that do not generate amplification curves are negative for *M. ulcerans* DNA (these samples should be formally reported as "*M. ulcerans* DNA not detected"). Again, a negative PCR result (for clinical specimens) should be accompanied with the comment: "This result may not exclude the presence of *M. ulcerans*. Routine culture is recommended in addition to this test."
4. Test samples that generate amplification curves in one replicate only should be assayed again. If, after testing a second time, the sample still generates an amplification curve in one replicate only, the result is considered inconclusive and a repeat DNA extract should be prepared.
5. Assay inhibition is considered to have occurred in a test sample if the IPC C_T value is >5 cycles higher than the IPC C_T value for the NTC. In such as case, the sample should be diluted and assayed again (see Notes 6 and 19).

4. Notes

1. To minimize contamination, separate equipment for DNA extraction, PCR mastermix preparation, and PCR template addition should be used, including biosafety cabinets, centrifuges, microcentrifuge tube racks, pipettes, laboratory gowns, gloves, reagents, etc. It is *CRUCIAL* that cultures of *M. ulcerans* are not processed using the same reagents or equipment as those used for DNA extraction from primary specimens.
2. Dividing reagents for DNA extraction and PCR into smaller aliquots (the volume will depend on the number and frequency of samples processed) prevents wastage of reagents in the event of contamination and avoids repeat freezing/thawing of reagents that are stored frozen.
3. We have found the Respiratory Specimen Preparation Kit used with the QIAamp DNA Mini Kit gives high DNA extraction efficiency from dry swabs, swabs in transport medium and fresh tissue and with minimal inhibition, but numerous alternative in-house or commercial methods are available and may be used instead.
4. Screw-capped tubes are preferable to flip-top tubes, which might "pop" open when tubes are incubated and spray tube contents.

5. To minimize contamination, all instruments (pipettes, laboratory benches, biosafety cabinets, microcentrifuge tube racks, etc.) should be cleaned thoroughly with 10% bleach or similar product as appropriate in between each extraction/PCR.
6. We have found the FastDNA Spin Kit for Soil to be excellent for extracting DNA from soil, water, biofilm, vegetation, animal feces, and other environmental samples with inhibitory substances, but alternative in-house or commercial methods are available and may be used instead. We find that the Mo Bio Power Soil™ Kit (MO BIO Laboratories, Inc., Carlsbad, USA) is superior to the FastDNA® Spin Kit for Soil for removing PCR inhibitors; however, it extracts 10–100-fold less DNA (unpublished data). If extracting DNA from insects (e.g., mosquitoes), we suggest using the regular FastDNA Kit (Qbiogene).
7. Negative controls should be included with every DNA extraction and PCR to monitor contamination at the DNA extraction and PCR stages. Although we do not include a positive DNA extraction control with every DNA extraction, we recommend including a positive extraction control (e.g., a piece of culture-confirmed tissue or a swab impregnated with a dilute suspension of *M. ulcerans* organisms) when a laboratory is implementing the method for the first time and when a new batch of reagents are being introduced. It is also recommended that reagent lot numbers are recorded for every DNA extraction and PCR so that any problems can be traced.
8. Wherever possible, leave gaps in between sample tubes when centrifuging or opening tubes in racks to add reagents to prevent cross-contamination.
9. This 60°C incubation step can be extended to, but should not exceed, 1 h.
10. At this point, the lysate can also be processed using an automated DNA extraction instrument. This is useful when the number of samples is large (i.e., greater than 24). We have used the Qiagen QIAxtractor (Qiagen) for 12 months and find it equivalent to the manual procedures described here, both in terms of DNA extraction efficiency and removal of inhibitors (unpublished data).
11. This 56°C incubation step can be extended to 2–3 h if required and may help to reduce inhibition in samples with a lot of clinical material.
12. Extending this elution incubation to 5 min may increase DNA yield.
13. This DNA extraction procedure for paraffin-embedded sections is the result of evaluation of several available methods and

is a modification of the protocol for tissue provided with the QIAamp DNA Mini Kit.

14. Serial paraffin sections should be cut from the paraffin block(s) using a fresh/sterile scalpel to prevent carryover of contaminating DNA.
15. This 56°C incubation step should not be less than 20 h or more than 48 h. Where there is a lot of tissue, a longer incubation period may increase DNA yield.
16. Sometimes, particularly when processing viscous samples, the lysate may not flow through the spin filter. If this occurs, extend the centrifuge time (i.e., 2–3 min) and/or remove the liquid remaining in the catch tube using a sterile transfer pipette.
17. To avoid over-dilution of the purified DNA, use the smallest amount of DES required to resuspend Binding Matrix pellet. Extending the incubation to 5 min may increase DNA yield.
18. If the steps summarized in Fig. 1 are not followed, the results of the test samples should be interpreted with caution at best or regarded as invalid at worst. Figure 1 outlines how to identify common problems and what should be done in each case.
19. Mild inhibition can usually be overcome at the PCR stage, either by diluting the DNA extract (e.g., by 1/2 to 1/10, depending on the level of inhibition) and repeating the PCR or by using a specialized real-time PCR mastermix, such as the TaqMan® Environmental Mastermix (Applied Biosystems).

References

1. MacCallum P, Tolhurst J, Buckle G, Sissons H (1948) A new mycobacterial infection in man. J Path Bacteriol 60:93–122
2. Fyfe JA, Lavender CJ, Johnson PD et al (2007) Development and application of two multiplex real-time PCR assays for the detection of *Mycobacterium ulcerans* in clinical and environmental samples. Appl Environ Microbiol 73:4733–4740
3. Phillips R, Horsfield C, Kuijper S et al (2005) Sensitivity of PCR targeting the IS*2404* insertion sequence of *Mycobacterium ulcerans* in an Assay using punch biopsy specimens for diagnosis of Buruli ulcer. J Clin Microbiol 43:3650–3656
4. Stinear T, Ross B, Davies J et al (1999) Identification and characterization of IS*2404* and IS*2606*: two distinct repeated sequences for detection of *Mycobacterium ulcerans* by PCR. J Clin Microbiol 37:1018–1023
5. Portaels F, Johnson P, Meyers WM, eds. *Buruli ulcer; Diagnosis of Mycobacterium ulcerans disease; a manual for health care providers.* 2001, World Health Organization: Geneva. Available at: http://www.who.int/buruli/information/diagnosis/en/index.html.
6. Portaels F, Meyers WM, Ablordey A et al (2008) First Cultivation and Characterization of *Mycobacterium ulcerans* from the Environment. PLoS Negl Trop Dis 2:e178
7. Rondini S, Mensah-Quainoo E, Troll H et al (2003) Development and application of real-time PCR assay for quantification of *Mycobacterium ulcerans* DNA. J Clin Microbiol 41:4231–4237
8. Ross B, Marino L, Oppedisano F et al (1997) Development of a PCR assay for rapid diagnosis of *Mycobacterium ulcerans* infection. J Clin Microbiol 35:1696–1700
9. Ross B, Johnson P, Oppedisano F et al (1997) Detection of *Mycobacterium ulcerans* in environmental samples during an outbreak of ulcerative disease. Appl Environ Microbiol 63:4135–4138

10. Williamson HR, Benbow ME, Nguyen KD et al (2008) Distribution of *Mycobacterium ulcerans* in Buruli Ulcer Endemic and Non-Endemic Aquatic Sites in Ghana. PLoS Negl Trop Dis 2:e205
11. Lavender CJ, Stinear TP, Johnson PD et al (2008) Evaluation of VNTR typing for the identification of *Mycobacterium ulcerans* in environmental samples from Victoria, Australia. FEMS Microbiol Lett 287:250–255

Chapter 14

Detection of *Bartonella* spp. DNA in Clinical Specimens Using an Internally Controlled Real-Time PCR Assay

Anneke M.C. Bergmans and John W.A. Rossen

Abstract

Bartonella henselae is the causative agent of cat-scratch disease (CSD), usually presenting itself as a self-limiting lymphadenopathy. In this chapter an internally controlled Taqman probe-based real-time PCR targeting the *groEL* gene of *Bartonella* spp. is described. This assay allows for the rapid, sensitive, and simple detection of *Bartonella* spp. in samples from CSD or endocarditis suspects, and it is suitable for implementation in the diagnostic microbiology laboratory.

Key words: Real-time PCR, *Bartonella*, Molecular diagnostics, Cat-scratch disease

1. Introduction

Bartonella species can cause cat-scratch disease in human and may cause endocarditis, trench fever, bacillary angiomatosis, Oroya fever, and verruga peruana (1–11). Bacteria within the genus *Bartonella* are microaerophilic, Gram-negative, fastidious, slow-growing organisms that belong to the class Alphaproteobacteria on the basis of their 16S rDNA sequences (12). Laboratory methods for the diagnosis of *Bartonella* infections include isolation of the organisms by culture, serological assays, histopathological examination, and molecular detection of *Bartonella* DNA in affected tissue (1, 12–14). Routine bacterial culture protocols do not usually allow detection of these organisms. Primary isolates are typically obtained after 12–14 days, although prolonged incubation periods of up to 45 days are sometimes necessary (15). The bacilli can be detected in tissue specimens with the Warthin-Starry silver stain, but the technique is difficult to perform and the result is not specific for *Bartonella*. Evaluation of serological tests, mainly by immunofluorescence assay or enzyme immunoassay have

Mark Wilks (ed.), *PCR Detection of Microbial Pathogens: Second Edition*, Methods in Molecular Biology, vol. 943,
DOI 10.1007/978-1-60327-353-4_14, © Springer Science+Business Media, LLC 2013

reported various sensitivities and specificities, depending on study population and definitions of CSD, as well as materials and techniques used (14, 16). Polymerase chain reaction (PCR) offers a rapid and specific means to detect the organism directly from clinical specimens. Therefore, PCR as a quick and more reliable diagnostic test is being increasingly used (1, 12, 13, 16–19). In this chapter an internally controlled Taqman probe-based real-time PCR is described using the *groEL* gene, (which codes for a major heat shock protein) as a target sequence for the detection of *Bartonella* spp. DNA in clinical specimens.

The real-time PCR theoretically detects the following *Bartonella* species: *B. henselae* (both the Houston-1 type (*Alu*I RFLP type A) and the Marseille type (*Alu*I RFLP type B) (20), *B. birtlesii, B. vinsonii subsp. vinsonii, B. vinsonii subsp. arupensis, and B. doshiae* (21). Of these *Bartonella* spp. only *B. henselae* has been found to infect human frequently. In addition, only one case has been reported in which *B. vinsonii subsp. arupensis* (normally isolated from mice) infected a human, resulting in blood culture-negative endocarditis (22). Therefore, we expect that in clinical practice, using this real-time PCR assay, only *B. henselae* will be detected in the vast majority of cases. In patients with suspected culture-negative endocarditis, it would be prudent to confirm positive real-time PCR results by PCR on a *B. henselae*-specific target or by DNA sequence analysis. The performance of the real-time PCR was evaluated against a conventional *B. henselae*-specific PCR-hybridization assay targeting the16S rRNA gene (1), and 100% agreement between the two assays was found: 29 (40%) of 73 clinical specimens from CSD suspects gave positive PCR results, and 44 specimens (60%) were negative in both PCR assays (21). Furthermore, the assay generated negative results with DNA from the following bacterial species: *Bartonella bacilliformis* (KC583), *Bartonella clarridgeiae, Bartonella quintana* (Oklahoma strain), *Bartonella tribocorum, Bartonella elisabethae, Bartonella alsatica, Bartonella taylorii, Mycobacterium tuberculosis, Mycobacterium avium, Mycobacterium kansasii, Mycobacterium intracellulare, Mycobacterium gordonae, Mycobacterium xenopi, Mycobacterium abscessus, Mycobacterium fortuitum, Mycobacterium malmoense, Mycobacterium africanum, Streptococcus pyogenes, Staphylococcus aureus* (e.g., ATCC 29213), *Escherichia coli* (e.g., ATCC 25922), *Shigella flexneri, Streptococcus pneumoniae* (ATCC49619), *Legionella pneumophila* (ATCC 33623), *Bordetella pertussis* (Tohama), *Bordetella parapertussis* (B24), *Mycoplasma pneumoniae* (ATCC15293), *Chlamydia pneumoniae* (ATCCVR1355), *Acinetobacter baumannii, Pseudomonas aeruginosa* (ATCC27853) (21). The bacterial isolates without culture collection numbers are clinical isolates.

The *Bartonella* real-time PCR assay is used successfully in daily practice in our microbiology laboratory in Tilburg, The Netherlands, for the detection of *Bartonella* spp. in clinical material, mostly from CSD and endocarditis suspects.

2. Materials

2.1. Collection and Total Nucleic Acid Extraction from Clinical Specimens

1. DNAse-/RNAse-free containers are used to collect and store clinical specimens prior to nucleic acid (NA) extraction.
2. MagNA Lyser (Roche) is used to lyse biopsies and pus aspirates prior to NA extraction, as the MagNA Pure LC extraction robot requires liquid input material.
3. Lysis/Binding Buffer (LBB; from the MagNA Pure LC Total Nucleic Acid Isolation Kits—see below) is used for lysis of biopsies and pus aspirates in the MagNA Lyser.
4. MagNA Lyser Green Beads (Roche) are 2 ml-reaction vessels prefilled with beads that are used for lysis of biopsies and pus aspirates in the MagNA Lyser.
5. MagNA Pure LC Total Nucleic Acid Isolation Kits (Roche) are used in combination with the MagNA Pure LC extraction robot (MPLC; Roche) and accessories for extraction of total nucleic acids (DNA/RNA) from clinical specimens (see Note 1).
6. Molecular biology grade water.
7. DNAse-/RNAse-free filter pipette tips are used to prevent contamination, and thus avoid false positives.
8. DNAse-/RNAse-free 2 ml-cups with screw cap (Sarstedt) are used for storage of DNA extracts from clinical specimens.
9. Disposable scalpels are needed to cut biopsies into small pieces.
10. Non-powdered gloves are used during all handlings (see Note 2).

2.2. Real-Time PCR

1. The *Bartonella* spp. Real-Time PCR was developed using Primer Express Software (Applied Biosystems, Foster City, CA). Conserved target regions in the *gro*EL genes of *Bartonella* spp. were identified using BLAST (www.ncbi.nlm.nih.gov/blast) (21). Sequences of the primers and probes used for the detection of *Bartonella* spp. and for the detection of the internal control are summarized in Table 1.
2. Desalted primers are ordered at a 200 nmol scale from Invitrogen Ltd (Paisley, UK). Upon arrival the freeze-dried primers are suspended in Tris–EDTA buffer (10 mM Tris–HCl, pH 8.0, 1 mM EDTA) to a concentration of 200 μM. Subsequently, primers are aliquoted and stored at −20°C (see Note 3).
3. HPLC-purified, fluorescently labeled probes are ordered at a 50,000 pmol scale from Applied Biosystems. Upon arrival the probes (100 μM in Tris–EDTA-buffer) are aliquoted and stored at −20°C (see Notes 3 and 4).

Table 1
Sequences of *Bartonella* spp. specific primers and probe

Name	Sequence
Bhe-forward	ACA GGC TAT TGT CCA AGA AGG TGT A
Bhe-reverse	TCA ACA GCA GCA TCG ATA CCA
Bhe-probe	FAM-AAA GCC GTT GCT GCA G-MGB
PhHV-forward	GGG CGA ATC ACA GAT TGA ATC
PhHV-reverse	GCG GTT CCA AAC GTA CCA A
PhHV-probe	VIC-TTT TTT ATG TGT CCG CCA CCA TCT GGA TC-TAMRA

MGB minor groove binding, *FAM* 6-carboxyfluorescein, *VIC* 2′-chloro-7′-phenyl-1,4-dichloro-6-carboxyfluorescein, *TAMRA* 6-carboxytetramethylrhodamine, *PhHV* Phocine Herpes Virus

4. TaqMan®Universal PCR Mastermix is obtained from Applied Biosystems and is stored at 4°C (see Note 5).
5. MicroAmp™ Optical 96-Well Reaction Plates with Barcode for the 7900 HT Real-Time PCR Systems are obtained from Applied Biosystems.
6. MicroAmp™ Fast Optical 96-Well Reaction Plates with Barcode, for the 7500 Fast Real-Time PCR System are obtained from Applied Biosystems.
7. MicroAmp™ Optical Adhesive Films are obtained from Applied Biosystems.
8. Non-powdered gloves are used during all handlings (see Note 2). In addition, filter tips are used to prevent contamination, and thus avoid false positives.

3. Methods

3.1. Collection and Total Nucleic Acid Extraction from Clinical Specimens

3.1.1. Collection and Storage of Clinical Specimens (Room II, see Note 6)

1. Collect pus aspirates, swabs, or biopsy material from lymph nodes, aortic valve, brain, bone marrow, intraocular fluids, or other clinical material, at a volume of approximately 200–250 μl (minimum 50 μl), or 5–10 mm^3.
2. Store the specimens in a DNAse-/RNAse-free container at 4°C prior to nucleic acid isolation, for a maximum of 1 week. If specimens are not processed within 1 week they can be stored at −70°C for more than 1 year.

3.1.2. Lysis of Clinical Material (Room II, see Note 6)

1. Biopsies and pus aspirates need to be lysed in a MagNA Lyser before DNA extraction in a MagNA Pure LC robot can be performed (see Note 7).
2. Lyse the specimens as follows:
   - Cut biopsies into small pieces of 1 mm^3, put them into a MagNA Lyser Green Beads Tube containing Green Beads and add 500 μl of LBB.

     or

   - Put 250 μl of pus aspirate into an MagNA Lyser Green beads Tube containing Green Beads and add 250 μl of LBB.
3. Fix the tubes in the MagNA Lyser rotor according to the manufacturer's instruction.
4. Fix the retention plate and the red screws and close the lid.
5. Adjust speed to 6,000 rpm and shake for 30 s.
6. Push the Start button.
7. When the program is finished, release the red screws and remove the retention plate.
8. Put the rotor containing the tubes in the MagNA Lyser cooling block (pre-cooled at 4°C).
9. Centrifuge the sample for 10 s to pellet the cell debris, and then proceed with nucleic acid preparation using the supernatant (see Note 8).

3.1.3. Nucleic Acid Isolation from Clinical Material (Room II, see Note 6)

1. Start the MPLC apparatus and the MPLC software (see Note 1).
2. Click on "Sample Ordering."
3. At "Sample Protocol" select "Total NA—Total NA Variable_elution_volume.blk."
4. Adjust "Sample Volume" to 200 μl and "Elution Volume" to 50 μl.
5. Fill in the kit lot number and the expiry date.
6. Fill in the sample numbers in cell A1. Include at least 1 negative extraction control. In a run of 32 extractions two negative extraction controls are included. A negative extraction control exists of 200 μl of water and is treated as a clinical sample (see Note 9).
7. Click on "Print Sample Order" to print the worklist.
8. Click on "Start Batch" and follow the instructions on the screen.
9. Make a working solution of Proteinase K (add 5 ml of elution buffer to a bottle of Proteinase K and shake on a vortex) and pipette it into the appropriate tub on the MPLC.

10. Shake the Magnetic Glass Particles (MGP) before use in order to obtain a homogeneous suspension and pipet them just before starting the run.
11. All other reagents are ready-to-use.
12. Fill the Reagent Tubes with the appropriate volumes of the different reagents (bottles 1–7) and close the Tubes with Tube Lids.
13. Pipette 5 μl of Internal Control (Phocine Herpesvirus—1 (PhHV) virus suspension 2×10^{-4}) for each clinical specimen in the appropriate well of the Sample Cartridge (see Note 10).
14. Pipette 200 μl liquid sample or 200 μl of a clinical specimen liquidized using the MagNA Lyser into the wells containing the Internal Control.
15. On the screen, click on all blocks that contain reagents/specimens/disposables, wait until the screen shows the "Heat Block Status PASS" sign, click on "Cover Lock" and "OK".
16. The run starts, and takes 1.30 h for 32 samples.
17. After completion of the run, for each specimen a green message box with "PASS" or a red message box with "FAIL" appears on the screen. Specimens with the "FAIL" message need to be re-extracted.
18. Click on "Close" (and on "Unlock Door" if necessary) and open the door.
19. The DNA extracts (eluates) are in the Sample Cartridge in the Cooling Block.
20. Pipette the eluates into DNAse-/RNAse-free 2 ml-tubes with screw cap lids.
21. DNA extracts can be stored at 4°C for up to 1 year before use in the real-time PCR (see Note 11).

3.2. Real-Time PCR

3.2.1. Preparation of Primer/Probe Mixtures (Room I, see Note 12)

When primers and probe are ordered for the first time (see Note 13), their optimal concentrations for their use in the real-time PCR need to be determined. For this purpose a primer/probe matrix is used. By independently varying forward and reverse primer concentrations, the concentrations that provide optimal assay performance can be identified. The primer concentrations used in the primer optimization matrix are shown in Table 2. For the Real-Time PCR assay, optimal performance is achieved by selecting the primer concentrations that provide the lowest cycle threshold (C_t) and highest normalized fluorescence values (ΔRn) for a fixed amount of target template.

Using the optimal primer concentrations defined by the primer optimization matrix ensures excellent assay performance when using a 250 nM probe concentration. However, a probe optimization experiment can prove useful to reduce assay running costs.

Table 2
Primer concentrations used in the primer optimization matrix

Primer (nM)	Forward 300	Forward 600	Forward 900
Reverse 300	300/300	300/600	300/900
Reverse 600	600/300	600/600	600/900
Reverse 900	900/300	900/600	900/900

In our lab a probe optimization experiment is performed in which the probe concentration is varied from 50 to 250 nM. We select the probe concentration that provides the lowest C_t value. It should be noted, however, that to ensure the best reproducibility, especially when wishing to detect low copy numbers of a target sequence, it is necessary to avoid probe limiting concentrations. By using a 250 nM concentration, probe limitation is avoided and large ΔRn values are ensured.

Having determined the optimal primers and probe concentrations, 10× concentrated pre-mixtures containing primers and probe are prepared and stored at 4°C for direct use within 1 month of preparation or at −20°C for long-term storage. These mixtures also contain the primers and probe for the detection of the internal control (PhHV) (Table 1 and see Note 10).

3.2.2. Preparation of Amplification Mix (Room I)

1. Prepare the amplification mix.

   For each sample to be tested:

   - 2.5 μl primer/probe mixture.
   - 12.5 μl TaqMan®Universal PCR Mastermix.
   - Prepare additional mix for the negative isolation control (NEG), the no template control (NTC), the high (HPC), and low (LPC) positive run controls (*see* **Note 14**).
2. Pipette 15 μl of the amplification mix into a well of the reaction plate. For the NTC add 10 μl of the water used for the preparation of the primer/probe mixtures into one of the wells containing amplification mix.
3. Gently seal the reaction plate with the optical adhesive film in such a way that the seal can still be removed.
4. Transfer the reaction plate to room II (see Note 6).

3.2.3. Adding the Samples (Room II)

1. Place the reaction plate into a PCR or laminar flow cabinet.
2. Gently remove the seal from the reaction plate and place it upside down in a corner of the cabinet.
3. Add 10 μl of each of the extracted samples (also containing the Internal Control) to separate wells containing the amplification mix.

4. Add 10 μl of the NEG, HPC, and LPC to separate wells containing the amplification mix (see Note 15).
5. Gently seal the reaction plate again with the previously used optical seal. Make sure that the seal is now attached firmly to the plate to prevent cross-contamination and evaporation.
6. Spin the plate in a plate centrifuge for 15 s at a maximum speed of 500 × *g*.
7. Clean work benches, centrifuge, and tube racks with sodium hypochlorite (approximately 0.4% of free chloride) after use.
8. Transfer the sealed reaction plate to room III (see Note 16).

3.2.4. Amplification

1. Put the plate into the Real-Time PCR machine (we use an Applied Biosystems 7900HT and/or 7500 Fast Real-Time PCR System).
2. The following protocol is used (see Note 17):
   - 2 min at 50°C (UNG incubation, *see* **Note 18**).
   - 10 min at 95°C (polymerase activation).
   - 45 cycles of 15 s at 95°C (denaturation) and 1 min at 60°C (annealing and extension).

4. Interpretation and Reporting of Results

The results are analyzed using the manufacturer's software. The threshold for determination of the C_t is set in the middle of the logarithmic phase of the amplification curve (see also the figure below).

Figure 1 shows a typical *Bartonella* real-time PCR amplification curve from clinical specimens. A result is unequivocally *positive* if:

- The signal of the clinical sample is above a (by the software) defined threshold.
- All negative controls are negative.

A result is unequivocally *negative* if:

- The signal of the clinical sample is below a (by the software) defined threshold.
- The run control is positive and is within defined limits (see Note 15).
- The internal control of the sample is positive and within defined limits (no inhibition of the amplification in the reaction).

Equivocal results

- Samples with equivocal results must be reported as such and a new sample requested.

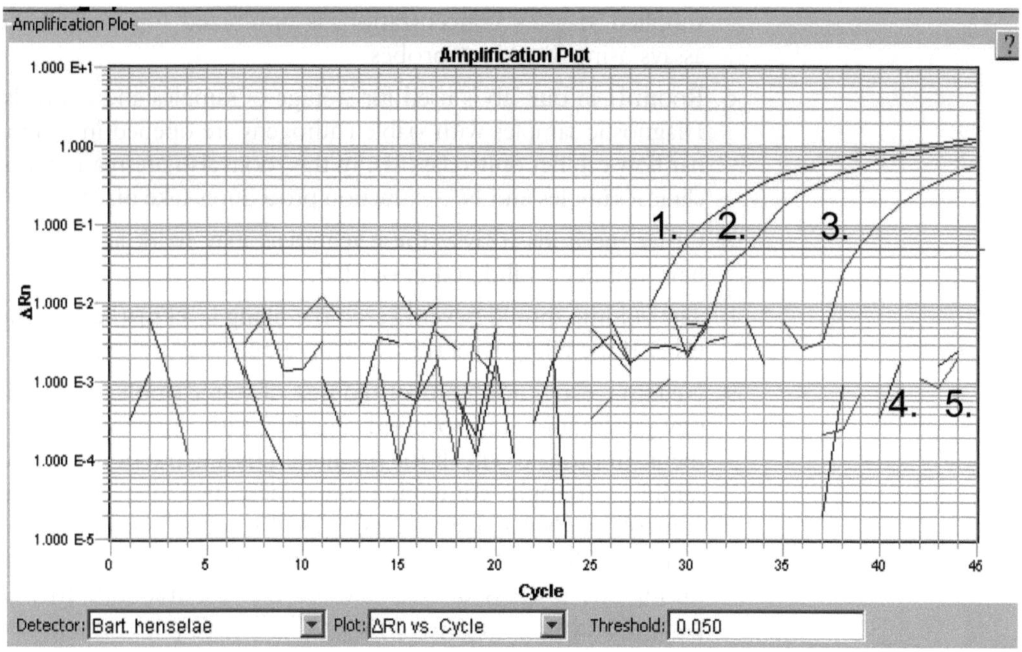

Fig. 1. Real-time PCR amplification curve resulting from PCR using primers and probe (depicted in Table 1) targeting *Bartonella* spp. 1, positive clinical sample; 2, high positive control; 3, low positive control; 4, negative control; 5, no template control.

5. Notes

1. For NA extraction in our lab the MagNA Pure LC is used. However, any alternative method to isolate DNA may work as well but was not validated in our lab.
2. Please note that powdered gloves should not be used as the powder negatively influences the PCR if present even in small amounts in the reaction tubes.
3. In our lab each aliquot of primer or probe contains enough to prepare primer/probe mixtures to be used in daily practice for about 1 year.
4. Probes are not allowed to go through more than four freeze-thaw cycles, are stored in the dark and during preparation exposure to direct light is avoided.
5. TaqMan® Universal PCR Master Mix conveniently combines AmpliTaq Gold® DNA Polymerase, AmpErase® UNG, dNTPs with dUTP, Passive Reference 1, and optimized buffer components in an easy-to-use premix. Proprietary buffer components and stabilizers are optimized to enhance reaction performance across a wide variety of cDNA or DNA sample types. The mix,

supplied at a 2× concentration, is optimized for 5′ nuclease assays using TaqMan® probes.

6. Room II in our lab is used for storage of samples and controls. Diagnostic samples with viable pathogens are opened in a laminar flow cabinet only (class II). In this room also pretreatment of clinical samples and NA-extraction procedures take place (23).
7. Other methods to lyse these materials may work as well but have not been validated in our lab.
8. The volume should never exceed the filling level marked on the tube. If the filling level is exceeded, sample material can escape.
9. Negative extraction controls consist of water (molecular biology grade) and are processed together with the clinical specimens, in order to check for adequate DNA extraction without sample-to-sample contamination. These negative extraction controls ought to give a negative PCR result in the *Bartonella* real-time PCR, otherwise contamination has been occurred during the process of lysis and/or DNA extraction. In this case, the extraction procedure of all clinical specimens in the run needs to be repeated. In addition, the negative extraction controls need to give a positive PCR result in the PhHV PCR well (Internal Control see Note 10), otherwise the DNA extraction was not successful, i.e., the DNA could have been lost during the extraction process, or the DNA extract contains inhibitory components for the real-time PCR.
10. As mentioned earlier an Internal Control (PhHV) is used to determine the extraction efficiency and purity of the DNA (24). We determined that this internal control can be detected in the same reaction as the target DNA without internal control primers and probe interfering with the *Bartonella* spp. primers and probe. Such interference may result in nonspecific signals and/or loss of sensitivity.
11. Proteinase K (Proteinase K (lyophilizate) recombinant, PCR Grade, Roche) once dissolved, can be stored for maximal 3 weeks at 4°C. Reagents in Tubes 1–3 and 6 can be sealed by Tube Lid Seals, and stored at room temperature until the next run.
12. Room I in our lab is restricted for qualified personnel only (no cleaning personnel). Here we also store stock reagents, primers, and probes. No clinical samples neither extracted DNA nor RNA are allowed in this room. Dedicated lab coats are used (23).
13. Reordered primers and probes are separately tested on arrival using a fixed amount of target template. This is performed in such a way that the new component is tested in a mixture containing already validated primers and/or probes and in parallel with a validated primer/probe mixture. Thus, if a reverse primer

is reordered it is tested in a mixture with an already validated forward primer and probe. C_t-values obtained using the "test" mixture and the previously validated primer/probe mixture should be comparable and within a range of 2× the standard deviation of the average C_t-value found for the fixed amount of target template. This average value is determined before introducing the diagnostic test into the daily practice by testing the fixed amount of target template for at least 20 times.

14. To make sure that there is adequate amplification mix, prepare an additional amount of mix equal to two extra reactions.
15. The HPC is a fixed amount of the target template that results in a C_t-value of approximately 30. The LPC is a 10× dilution of the HPC. Both controls are used to monitor the efficiency of the amplification. C_t-values obtained for both controls are recorded in a Stewart plot. On a regular base this plot is checked to monitor possible changes in amplification efficiency in time. C_t-values obtained for the HPC and LPC should be within a range of 2× the standard deviation of the average C_t-value found for these controls. The average value is determined for each new batch of controls before introducing the diagnostic test into the daily practice by determining C_t values for the HPC and LPC at least 20 times.
16. Room III in our lab is used for amplification and analysis of clinical samples and houses the Real-Time PCR equipment.
17. In our lab all assays are designed in such a way that they are running under identical Real-Time PCR conditions, in order to allow different assays to be performed in the same plate at the same time.
18. AmpErase® uracil-*N*-glycosylase (UNG) is added to the reaction to prevent the reamplification of carryover PCR products by removing any uracil incorporated into amplicons. This is why dUTP is used rather than dTTP in the PCR reaction. UNG is inactive above 55°C (25). Both UNG and dUTP are present in the TaqMan® Universal PCR Master Mix we use.

References

1. Bergmans AMC, Groothedde JW, Schellekens JFP, van Embden JDA, Ossewaarde JM, Schouls LM (1995) Etiology of cat scratch disease: comparison of polymerase chain reaction detection of *Bartonella* (formerly *Rochalimaea*) and *Afipia felis* DNA with serology and skin tests. J Infect Dis 171:916–923
2. Dolan M, Wong M, Regnery R et al (1993) Syndrome of *Rochalimaea henselae* adenitis suggesting cat scratch disease. Ann Intern Med 118:331–336
3. Regnery R, Olson J, Perkins B, Bibb W (1992) Serological response to "*Rochalimaea henselae*" antigen in suspected cat scratch disease. Lancet 339:1443–1445
4. Varela G, Vinson JW, Molina-Pasquel C (1969) Trench fever II: Propagation of *Rickettsia quintana* on cell-free medium from the blood of two patients. Am J Trop Med Hyg 18:708–712
5. Koehler J, Quinn F, Berger T, LeBoit T, Tappero J (1992) Isolation of *Rochalimaea* species from cutaneous and osseous lesions of

bacillary angiomatosis. N Engl J Med 327: 1625–1631
6. Daly SJ (1995) The 10 most common questions about *Bartonella (Rochalimaea)* species. Infect Dis Clin Practice 4:359–364
7. Garcia-Caceres U, Garcia FU (1991) Bartonellosis. An immunodepressive disease and the life of Daniel Alcides Carrion. Am J Clin Pathol 95:S58–S66
8. Daly J, Worthington M, Brenner D et al (1993) *Rochalimaea elizabethae* sp. nov. isolated from a patient with endocarditis. J Clin Microbiol 31:872–881
9. Brenner DJ, Oconnor SP, Winkler HH, Steigerwalt AG (1993) Proposals to unify the genera *Bartonella* and *Rochalimaea*, with descriptions of *Bartonella quintana* comb. nov., *Bartonella vinsonii* comb. nov., *Bartonella henselae* comb. nov., and *Bartonella elizabethae* comb. nov, and to remove the family Bartonellaceae from the order Rickettsiales. Int J Syst Bacteriol 43:777–786
10. Hadfield T, Warren R, Kass M, Brun E, Levy C (1993) Endocarditis caused by *Rochalimaea henselae*. Hum Pathol 24:1140–1141
11. Spach D, Callis K, Paauw D et al (1993) Endocarditis caused by *Rochalimaea quintana* in a patient infected with human immunodeficiency virus. J Clin Microbiol 31:692–694
12. Anderson BE, Neuman MA (1997) *Bartonella* spp. as emerging human pathogens. Clin Microbiol Rev 10:203–219
13. Sander A, Penno S (1999) Semiquantitative species-specific detection of *Bartonella henselae* and *Bartonella quintana* by PCR-enzyme immunoassay. J Clin Microbiol 37: 3097–3101
14. Sander A, Berner R, Ruess M (2001) Serodiagnosis of cat scratch disease: response to *Bartonella henselae* in children and a review of diagnostic methods. Eur J Clin Microbiol Infect Dis 20:392–401
15. Maurin M, Roux V, Stein A, Ferrier F, Viraben R, Raoult D (1994) Isolation and characterization by immunofluorescence, sodium dodecyl sulfate-polyacrylamide gel electrophoresis, Western blot, restriction-fragment length polymorphism PCR, 16 S rDNA gene sequencing, and pulsed-field gel electrophoresis of *Rochalimaea quintana* from a patient with bacillary angiomatosis. J Clin Microbiol 32: 1166–1171
16. Bergmans AM, Peeters MF, Schellekens JF, Vos MC, Sabbe LJ, Ossewaarde JM, Verbakel H, Hooft HJ, Schouls LM (1997) Pitfalls and fallacies of cat scratch disease serology: evaluation of *Bartonella henselae*-based indirect fluorescence assay and enzyme-linked immunoassay. J Clin Microbiol 35:1931–1937
17. Avidor B, Kletter Y, Abulafia S, Golan Y, Ephros M, Giladi M (1997) Molecular diagnosis of cat scratch disease: a two-step approach. J Clin Microbiol 35:1924–1930
18. Avidor B, Varon M, Marmor S, Lifschitz-Mercer B, Kletter Y, Ephros M, Giladi M (2001) DNA amplification for the diagnosis of cat-scratch disease in small-quantity clinical specimens. Am J Clin Pathol 115: 900–909
19. Hansmann Y, DeMartino S, Piemont Y, Meyer N, Mariet P, Heller R, Christmann D, Jaulhac B (2005) Diagnosis of cat scratch disease with detection of *Bartonella henselae* by PCR: a study of patients with lymph node enlargement. J Clin Microbiol 43:3800–3806
20. Bergmans AMC, Schellekens JF, van Embden JD, Schouls LM (1996) Predominance of two *Bartonella henselae* variants among cat-scratch disease patients in the Netherlands. J Clin Microbiol 34:254–260
21. Diederen BMW, Vermeulen MJ, Verbakel H, van der Zee A, Bergmans AMC, Peeters MF (2007) Evaluation of an internally controlled real-time polymerase chain reaction assay targeting the *gro*EL gene for the detection of *Bartonella* spp. Eur J Clin Microbiol Infect Dis 26:629–633
22. Fenollar F, Sire S, Raoult D (2005) *Bartonella vinsonii* subsp. *arupensis* as an agent of blood culture-negative endocarditis in a human. J Clin Microbiol 43:945–947
23. Noordhoek GT, Ieven M, Roosendaal R. on behalf of the Working Group on Molecular Diagnostics of Infectious Diseases (WMDI) (2007) Basic requirements for reliable performance of nucleic acid amplification based methods for diagnosis of infectious diseases. http://www.wmdi.nl/pages/documents/Guidelines%20MDI_definitief2_%20maart2007.pdf
24. Niesters HG (2001) Quantification of viral load using real-time amplification techniques. Methods 25:419–429
25. Longo MC, Berninger MS, Hartley JL (1990) Use of uracil DNA glycosylase to control carry-over contamination in polymerase chain reactions. Gene 93:125–128

Chapter 15

Simultaneous Direct Identification of Genital Microorganisms in Voided Urine Using Multiplex PCR-Based Reverse Line Blot Assays

Michelle L. McKechnie, Fanrong Kong, and Gwendolyn L. Gilbert

Abstract

Our aim was to develop and evaluate sensitive methods that would allow simultaneous direct identification of multiple potential pathogens in clinical specimens for diagnosis and epidemiological studies, using a multiplex PCR-based reverse line blot assay. We have previously developed assays suitable for detection of bacterial respiratory and systemic pathogens. In this chapter we describe, in detail, a method developed to identify 14 genital microorganisms, for use in epidemiological studies of genital infection or colonization, using first voided urine specimens. The 14 urogenital pathogens or putative pathogens studied were *Trichomonas vaginalis, Streptococcus pneumoniae, Neisseria gonorrhoeae, N. meningitidis, Chlamydia trachomatis, Ureaplasma parvum, U. urealyticum, Mycoplasma hominis, M. genitalium, Gardnerella vaginalis, Haemophilus influenzae*, herpes simplex virus 1 and 2, and adenovirus. Two species specific primer pairs and probes were designed for each target. The method was validated using a reference strain or a well-characterized clinical isolate of each target organism. In a clinical study among men attending sexual health clinics in Sydney, we used the assay to compare rates of detection of the 14 organisms in men with urethritis with those in asymptomatic controls and found the method to be sensitive, specific, convenient, and relatively inexpensive.

Key words: Multiplex PCR-based reverse line blot, Genital microorganisms, First void urine specimens

1. Introduction

Diagnosis of infectious disease syndromes for which there are multiple possible causes—such as pneumonia, meningitis, and genital syndromes (urethritis/cervicitis)—usually involves a combination of various types of antigen detection, culture, nucleic acid amplification, and serological tests, with great variation in sensitivity, specificity, cost, turnaround times, and availability. Specimen

Mark Wilks (ed.), *PCR Detection of Microbial Pathogens: Second Edition*, Methods in Molecular Biology, vol. 943, DOI 10.1007/978-1-60327-353-4_15, © Springer Science+Business Media, LLC 2013

requirements for optimal diagnostic sensitivity often involve relatively invasive procedures, such as collection of a broncho-alveolar lavage or urethral/cervical swabs, before treatment is started, which is often impracticable. Therefore, a specific diagnosis is often not made or is delayed, which may result in suboptimal management, inadequate (if any) epidemiological investigation, and confusion about the roles of some pathogens.

The high level of sensitivity of nucleic acid amplification tests, such as PCR, allows the use of less invasive specimen types, including first voided urine specimens or self-collected vaginal swabs that are unsuitable for less sensitive methods such as culture and antigen tests (1). Multiplex PCR (mPCR) is a variant of PCR in which two or more target sequences can be amplified by including more than one pair of primers in the same reaction (2). We have developed an mPCR system, which allows amplification of multiple targets, simultaneously. Primers are designed with similar melting temperatures (T_m) and amplicon sizes, allowing amplification of target DNA under similar PCR conditions, without significant interference or loss of sensitivity. Amplicons are identified by hybridization to probes fixed to a membrane in a macroarray (3). It allows the simultaneous detection, identification, and typing of multiple microorganisms in clinical specimens. It is a convenient way to identify up to 43 targets (which are labeled onto the membrane) in 43 individual specimens simultaneously in a single reaction. (Note: The Miniblotter has 45 lanes, and the first and last lanes are not used, leaving 43 lanes to initially label the membrane with probes and then run specimens at right angles.) It is more flexible, similar to perform, and less expensive than DNA microarray. The number of targets is adequate for most clinical and epidemiological applications and much more practical than conventional mPCR.

We have previously developed and evaluated multiplex PCR-based reverse line blot (mPCR/RLB) assays for detection of bacterial respiratory and systemic pathogens (4, 5). In this chapter, we describe, in detail, the development of an mPCR/RLB assay which can identify 14 genital microorganisms in first voided urine specimens, which is suitable for epidemiological studies of urethritis and other genital syndromes. In particular it is suitable for use in studies designed to evaluate the roles of a number of putative genital pathogens or potential pathogenic combinations of pathogens, which are difficult to identify by conventional methods. The results of our preliminary evaluation of this method and the epidemiological study for which it was used have been (6) or will be (7) published elsewhere.

The equipment used includes a "miniblotter," which consists of two plastic blocks and nylon membrane fixed between them. Using channels in the blocks, oligonucleotide probes, which have been synthesized with a 5′ terminal amino group, are covalently linked to the carboxyl groups in the membrane, in parallel rows. Biotin-labeled PCR products are applied to the membrane at right angles to the probes and incubated to allow hybridization to the corresponding probe(s).

Hybridization is identified by addition of peroxidase-labeled streptavidin, which reacts with biotin-labeled primers to produce enhanced chemiluminescence (ECL) after addition of ECL detection reagents (hydrogen peroxide and luminal), which is detected on a light-sensitive film.

Sexually transmitted infections (STIs) are a major global health problem. Worldwide, an estimated 340 million cases of curable STIs, including chlamydial infection, gonorrhoea, trichomoniasis, and syphilis, occur annually and their incidence is increasing in many parts of the world (8). In developing countries, their complications rank in the top five disease categories for which adults seek health care (8). Many STIs cause asymptomatic infection; for example, up to 70% of men and women with gonococcal and/or chlamydial infections are symptom free (8), which creates the potential for unrecognized transmission with significant implications for both individual and population health.

Urethritis is characterized by discharge and dysuria (9) and is broadly classified as nongonococcal (NGU) or gonococcal. It occurs in both men and women, but is often unrecognized in women. Acute NGU is one of the commonest STIs affecting heterosexual men, yet a specific pathogen—most commonly *Chlamydia trachomatis*—is only identified in 50–70% of cases (10). PID is an important complication of STI in women; *C. trachomatis* and *N. gonorrhoeae* are commonly implicated but the cause is often unknown. Bacterial vaginosis is the commonest cause of vaginal discharge and is associated with both recognized STIs and other genital syndromes (11, 12). Additional epidemiological studies are needed to determine the significance of organisms, other than recognized genital pathogens, in urethral and vaginal syndromes (10, 13–15). In particular, the pathogenic roles, if any, of the two recently defined human *Ureaplasma* species (16)—*U. urealyticum* (previously *U. urealyticum* biovar 2) and *U. parvum* (previously *U. urealyticum* biovar 1)—and several other genital (17–19) and respiratory pathogens (20–23) in NGU are unclear.

We describe the development and evaluation of an mPCR/RLB assay (3) designed to detect the presence of any of 14 recognized and potential genital pathogens in urine specimens, for use in clinical and epidemiological studies of genital infections.

2. Materials

2.1. DNA Extraction

1. Roche COBAS® Amplicor extraction kit (Roche Diagnostics Australia Pty. Limited Systems, Australia) was used according to the manufacturer's instructions.

2.2. Multiplex PCR

1. Deoxynucleoside triphosphates (dNTP) set: 0.125 mM of each dNTP (Roche). Store at –20°C.

2. Molecular biology grade water.
3. *Taq* PCR core Kit: Kit contains 10× buffer; $MgCl_2$ 25 mM, and Hotstar *Taq* polymerase 5 U/µl (Qiagen, Valencia CA, USA). Store at −20°C.

2.3. Reverse Line Blot Hybridization Assay

2.3.1. Preparation of RLB Membrane

1. 10% (w/v) Sodium dodecyl sulfate (SDS): Take care when preparing SDS stock solution from powder, which is corrosive and causes irritation to the skin, eyes, and respiratory tract—use respiratory protection, protective clothing, and equipment. Store at room temperature.
2. 20× Saline sodium phosphate (SSP)–ethylenediaminetetraacetic acid (EDTA) buffer (SSPE) (Astral Scientific Pty. Ltd., Carringbah, Australia). Store at room temperature.
3. 0.5 M $NaHCO_3$: Store at room temperature.
4. 16% *N*-(3-dimethylaminopropyl)-*N*-ethylcarbodiimide hydrochloride (EDAC): Store at room temperature.
5. Support cushions for Miniblotter (Immunetics Inc., Boston, Massachusetts, USA).
6. 0.1 M NaOH. Store at room temperature.
7. 2× SSPE/0.1% SDS: Freshly prepared and pre-warmed to 60°C.
8. 20 mM EDTA. Store at room temperature.

2.3.2. Hybridization and Detection of mPCR Products

1. 2× SSPE/0.1% SDS: Freshly prepared and pre-warmed to 60°C.
2. Miniblotter (Immunetics Inc., Boston, Massachusetts, USA).
3. Hybridization oven (Model: Shake "N" Stack, Thermo Scientific, USA), rolling bottle (Thermo Scientific, USA), and nylon separating mesh.
4. 2× SSPE/0.5% SDS: Freshly prepared and pre-warmed to 60°C.
5. Streptavidin–peroxidase conjugate (Roche Diagnostics Australia PTY. Limited Systems, Australia).
6. 2× SSPE/0.5% SDS: Freshly prepared and pre-warmed to 42°C.
7. 2× SSPE: Freshly prepared and stored at room temperature.
8. ECL™ Solution (Amersham Biosciences, Buckinghamshire, UK).
9. Film cartridge.
10. Rocking platform (Model: Belly Dancer, Pegasus Scientific Inc., Rockville MD, USA).
11. X-ray film exposure cassette (Sigma-Aldrich, Sydney, Australia).
12. X-ray film developer.

2.3.3. Regeneration of RLB Membrane

1. 1% SDS: Freshly prepared and pre-warmed to 80°C.
2. 20 mM EDTA.

3. Methods

3.1. Primer and Probe Design

Primers and probes were designed and selected from previously published primer and probe sequences. Sequence databases Biomanager, Sigma-Genosys, GenBank (USA), and Entrez Pubmed were used in the selection and design as follows:

1. *Biomanager*—https://biomanger.info/
   Provided by the Australian National Genomic Information Service (ANGIS). It includes approximately 100 programs for sequence editing, search, comparison, multiple sequence alignment, phylogenetic analysis, and primer/probe design. Sequences and data can also be stored in this program.
2. *Sigma-Genosys*—http://row.sigma-genosys.eu.com
   This program allows assessment of oligonucleotide characteristics (melting temperature [T_m], secondary structure, primer dimer formation, GC content, and molecular weight). Using this program the physical characteristics of primer and probes were modified to allow optimal performance under the same PCR conditions: length 18–30 bp; T_m 58–65°C; secondary structure moderate, weak, or no structure; no dimer formation; and a predicted amplicon size of 80–400 bp.
3. *GenBank (USA)*—http://www.ncbi.nlm.nih.gov
   This program provides access to relevant sequences and allows BLASTn searches of primer/probe sequences against deposited sequences to confirm specificity. This program, given a DNA query, returns the most similar DNA sequence from the DNA database specified by the user.
4. *Entrez-Pubmed*—http://www.ncbi.nlm.nih.gov/entrez/query.fcgi?db=Pubmed
   This program enables literature searches and provides access to many full-text articles. It also allows selection of the best published primers and probes for modification.

3.1.1. Oligonucleotide Primers

Thirteen primer pairs (one antisense "Ab" and one sense "Sb") were used to amplify species-specific target sequences from pathogens of interest. Most of the primers were selected from published papers and modified to match the desired characteristics. The adenovirus primers used for the mPCR were specifically chosen to have "degenerative" characteristics. Degenerative primers make it possible to amplify related, but distinct, nucleic acid sequences.

Primers were purchased and synthesized from two companies, Sigma-Aldrich Australia and AUGCT in China. All primers were labeled with biotin at the 5′ end.

3.1.2. Probes

For every target region two probes, one sense (Sp) and one antisense (Ap), were used for hybridization. The probes were designed to have the same characteristics as the primers. All probes were labeled with amine at the 5′ end to allow coupling to the RLB membrane.

The HSV antisense (Ap) probe used for the RLB was a common probe which detects both HSV types 1 and 2; they were differentiated by type-specific sense probes, HSV1-Sp and HSV2-Sp.

Primers and probes used for the mPCR/RLB are shown in Tables 1 and 2.

3.1.3. Primer and Probe Preparation

1. On receipt from the manufacturer, centrifuge the primers and probes at 13,200 rpm (29,272 × *g*) and throughout for 30 s before opening.
2. Add molecular biology grade water to each tube to a concentration of 100 pmol/μl.
3. Separate stock and working solutions to avoid possible contamination or loss of primer/probe labels associated with frequent freezing/thawing of stock solution. Store solutions at −20°C and working solutions, for a maximum of 4 weeks, at 4°C. Stock solutions can be stored for several years.

3.1.4. DNA Extraction

The Roche COBAS® AMPLICOR CT/NG specimen preparation kit was used according to the manufacturer's instructions (Roche Diagnostics Australia PTY. Limited Systems, Australia).

1. Vortex urine specimens thoroughly for 3–10 s. Transfer 500 μl of each specimen to a tube containing 500 μl of "CT/NG Urine Wash."
2. Incubate the specimens at 37°C in a heating block for 15 min.
3. After incubation centrifuge at 13,200 rpm for 15 min.
4. Remove the supernatant and add 250 ml of "CT/NG LYS" solution to each tube.
5. Vortex the tubes briefly, and then incubate at room temperature for 15 min.
6. After incubation, add 250 μl of "CT/NG DIL" to each tube and mix by vortexing.
7. Finally, centrifuge the tubes for 10 min at 13,200 rpm and store at 4°C, if PCR is to be performed within 24 h, or −70°C until further analysis.

Table 1
Primers and probes used for the mPCR–RLB

Primer/Probe Name[a]	Gene Target	GenBank Number[b]	Primer–Probe Sequence (5′-3′)[c]	T_m (°C)[d]	Number of bases	References[e]
CT24b	*C. trachomatis* cryptic plasmid	X06707	840GGG ATT CCT GTA ACA ACA AGT CAG G864	67	25	Unpublished preliminary study
CTS1p			865TTG CGC ATA ATT TTA GGC TTG885	64	21	
CTA2p			1021ACA CTT TGT CTC GAT GAA AGA CA999	63	23	
CT27b			1047CCT CTT CCC CAG AAC AAT AAG AAC AC1022	67	26	
NGpSb	*N. gonorrhoeae* cryptic plasmid	M310316	3249TGC TGT TTC AAG TCG TCC AG3268	64	20	
NGpAp			3317GAT AGT CAT AGC AGG GCT GTT C3296	62	22	
NGpSp			3452CCG TAA CGT CTC TAA GTC TGC TT3474	62	23	
NGpAb			3503CGA AGC CGC CAG CAT AGA GC3484	71	20	
NG16Sb	*N. gonorrhoeae* (16 S rRNA *and* 23 S rRNA)	AF223396	404CCA AAA CTT AAC AAA TGA AAG CAA G428	63	24	This study
NG1S			453TGA TTT GCG AAG TAG AAT AAC G474	61	22	
NG2A2			456ATC AAA ATA AGC TGC TAA AAA CAG433	59	24	
NGITSAb			490TGT TAA AGA TCG ATG CGT CGT472	64	21	
MgPa-Sb	*M. genitalium* (*MgPa*)	M31431	1420GAG AAA TAC CTT GAT GGT CAG CAA1444	65	24	(24)
MgPAa-Ap			1463TAT CAT ACC TTC TGA TTG CAA AGT1445	60	24	
MgPa-Sp			1473CGG TAG AGC TTT ATA TGA TAT TAA CTT AGC1503	61	30	This study
MgPa-Ab			1492GTT AAT ATC ATA TAA AGC TCT ACC GTT GTT ATC1530	63	33	(24)
UU-Sb	*U. urealyticum* (*ureB*)	AF085729	640GAT CAC ATT TCC ACT TAT TTG AAA CA665	64	26	(25)
UU-Ap			705CTT CAT TTC CTT TTT CAT CAA AAA ATA C678	63	28	Unpublished preliminary study
UU-Sp			690AAA AAG GAA ATG AAG ATA AAG AAC G714	61	25	(25)
UU-Ab			739AAA CGA CGT CCA TAA GCA ACT TTA716	64	24	(25)
UP-Sb	*U. parvum* (*ureB*)	AF085731	637GAT CAC ATT TTC ACT TGT TTG AAG TG662	64	26	

(continued)

Table 1 (continued)

Primer/Probe Name[a]	Gene Target	GenBank Number[b]	Primer–Probe Sequence (5′-3′)[c]	T_m (°C)[d]	Number of bases	References[e]
UP-Ap			702CTT CAT TTC CTT TTT CAT CAA AAA ATA C675	63	28	Unpublished preliminary study
UP-Sp			688AAA AAG GAA ATG AAG ATA AAG AAC G712	61	25	Modified from reference (25)
UP-Ab			735AAC GTC GTC CAT AAG CAA CTT TG713	66	23	(25)
TV-Sb	*T. vaginalis* (*btub*)	L05468	850CAT TGA TAA CGA AGC TCT TTA CGA T874	63	25	(14)
TV-Ap			904TGT TGT GAG CTT GAG TGT ACG G883	65	22	Unpublished preliminary study
TV-Sp			916CGA TCT TAA CCA CCT TGT TTC C945	63	22	Modified from reference (14)
TV-Ab			962CGC ATG TTG TGC CGG ACA945	71	18	(14)
GV-Sb	*G. vaginalis* (*intergenic transcribed sequence* (*ITS*)	L08167	360GCT TTT ACT GGT GTA TCA CTG TAA GG385	63	25	(26)
GV-Ap GV-Sp			416TCC TGT CTA CCA AGG CAT CC397 632CGT GTG ATA ACC GTC AGG TG651	64 64	19 19	Unpublished preliminary study
GV-Ab			695CCG TCA CAG GCT GAA CAG T677	64	18	(26)
HSV-Sb	HSV 1 (*gD*)	L09242	460ATC CGA ACG CAG CCC CGC TG479	77	19	(27)
HSV-Ap			516ATC CTC GCT GAC GGC G501	67	15	
HSV1-Sp			545CGT TTG AGA CCG CCG GCA562	73	17	Unpublished preliminary study
HSV-Ab			601TCT CCG TCC AGT CGT TTA TCT TC579	65	22	(27)

HSV-Sb	HSV 2 (*gD*)	L09242	487ATC CGA ACG CAG CCC CGC TG506	77	19	
HSV-Ap			543ATC CTC GCT GAC GGC G528	67	15	
HSV2-Sp			572CCT TCG AGA CCG CGG GTA589	68	17	Unpublished preliminary study
HSV-Ab			628TCT CCG TCC AGT CGT TTA TCT TC606	65	22	(27)
AdVdeSb	Adenovirus (*hexon*)		GCC SCA RTG GKC WTA CAT GCA CAT C	69	25	Modified from reference (28)
AdVdeAp			CCY ACR GCC AGI GTR WAI CGM RCY TTG TA	68	32	
AdVdeSp			GCC CGY GCM ACI GAI ACS TAC TTC	64	24	
AdVdeAb			CAG CAC SCC ICG RAT GTC AAA	63	21	
NM-Sb	*N. meningiditis* (*porA*)	AY319969	929GCT TCG GTA ATG CAG TTC CA948	64	19	(29)
NM-Ap			1007GTA TTT TCG CCT TTT TTA CCG CG985	68	22	(29)
NM-Sp			961GCC CAT GGT TTC GAC TTT ATC937	64	24	Unpublished preliminary study
NM-Ab			1055CGT TTG GAA AAA TCA TAA TCA ACG1032	65	23	(29)
HigyrBSb	*H. influenzae* (*gyrB*)	U32738	5926GAA GCA CAG TCA TAA TAA CTT CTG CT5951	63	25	(30)
HigyrBAp			6028GAT GAT AAT TCT GTA TCG GTG CAA6005	64	23	
HigyrBSp			5974GAA TAT CCA CAG GAA TCC CG5993	64	19	
HigyrBAb			6159AGC GTC CTG GTA TGT ATA TCG G6138	64	15	
lytAS2b	*S. pneumoniae* (*lyt*A)	M13812	681CAA CCG TAC AGA ATG AAG CGG701	66	20	(31)
lytAAp			721GTC TTT CCG CCA GTG ATA AT702	61	19	
lytASp			955AGG GAG TTT AGC TGG AAT TAA AA977	61	22	
lytAA1b			999TTA TTC GTG CAA TAC TCG TGC G978	67	21	

S=G+C, R=A+G, K=G+T, W=A+T, Y=C+T, M=A+C, I=Inosine

[a]The suffix b indicates biotin-labeled primer and p indicates amine-labeled probe. A indicates antisense and S indicates sense

[b]A number assigned to a nucleotide sequence by the National Center for Biotechnology Information (NCBI)

[c]Numbers represent the base positions at which the primer/probe sequence starts and finishes (starting at point 1 of the corresponding gene sequence in GenBank)

[d]Melting temperature values provided by primer synthesizer (Sigma-Aldrich)

[e]mod=modified. Primers were modified from the published primers

Table 2
The mPCR master mix

Reaction component	Stock concentration	Tube volume (μl)	Final concentration
PCR buffer (15 mM $MgCl_2$)	10×	2.5	1 × 1.5 mM
$MgCl_2$	25 mM	3.0	3.0 mM
dNTP	2.5 mM	1.25	0.125 mM
Primer Sb[a]	50 μM	0.075 × 15 = 1.125	0.25 μM
Primer Ab[b]	50 μM	0.075 × 15 = 1.125	0.25 μM
Hotstar *Taq* (5 U/μl) (Qiagen™)	5 U/μl	0.2	1 U
DNA template	–	10.0	–
Water	–	5.8	–
Total		25	

[a] *Sb* sense primer
[b] *Ab* antisense primer

Timing: Depending on the number of specimens, 1–4 h. For example a run of 43 specimens will take approximately 2–3 h to complete extraction.

3.2. Multiplex PCR

1. Prepare the mPCR master mix for a 25 μl system on ice, as shown in Table 2. Prepare enough for the required number of samples, plus approximately 10% extra.
2. Mix the mPCR master mix well by vortexing and spin in a microcentrifuge. It is important to ensure that the master mix is thoroughly mixed and spun before aliquoting. *Taq* is usually stored in a glycerol-containing buffer and so is heavier than other reagents; without thorough mixing, it may not be homogeneously distributed (see Note 1).
3. Aliquot 15 μl of the mPCR master mix into the required number of PCR tubes.
4. Add 10 μl of DNA template to each PCR tube in the Biological Safety Cabinet Class II.
5. Mix the contents of the tubes by spinning for 10 s in a microcentrifuge.
6. Transfer the PCR tubes to a Mastercycler gradient thermal cycler with the following thermal profile program:

   1 cycle of initial denaturation for 15 min at 95°C

   40 cycles of denaturation at 94°C for 30 s, followed by

   40 cycles of annealing at 60°C for 30 s and extension at 72°C for 90 s

1 cycle of final extension at 72°C for 10 min

PCR holding temperature 4 or 22°C

mPCR products can be stored in a refrigerator at 4°C for a period of up to 2 weeks prior to RLB assay. Ideally, mPCR products should not be frozen, but if a longer period of storage is required, they can be frozen at −70°C and thawed once only.

Timing: Approximately 3.5 h if tubes and primers have been prepared to set up and run the mPCR.

3.3. Preparation of RLB Assay Membrane

1. Make all required buffers and solutions and pre-warm to the appropriate temperatures. Turn on all incubators before starting the procedure.
2. Dilute oligonucleotide probes to their optimal concentrations, ranging from 0.6 to 5.4 pmol/μl in 180 μl of 0.5 M $NaHCO_3$ (see Table 3).
3. Cut a Biodyne C membrane (Pall Life Sciences) to a size of 15 × 15 cm and mark one edge with a pencil line to indicate the parallel direction of the probes. Label the membrane with relevant information: type of assay, membrane number, person performing the assay, and date of assay.
4. Incubate the membrane in a sealed plastic bag with 18 ml of 16% EDAC and place in a gentle rocker at room temperature for 10 min.
5. Remove the membrane from the plastic bag and briefly wash with 250 ml of distilled water.
6. Place the membrane onto a support cushion in the Miniblotter such that the slots are parallel and remove excess fluid from the slots by aspiration.
7. Add 150 μl of each diluted oligonucleotide probe to the lanes on the Miniblotter. Exclude the first and last lanes to ensure that the RLB results are within the reading frame. It is important to avoid the formation of air bubbles in the slots as this can cause loss of hybridization signal. This can be best done by sucking the solution up and down while holding the pipette in the lane if bubbles are present.
8. Add 150 μl of 0.5 M $NaHCO_3$ to the first and last lanes of the membrane.
9. Incubate for 5 min at room temperature. Avoid moving or rocking the miniblotter during this incubation.
10. Remove the oligonucleotide/0.5 M $NaHCO_3$ solution from the Miniblotter by aspiration.
11. Remove the membrane from the Miniblotter and inactivate by incubation in 250 ml of 0.1 M NaOH at room temperature for 9 min on a gentle rocker. It is critical that the incubation is no longer than 10 min.

Table 3
Oligonucleotide probe concentrations for RLB membrane labeling

Target organism	Probe name	Probe concentration pmol/μl[a]
C. trachomatis	CTA2p	1.8
	CTS1p	1.8
N. gonorrhoeae	NGpAp	1.8
	NGpSp	1.8
	NG2A3	0.6
	NGIS	0.6
M. genitalium	MgPaAp	1.8
	MgPaSp	1.8
U. urealyticum	UUAp	1.8
	UUSp	1.8
U. parvum	UPAp	1.8
	UPSp	1.8
T. vaginalis	TVAp	1.8
	TVSp	1.8
M. hominis	MHAp	1.8
	MHSp	1.8
G. vaginalis	GVAp	1.8
	GVSp	1.8
HSV	HSVAp	1.8
	HSV1Sp	1.8
	HSV2Sp	3.2
Adenovirus	AdVdeAp	1.8
	AdVdeSp	1.8
N. meningitidis	NMAp	1.8
	NMSp	1.8
H. influenzae	HigyrBAp	1.8
	HigyrBSp	1.8
S. pneumoniae	LytAAp	5.4
	LytASp	5.4

[a]Higher concentrations of HSV2Sp, LytASp, and LytAAp are needed due to poor RLB results at 1.8 pmol/μl

12. Incubate the membrane in 250 ml of pre-warmed 2× SSPE/0.1% SDS solution at 60°C for 5 min in the oven.
13. Incubate in 250 ml of 20 mM EDTA at room temperature for 20 min.
14. Store the membrane in 20 mM EDTA in a sealed plastic bag at 4°C until use. Make sure that the membranes are properly sealed and stored as dehydration will render them useless for further hybridization assays.

Time required: 3–4 h.

3.4. Hybridization and Detection of mPCR Products

1. Place 43×500 μl PCR tubes in a rack and add 150 μl of 2× SSPE/0.1% SDS, pre-warmed to 60°C to each tube.
2. Add 20 μl of PCR product to each tube and place in boiling water for 10 min.
3. Cool the tubes immediately on ice for 5–6 min.
4. Wash the membrane in pre-warmed 2× SSPE/0.1% SDS for 5 min in a 60°C oven, rocking at low speed.
5. Clean the Miniblotter with 70% alcohol and attach the membrane.
6. Add 150 μl of 2× SSPE/0.1% SDS to the first and last slots of the Miniblotter, and then add 143 μl of each diluted denatured PCR product into alternate slots.
7. Hybridize the membrane at 60°C for 1 h on a horizontal surface. Avoid excessive moving or rocking of the membrane during this incubation.
8. After hybridization, remove the samples from the Miniblotter by aspiration.
9. Wash the membrane twice in pre-warmed 2× SSPE/0.5% SDS in a 60°C oven for 10 min, rocking at low speed.
10. Moisten a piece of nylon mesh with 2× SSPE/0.5% SDS.
11. Add 3 μl of (500 U/ml) streptavidin–peroxidase conjugate (26) to 15 ml of 2× SSPE/0.5% SDS, which has been previously prepared and pre-warmed to 42°C.
12. Place the membrane onto the nylon mesh, roll it up, and place it into a rolling bottle. Unroll them inside the bottle and gently press the membrane and nylon mesh to the wall of the bottle using a sterilized glass pipette so that it adheres to the glass. The membrane should be on the inside of the rolled nylon mesh.
13. Next, add streptavidin–peroxidase conjugate to the bottle, before incubating it in a rolling chamber at 42°C for 60 min.
14. After incubation, remove the membrane from the bottle and wash the membrane twice in 250 ml of 2× SSPE/0.5% SDS in a 42°C oven for 10 min whilst rocking.

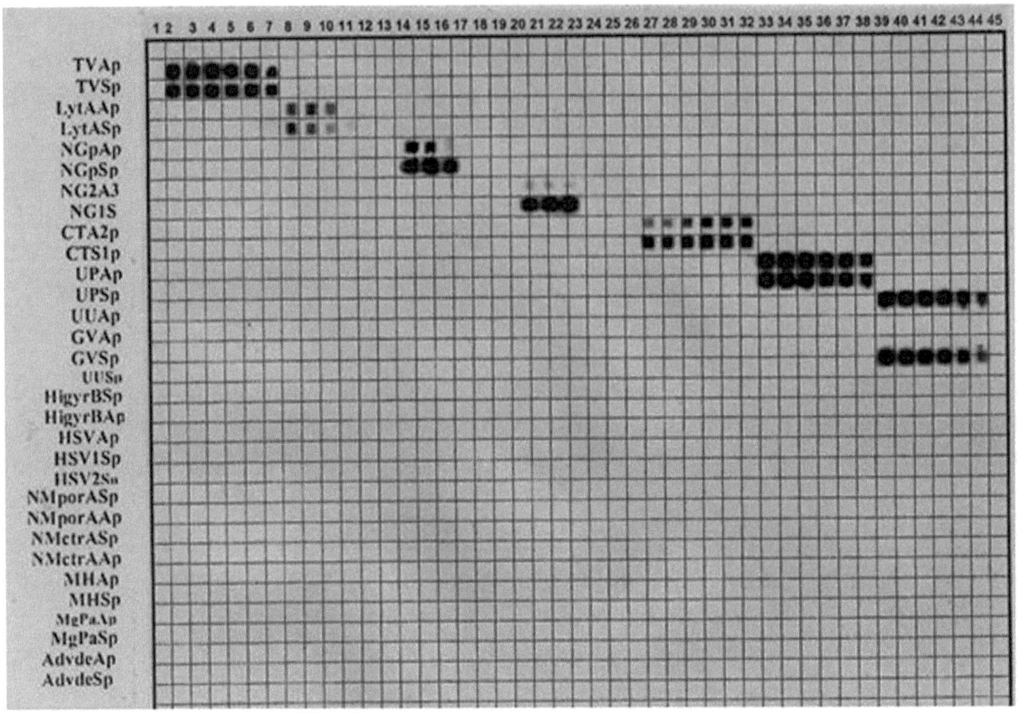

Fig. 1. Tenfold dilution series of reference strains. Lanes 2–7 *T. vaginalis*; Lanes 8–13 *S. pneumoniae*; Lanes 14–19 *N. gonorrhoeae* (plasmid); Lanes 20–26 *N. gonorrhoeae* (ITS); Lanes 27–32 *C. trachomatis*; Lanes 33–38 *U. parvum*; Lanes 39–44 *U. urealyticum*.

15. Prepare the film cartridge by cutting the plastic transparency sheets to size.

16. Wash the membrane twice in 250 ml 2× SSPE at room temperature, rocking at low speed for 5 min on a rocking platform.

17. Prepare the ECL™ solution (Amersham) (7.5 ml of detection reagent 1 and 7.5 ml of detection reagent 2) and add to the membrane. Gently rock for 2 min by hand to cover the membrane.

18. Finally place the membrane in the film cartridge and cover with the plastic transparency sheet.

19. Expose the membrane for 15 min to light-sensitive film in a darkroom. If the signal is too weak or strong, the membrane can be used again directly to expose another film for a longer or a shorter period (see Notes 2 and 4). Anticipated results are that visible dark signals, compared to the background, will appear on the grid, in positions corresponding with positive samples (Fig. 1). There are numerous reasons for the signal being too faint or too strong. These are listed with suggestions for remedying the problem (see Notes 2–4).

3.5. Regeneration of RLB Membrane

1. To allow reuse of the RLB membrane, wash it twice in 1% SDS at 80°C for 30 min whilst rocking.
2. Wash the membrane in 20 mM EDTA at room temperature for 15 min on the rocking platform.
3. Seal the membrane in a plastic bag with 20 mM EDTA at 4°C until next use.

Timing: Approximately 6 h.

4. Notes

Major problems or faults that can occur with this technique, possible explanations, and suggested solutions are listed below and discussed in more detail elsewhere (3).

1. Faint or no mPCR products are obtained. As with any PCR, one or more reagents for PCR master mix may have been omitted. In contrast to single/uniplex PCR however, increasing the overall primer concentration in mPCR may lead to primer dimer formation and adversely affect the results. A final concentration of 0.04–0.8 μM for each primer is recommended. We prefer to start with concentrations of 0.25 μM; concentrations of primer pairs, whose products give weak signals, can be increased, if necessary.
2. Inadequate binding of probes to the membrane can interfere with hybridization, check membrane preparation procedures, and in particular ensure that the EDAC is freshly prepared.
3. The mPCR was successful, but there are no hybridization signals or signals are faint. Errors in the preparation of the buffers are a major cause of unsuccessful RLB. Remake the buffers and repeat. Primers and/or probes may have been inefficiently labeled (with biotin or amine, respectively) during synthesis or labels have been degraded by incorrect use or storage. Resynthesize primers and/or probes and store them properly. Faint signals may be due to low concentrations of probes or mPCR products: repeat the hybridization with increased probe and mPCR product concentration. Incorrect hybridization temperatures (T_h) may have been used. Lowering the T_h may improve the signal intensity, especially when the probe region contains mutations.
4. There are strong background or hybridization signals. Incorrect T_h and stripping temperature may have been used: check hybridization and stripping procedures; increase T_h and incubation temperature. The probe concentrations may have been too high: lower the probe concentrations and repeat. The

Miniblotter may not have been cleaned properly; therefore thoroughly clean the Miniblotter using a dedicated brush and soak the apparatus, preferably overnight, in a soap solution, e.g., Extran (Merck).

Acknowledgments

Victor Weixiong provided some of the primer/probe designs. Dr Neisha Jeoffreys provided valuable advice and help with troubleshooting of initial problems, false positive results, and excessive background signals.

References

1. Chen MY, Donovan B (2005) Changes in testing methods for genital *Chlamydia trachomatis* in New South Wales, Australia. Sexual Health 2:251–253
2. Markoulatos P, Siafakas N, Moncany M (2002) Multiplex polymerase chain reaction: a practical approach. J Clin Lab Anal 16:47–51
3. Kong F, Gilbert GL (2006) Multiplex PCR-based reverse line blot hybridization assay (mPCR/RLB)—a practical epidemiological and diagnostic tool. Nat Protoc 1:2688–2680
4. Wang Y, Kong F, Yang Y, Gilbert GL (2008) A mulitplex PCR-based reverse line blot hybridisation (mPCR/RLB) assay for detection of bacterial respiratory pathogens in children with pneumonia. Pediatr Pulmonol 43:150–159
5. Wang Y, Kong F, Gilbert GL, Brown M, Gao W, Yu S, Yang Y (2008) Use of a multiplex PCR-based reverse line blot (mPCR/RLB) hybridisation assay for the rapid identifcation of bacterial pathogens. Clin Microbiol Infect Dis 14:155–160
6. Mckechnie ML, Hillman RJ, Couldwell D, Kong F, Freedman EV, Wang H, Gilbert GL (2009) Simultaneous identification of 14 genital microorgranisms in urine by use of a mPCR-based reverse line blot assay. J Clin Microbiol 47:1871–1872
7. Couldwell D, Gidding HF, Freedman EV, McKechnie ML, Biggs K, Sintchenko V, Gilbert GL (2010) *Ureaplasma urealyticum* is significantly associated with key symptoms and signs of non-gonococcal urethritis in heterosexual Sydney men. Int J STD AIDS 21:337–341
8. World Health Organization. (2007) Fact sheet: Sexually transmitted infections. 110:1-4.
9. Scholes D, Stergachis A, Heidrich FE, Andrilla H, Holmes KK, Stamm WE (1996) Prevention of pelvic inflammatory disease by screening for cervical chlamydial infection. N Engl J Med 334:1362–1366
10. Bradshaw CS, Tabrizi NS, Read RHT, Garland SM, Hopkins AC, Moss LL, Fairley KC (2006) Etiologies of nongonococcal urethritis: bacteria, viruses, and the association with orogenital exposure. J Infect Dis 193:336–345
11. Bailey JV, Farquhar C, Owen C, Mangtani P (2004) Sexually transmitted infections in women who have sex with women. Sex Transm Infect 80:244–246
12. Keane FEA, Thomas BJ, Whitaker L, Renton A, Taylor-Robinson D (1997) An association between non-gonococcal urethritis and bacterial vaginosis and the implications for patients and their sexual partners. Genitourin Med 73:373–377
13. Hagman M, Forslin L, Moi H, Danielsson D (1991) Neisseria meningitidis in specimens from urogenital sites. Sex Transm Dis 18:228–232
14. Hardick J, Yang S, Lin S, Duncan D, Gaydos C (2003) Use of the Roche lightcycler instrument in a real-time PCR for *Trichomonas vaginalis* in urine samples from females and males. J Clin Microbiol 41:5619–5622
15. Hartmann AA, Elsner P (1988) Urethritis caused by *Neisseria meningitidis* Group B: A case report. Sex Transm Dis 15:150–151
16. Deguchi T, Yoshida T, Miyazawa T, Yasuda M, Tamaki M, Ishiko H, Maeda S (2004) Association of *Ureaplasma urealyticum* (Biovar 2) with nongonococcal urethritis. Sex Transm Dis 31:192–195
17. Povlsen K, Bjornelius E, Lidbrink P, Lind I (2002) Relationship of *Ureaplasma urealyticum* Biovar 2 to nongonococcal urethritis. Eur J Clin Microbiol Infect Dis 21:97–101

18. Taylor-Robinson D (2007) The role of mycoplasmas in pregnancy outcomes. Best Pract Res Clin Obster Gynaecol 21:425–438
19. Taylor-Robinson D, McCormak WM (1980) The genital mycoplasmas (first of two parts). N Engl J Med 302:1003–1010
20. Koroglu M, Yakupogullari Y, Aydogan F (2007) A case of urethritis due to *Streptococcus pneumoniae*. Sex Transm Dis 34:1040
21. Noble RC (1985) Colonisation of the urethra with *Streptococcus pneumoniae*: a case report. Genitourin Med 61:345–346
22. Quentin R, Musser JM, Mellouet M, Sizaret P-Y, Selander RK, Goudeau A (1989) Typing of urogenital, maternal, and neonatal isolates of *Haemophilius influenzae* and *Haemophilius parainfluenzae* in correlation with clinical source of isolation and evidence for a genital specificity of *H. influenzae* biotype IV. J Clin Microbiol 27:2286–2294
23. Sturm AW (1986) *Haemophilius influenzae* and *Haemopilius parainfluenzae* in Nongonococcal urethritis. J Infect Dis 153:165–167
24. Jensen JS, Borre MB, Dohn B (2003) Detection of *Mycoplasma genitalium* by PCR amplification of the 16 S rRNA gene. J Clin Microbiol 41(1):261–266
25. Mallard K, Schopfer K, Bodmer T (2005) Development of real-time PCR for the differential detection and quantification of Ureaplasma urealyticum and Ureaplasma parvum. J Microbiol Methods 60(1):13–19
26. Zariffard MR, Saifuddin M, Sha BE, Spear GT (2002) Detection of bacterial vaginosis-related organisms by real-time PCR for Lactobacilli, *Gardnerella vaginalis* and *Mycoplasma hominis*. FEMS Immunol Med Microbiol 34(4):277–281
27. Stevenson J, Hymas W, Hillyard D (2005) Effect of sequence polymorphisms on performance of two real-time PCR assays for detection of herpes simplex virus. J Clin Microbiol 43(5):2391–2398
28. Allard A, Albinsson B, Wadell G (2001) Rapid typing of human adenoviruses by a general PCR combined with restriction endonuclease analysis. J Clin Microbiol 39(2):498–505
29. Jordens JZ, Heckels JE (2005) A novel porA-based real-time PCR for detection of meningococcal carriage. J Med Microbiol 54(5): 463–466
30. Roth SB, Jalava J, Ruuskanen O, Ruohola A, Nikkari S (2004) Use of an oligonucleotide array for laboratory diagnosis of bacteria responsible for acute upper respiratory infections. J Clin Microbiol 42(9):4268–4274
31. Llull D, López R, García E (2006) Characteristic signatures of the lytA gene provide a basis for rapid and reliable diagnosis of *Streptococcus pneumoniae* infections. J Clin Microbiol 44(4): 1250–1256

Chapter 16

Diagnosis of *Clostridium difficile* Infection Using Real-Time PCR

Renate Johanna van den Berg, Dennis Bakker, and Ed J. Kuijper

Abstract

Clostridium difficile is known to cause antibiotic-associated diarrhea and pseudomembranous colitis. Toxinogenic strains of the bacterium produce toxins A (TcdA) and B (TcdB), which are associated with the pathogenicity. The standard methods for diagnosis of *C. difficile* infection include the cell cytotoxicity assay and the culture of a toxinogenic strain. Due to the long turnaround time of these methods, more rapid methods are preferred. Enzyme immunoassays are fast, but lack sensitivity. Therefore, real-time PCR methods have been developed.

The real-time PCR described in this chapter detects *tcd*B, the gene coding for toxin B. Since toxin A-negative, toxin B-positive strains have been reported to cause disease as well, these strains can also be detected by this method which uses an automated STAR-MagnaPure method for the optimum isolation of DNA from feces. An internal control is included as well to control for inhibition of the PCR method.

Key words: *Clostridium difficile*, Real-time PCR, Toxin B, Pathogenicity locus, DNA isolation from feces

1. Introduction

Symptoms of *Clostridium difficile* infection (CDI) can vary from mild diarrhea to pseudomembranous colitis. In more than 85% of CDI, antibiotic treatment preceded the onset of symptoms. The pathogenicity of *C. difficile* is associated with the production of two toxins: toxin A (TcdA) and toxin B (TcdB) (1, 2). It has been suggested that both toxins act synergistically (3, 4), although some reports mention CDI by strains that are TcdA negative, TcdB positive (5–8). These strains have a deletion in the *tcd*A gene, and are therefore unable to produce normal TcdA. The role of the binary toxin is less clear although a recent publication suggested that it increases

Mark Wilks (ed.), *PCR Detection of Microbial Pathogens: Second Edition*, Methods in Molecular Biology, vol. 943,
DOI 10.1007/978-1-60327-353-4_16, © Springer Science+Business Media, LLC 2013

binding of *C. difficle* to intestinal epithelial (9–11). The genes for TcdA and TcdB, together with the toxin-regulating genes *tcd*C and *tcd*R, are located on a pathogenicity locus, which is absent in nontoxinogenic, and therefore nonpathogenic, strains.

The diagnosis of CDI requires the detection of toxin-producing *Clostridium difficile* or its products in diarrheal samples. In the past, diagnosis of CDI was established by a positive cell cytotoxicity assay, or by positive culture of a toxinogenic strain. The cytotoxicity is mainly dependent on the action of TcdB. Due to the long turnaround time of these assays, other methods have been developed, such as enzyme immune-assays for detection of the toxins and PCR for the detection of toxin genes. However, the positive predictive values are too low to be used as the single detection method for CDI (12): http://www.pasa.nhs.uk/pasa/Doc.aspx?Path = %5bMN%5d%5bSP%5d/NHSprocurement/CEP/CEP08054.pdf. Real-time PCR for *C. difficile* is one of the preferred methods, due to its fast turnaround time, where results can be obtained within a day.

Thus far, several methods have been described on the application of real-time PCR to detect *tcd*B for CDI (13–16), which makes it applicable for the detection of TcdA-negative, TcdB-positive strains, as well as all other toxinogenic strains of *C. difficile*. In addition several companies have also developed first-generation assays detecting *tcd*B, such as the BD GeneOhm(TM) Cdiff Assay (BD GeneOhm) (17) and the GeneXpert *C. difficile* Assay (Cepheid) Biomerieux (18). The latter also detects binary toxin genes and deletions in *tcd*C. The principle of the PCR that we developed has been published by van den Berg et al. (16), but the sensitivity was lower than that of culture and adaptations have been made since to improve the sensitivity. These adaptations are included in this protocol and resulted in a sensitivity of 10^3 CFU/gram feces. The real-time PCR detects *tcd*B in DNA isolated from fecal samples, which makes the DNA isolation protocol of high importance for correct detection of the disease-causing agent. An internal control is included in the PCR as well to verify correct DNA isolation and PCR performance (19).

2. Materials

2.1. Obtaining Clinical Samples

The most common criteria applied to select fecal samples for CDI testing include loose or watery stools from hospitalized patients, diarrheal samples from patients above 65 years of age, stools from patients with previous antibiotic therapy, and all stools from nosocomial (development of diarrhea after 3 days of hospitalization) diarrhea (20, 21). In the Netherlands, we advice testing feces samples of all hospitalized patients who develop diarrhea 3 days after hospitalization, irrespective of the physician's request. Application of this algorithm resulted in 24% increase of CDI diagnosed patients (12). In a European surveillance study, 23% of the laboratories mention CDI to be diagnosed on feces samples from patients nursed at specific departments, i.e., oncology, hematology, intensive care, or gastroenterology (20). Interestingly, we have found an unexpected high incidence of CDI among diarrheal patients who attended a general practitioner (Bauer MP, Veenendaal D, Verhoef L, Bloembergen P, van Dissel JT, Kuijper EJ. Clinical and microbiological characteristics of community-onset *Clostridium difficile* infection in the Netherlands. Clin Microbiol Infect. 2009 Dec;15(12):1087–1092).

Consequently, we advice testing all diarrheal feces samples of patients for CDI, provided that common enteropathogens are absent and the patient is above 2 years of age.

2.2. DNA Isolation from Feces Samples

1. The consistency of the fecal samples would ideally be watery, loose, or unformed (taking the shape of the container); solid samples should be refused (see Notes 1–3).
2. Stool Transport and Recovery buffer STAR (Roche) (see Notes 4 and 5).
3. Chloroform p.A., stored in a safety cabinet.
4. MagNA Pure LC DNA Isolation Kit III (bacteria, fungi) (Roche). Store at room temperature. This kit, together with the LC system, is used to lyse, wash, and elute DNA form fecal samples by magnetic glass particles.
5. Isolated DNA can be stored at –20°C.
6. MagNA Pure LC system (Roche).

2.3. *Clostridium difficile tcdB*-Specific Primers and Probes

Primers and probes were selected from the non-repeat region from a known *tcd*B sequence (accession no. X53138 (1)).

1. The sequence of the forward primer is 5′-GAAAGTCCAAG TTTACGCTCAAT-3′. A stock of 500 μM is prepared and stored at –20°C. For short-term use, a stock of 10 μM is prepared and stored at 4°C. The working stock can be stored for up to 1 year.
2. The sequence of the reverse primer is 5′-GCTGCACCTA AACTTACACCA-3′. A stock of 500 μM is prepared and stored at –20°C. For short-term use, a stock of 10 μM is prepared and stored at 4°C. The working stock can be stored for up to 1 year.
3. The sequence of the probe is 5′-ACAGATGCAGCCAAAGT TGTTGAATT-3′, labeled with 6-carboxyfluorescein (FAM) at the 5′-end, and labeled with BHQ-1 at the 3′-end. Reconstitute to a concentration of 50 μM and store at –20°C. A working solution of 5 μM should be prepared and stored at 4°C for up to 1 year.

2.4. Real-Time PCR

1. HotStar Taq DNA Polymerase (Qiagen).
2. $MgCl_2$ solution (1 M). Dilute 1.25 ml in 48.75 ml molecular biology grade distilled water. Store the 25 mM stock in aliquots of 1 ml at 4°C.
3. Thermo fast PCR-plate (AB-gene) or CFX96-well PCR plate (BioRad).
4. Optical sealing tape (BioRad).
5. Optical Compression Pad for improved film sealing of 96-well plates in Opticon systems (BioRad).
6. CFX96 real-time PCR detection system (BioRad).

2.5. Internal Control

1. The sequence of the forward primer is 5′- GGGCGAATCAC AGATTGAATC -3′. A stock of 500 μM is prepared and stored at −20°C. For short-term use, a stock of 50 μM should be prepared and stored at 4°C. The working stock can be stored for up to 1 year.
2. The sequence of the reverse primer is 5′- GCGGTTCCAA ACGTACCAA -3′. A stock of 500 μM is prepared and stored at −20°C. For short-term use, a stock of 50 μM should be prepared and stored at 4°C. The working stock can be stored for up to 1 year.
3. The sequence of the probe is 5′- TTTTTATGTGTCCG CCACCATCTGGATC -3′, labeled with CY5 at the 5′-end, and labeled with BHQ2 at the 3′-end. A stock of 50 μM is prepared and stored at −20°C. For short-term use, a stock of 5 μM should be prepared and stored at 4°C. The working stock can be stored for up to 1 year.
4. Phocine Herpes Virus (PhHV) is diluted 1:10 (10^{-1} to 10^{-7}) in Eagle's Minimal Essential Medium (EMEM) + 10% Fetal Calf Serum (FCS). DNA is isolated using the QIAamp DNA blood mini kit (Qiagen). Perform a quantitative real-time PCR on this DNA and select the dilution that shows a C_t-value of 32–35. This dilution is then stored at −20°C.
5. For each sample isolation, use 10 μl PhHV of the above-mentioned dilution in the lysis buffer of the MagNA Pure isolation kit (see Notes 6 and 7).

3. Methods

3.1. DNA Isolation from Feces Samples

1. One part of feces is mixed with three parts of STAR buffer (see Note 8).
2. The STAR–fecal mixture is shortly vortexed, and 0.1× volume of chloroform is added (in an exhaust protective cabinet).

3. This mixture is centrifuged for 1 min at 1,000 × *g*.
4. 100 μl of the supernatant is then added to a mixture with 130 μl lysis buffer and 20 μl proteinase K, which are ingredients of the MagNA Pure LC DNA Isolation Kit III.
5. Heat this mixture for 10 min at 65°C, and subsequently for 10 min at 95°C.
6. Let the mixture cool down, after which 100 μl of each sample is used for the automated isolation using the MagNA Pure LC system; select the protocol "DNA bacteria III" (see Note 9).
7. Use the buffers and ingredients of the MagnaPure LC DNA isolation Kit III (remember to use the lysis buffer with PhHV).

3.2. Real-Time PCR

1. The amplification mixture per PCR reaction contains 25 μl of HotStar Taq DNA Polymerase, to which 0.4 μl of forward primer (10 μM), 0.4 μl of reverse primer (10uM), 7 μl of MgCl2 (25 mM), 1 μl of probe (5 μM), 0.3 μl of both PhHV-primers (50 μM), and 0.5 μl PhHV-Taqman probe (5 μM) are added. Molecular biology grade distilled water is added to reach a final volume of 40 μl.
2. The amplification mixture can now be divided over the appropriate number of wells in the PCR plate, and cover the plate with a seal.
3. Add 10 μl of the isolated DNA to reach a final volume of 50 μl (see Note 10).
4. The plate is then covered by an optical sealing tape, and after firm attachment using the optical compression pad; the plate is briefly centrifuged to deposit the contents of the wells (up to 400 × *g*) and then inserted into the CFX96 real-time PCR detection system.
5. The PCR protocol used consists of the following steps: 15 min at 95°C to activate the polymerase, subsequently 50 cycles of 30 s at 94°C for denaturation of the DNA, 30 seconds at 50°C for annealing of the primers, and 30 s at 72°C for elongation of the translated DNA. This protocol can be added to the system as a .prcl file (see Note 11).
6. The size of the generated fragment is 177 bp.

3.3. The CFX96 Real-Time PCR System

1. Turn on the system.
2. Turn on the attached computer.
3. Start the BioRad CFX manager software.
4. Click the start-up wizard.
5. Click on create experiment and select your PCR protocol.
6. Put the PCR plate in the correct orientation into the system and press "Start run."

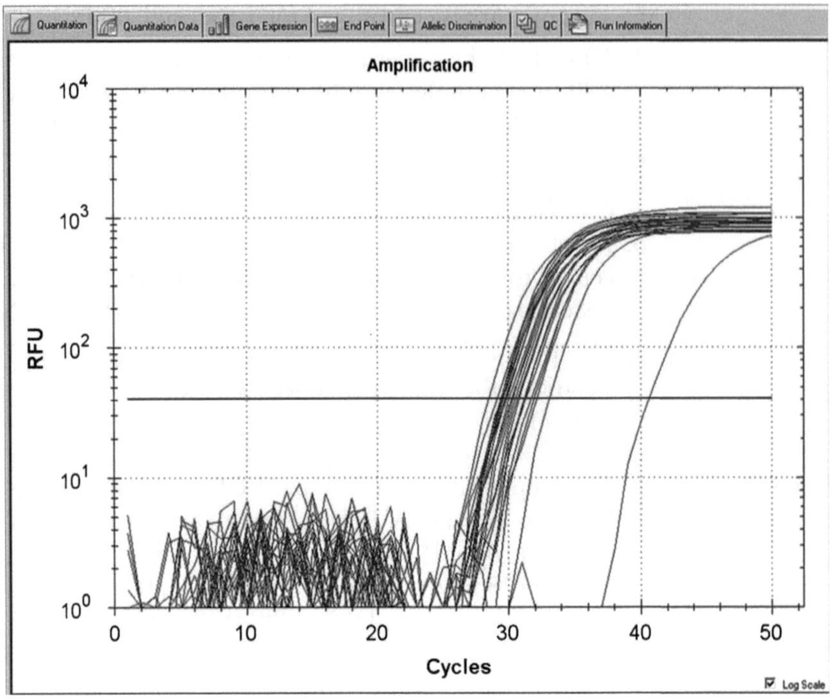

Fig. 1. On screen quantitation curve of 26 samples, of which six samples were positive.

7. The window "save optical data file" is opened; give a filename and click "save."
8. A new window appears: "run time central." The measurements can be followed real time in this screen (see Notes 12 and 13). The quantitation curve in the Data Analysis screen is shown in Fig. 1.
9. After the run, save the .pcrd file on a suitable location.

3.4. Analysis

1. Open the CFX96 manager Software, click the tab "file," and select open data file. Select your .pcrd file, and double-click. The section "data analysis" is now opened.
2. Select the wells you want to analyze by clicking "Analyze wells."
3. Select Cy5 as the fluorophore you want to analyze.
4. The baseline threshold and crossing threshold is auto calculated.
5. To adjust (per sample) the baseline threshold, click with the right mouse pad on your curve and select "settings baseline threshold." A new window opens, where you can select "user defined," select the well, and edit "baseline begin and/or baseline end." Click "ok" and "ok." The new background has now been calculated.

6. To correct the crossing threshold, move the threshold line with your mouse to correct the "threshold position." This line should be immediately above the background.
7. Now look at the C_t-values of the internal control. C_t-value should be between 32 and 35 cycles (see Note 14). If this is correct, select "quantitation data," and a table with results will pop-up. Now print.
8. Repeat steps 3–6 for the FAM fluorophore.
9. All samples above the threshold can be interpreted as positive (see Notes 15 and 16). Select "quantitation data," and a table with results will pop-up. Now print.
10. Save the edited file by clicking "file," "save," "save," and "yes".

4. Notes

1. Pre-incubation of the fecal samples for 24–48 h in broths with different concentrations of cefoxitin, cycloserine, taurocholate, and lysozyme did not have any effect on the sensitivity of the real-time PCR (D. Bakker, unpublished observation). Clearly, inhibition of normal fecal flora and simultaneously growth stimulation of *C. difficile* are difficult to achieve.
2. Samples sent by mail should preferably be delivered by courier or frozen and sent on dry ice.
3. If fecal samples are also used for other methods than PCR, and fecal samples need to be stored for a period of <5 days, it is advised to store samples at 4°C (22, 23). Samples for outbreak investigation sometimes need to be stored for a long time (>5 days); therefore it is advised to store samples at 4°C or −20° for this purpose for a maximum of 52 days (23, 24).
4. If the buffer is crystallized, incubate the buffer for 10 min at 37°C to redissolve.
5. After centrifugation, part of the supernatant can be stored at −20°C for later use.
6. 1 μl PhHV per test of a 10-times lower dilution can also be used: for example if the dilution of PhHV yielding a C_t value of 32–35 cycles dilution is 10^{-4}, then prepare a 10^{-3} dilution and use that one to add 1 μl PhHV to the lysis buffer.
7. When testing two negative controls in a real-time PCR, the C_t-values of the internal control in these wells should still be between 32 and 35 cycles.

8. As well as patient samples, always include a negative and a positive control. If you are testing more than 10 samples in a run, include an extra negative control halfway through the run.
9. If the laboratory does not own a MagNA Pure system, the manual QIAamp DNA blood mini kit (Qiagen) can be used for DNA isolation, although the automated method is preferred due to its short hands-on time. The methods show similar results in sensitivity (16).
10. Always pipette the DNA from the patient samples and negative controls, before adding the positive control.
11. Since the analysis of the product is performed in real time, no extended elongation time at the end of the cycles is necessary.
12. Do not use the computer during runs, to prevent crashes.
13. Do not remove the lid from the running system, because the computer will stop the program and all data will be lost.
14. The C_t-value of the internal control should be equal to the average C_t-value of the PhHV in the negative controls ±3 cycles. Normally this is between 32 and 35 cycles. If this is not the case, the sample is inhibited and the PCR needs to be repeated. When you repeat the PCR, also test a tenfold dilution of the initial DNA sample. If after this repeat the test is still inhibited, the result should be recorded as "inhibited."
15. The results can only be interpreted if the sample is not inhibited, and the C_t value(s) of the internal control from negative control(s) is present.
16. To use the real-time PCR quantitatively, include a standard dilution series of a positive control of which the CFU is known. After running the PCR, look at the standard curve. Check the correlation coefficient, PCR efficiency (E), and slope. The values can be corrected by changing the "threshold position." Now select "reports," "2 PCR Quant data," "sort data by," "well," and "print."

Acknowledgments

We would like to thank Ingrid Sanders for her help in optimizing the DNA isolation protocol and real-time PCR conditions. We also would like to thank the department of Clinical Microbiology Laboratory for providing the correct protocols for the use of the MagnaPure and CFX96 systems.

References

1. Barroso LA, Wang SZ, Phelps CJ, Johnson JL, Wilkins TD (1990) Nucleotide sequence of *Clostridium difficile* toxin B gene. Nucleic Acids Res 18:4004
2. Dove CH, Wang SZ, Price SB, Phelps CJ, Lyerly DM, Wilkins TD et al (1990) Molecular characterization of the *Clostridium difficile* toxin A gene. Infect Immun 58:480–488
3. Lyerly DM, Lockwood DE, Richardson SH, Wilkins TD (1982) Biological activities of toxins A and B of *Clostridium difficile*. Infect Immun 35:1147–1150
4. Lyerly DM, Saum KE, MacDonald DK, Wilkins TD (1985) Effects of *Clostridium difficile* toxins given intragastrically to animals. Infect Immun 47:349–352
5. Kato H, Kato N, Watanabe K, Iwai N, Nakamura H, Yamamoto T et al (1998) Identification of toxin A-negative, toxin B-positive *Clostridium difficile* by PCR. J Clin Microbiol 36:2178–2182
6. Kuijper EJ, de Weerdt J, Kato H, Kato N, van Dam AP, van der Vorm ER, Weel J, van Rheenen C, Dankert J (2001) Nosocomial outbreak of *Clostridium difficile*-associated diarrhoea due to a clindamycin-resistant enterotoxin A-negative strain. Eur J Clin Microbiol Infect Dis 20:528–534
7. van den Berg RJ, Claas EC, Oyib DH, Klaassen CH, Dijkshoorn L, Brazier JS et al (2004) Characterization of toxin A-negative, toxin B-positive *Clostridium difficile* isolates from outbreaks in different countries by amplified fragment length polymorphism and PCR ribotyping. J Clin Microbiol 42: 1035–1041
8. Goorhuis A, Legaria MC, van den Berg RJ, Harmanus C, Klaassen CH, Brazier JS, Lumelsky G, Kuijper EJ (2009) Application of multiple-locus variable-number tandem-repeat analysis to determine clonal spread of toxin A-negative *Clostridium difficile* in a general hospital in Buenos Aires, Argentina. Clin Microbiol Infect 15(12):1080–1086
9. Geric B, Carman RJ, Rupnik M, Genheimer CW, Sambol SP, Lyerly DM et al (2006) Binary toxin-producing, large clostridial toxin-negative *Clostridium difficile* strains are enterotoxic but do not cause disease in hamsters. J Infect Dis 193:1143–1150
10. Perelle S, Gibert M, Bourlioux P, Corthier G, Popoff MR (1997) Production of a complete binary toxin (actin-specific ADP-ribosyltransferase) by *Clostridium difficile* CD196. Infect Immun 65:1402–1407
11. Schwan C, Stecher B, Tzivelekidis T, van Ham M, Rohde M, Hardt WD, Wehland J, Aktories K (2009) *Clostridium difficile* toxin CDT induces formation of microtubule-based protrusions and increases adherence of bacteria. PLoS Pathog 5(10):e1000626
12. van den Berg RJ, Vaessen N, Endtz HP, Schulin T, van der Vorm ER, Kuijper EJ (2007) Evaluation of real-time PCR and conventional diagnostic methods for the detection of *Clostridium difficile*-associated diarrhoea in a prospective multicentre study. J Med Microbiol 56:36–42
13. Belanger SD, Boissinot M, Clairoux N, Picard FJ, Bergeron MG (2003) Rapid detection of *Clostridium difficile* in feces by real-time PCR. J Clin Microbiol 41:730–734
14. Peterson LR, Manson RU, Paule SM, Hacek DM, Robicsek A, Jr Thomson RB et al (2007) Detection of toxigenic *Clostridium difficile* in stool samples by real-time polymerase chain reaction for the diagnosis of *C. difficile*-associated diarrhea. Clin Infect Dis 45:1152–1160
15. Sloan LM, Duresko BJ, Gustafson DR, Rosenblatt JE (2008) Comparison of real-time PCR for detection of the tcdC gene with four toxin immunoassays and culture in diagnosis of *Clostridium difficile* infection. J Clin Microbiol 46:1996–2001
16. van den Berg RJ, Kuijper EJ, van Coppenraet LE, Claas EC (2006) Rapid diagnosis of toxinogenic *Clostridium difficile* in faecal samples with internally controlled real-time PCR. Clin Microbiol Infect 12:184–186
17. Stamper PD, Alcabasa R, Aird D, Babiker W, Wehrlin J, Ikpeama I et al (2009) Comparison of a commercial real-time PCR assay for tcdB detection to a cell culture cytotoxicity assay and toxigenic culture for direct detection of toxin-producing *Clostridium difficile* in clinical samples. J Clin Microbiol 47:373–378
18. Huang H, Weintraub A, Fang H, Nord CE (2009) Comparison of a commercial multiplex real-time PCR to the cell cytotoxicity neutralization assay for diagnosis of *clostridium difficile* infections. J Clin Microbiol 47(11): 3729–3731
19. Niesters HG (2002) Clinical virology in real time. J Clin Virol 25(Suppl 3):S3–S12
20. Barbut F, Delmee M, Brazier JS, Petit JC, Poxton IR, Rupnik M et al (2003) A European survey of diagnostic methods and testing protocols for *Clostridium difficile*. Clin Microbiol Infect 9:989–996

21. Johnson S, Gerding DN (1998) *Clostridium difficile*–associated diarrhea. Clin Infect Dis 26:1027–1034
22. Freeman J, Wilcox MH (2003) The effects of storage conditions on viability of *Clostridium difficile* vegetative cells and spores and toxin activity in human faeces. J Clin Pathol 56: 126–128
23. Wilcox MH (1998) *Clostridium difficile* infection: appendix. J Antimicrob Chemother 41(Suppl C):71–72
24. Brazier JS (1998) The diagnosis of *Clostridium difficile*-associated disease. J Antimicrob Chemother 41(Suppl C):29–40

Chapter 17

Detection of Pathogenic *Leptospira* spp. Through Real-Time PCR (qPCR) Targeting the LipL32 Gene

Robyn Anne Stoddard

Abstract

Rapid diagnosis of leptospirosis, through culture and/or serology, can be difficult without proper expertise and is often delayed due to the length of time required to obtain results. Polymerase chain reaction (PCR), more specifically the real-time detection of the amplified PCR product, is a methodology that can provide a diagnosis in a timelier manner compared to culture and serology. There are a limited number of real-time PCR (qPCR) assays for detecting *Leptospira* and not all of these assays are able to distinguish pathogenic from nonpathogenic species. In addition, there are a variety of probe technologies and qPCR instruments that are utilized with these assays. This chapter presents a qPCR assay that targets *lipL32*, a gene which is present only in pathogenic *Leptospira* spp. This assay utilizes a TaqMan probe and instructions for use on either the Lightcycler 1.2 (Roche Diagnostics, Indianapolis, IN) or the ABI 7500 (Applied Biosystems, Foster City, CA) are provided.

Key words: *Leptospira*, Leptospirosis, Real-time PCR, TaqMan, Diagnosis, LipL32

1. Introduction

Leptospirosis is one of the main causes of acute febrile illness and is presumed to be the most widespread zoonotic disease in the world (1). There are 18 different *Leptospira* species including more than 200 serovars, but not all are considered to be pathogenic (2). The clinical presentation in humans is difficult to distinguish from dengue, malaria, influenza, and many other diseases characterized by fever, headache, and myalgia (1). Definitive diagnosis of leptospirosis in humans is made by demonstration of the organism using dark-field microscopy, isolation of the bacteria from tissue or body fluids, or through use of the microscopic agglutination test (MAT) for serology to detect a fourfold increase in titer (1). However,

Mark Wilks (ed.), *PCR Detection of Microbial Pathogens: Second Edition*, Methods in Molecular Biology, vol. 943,
DOI 10.1007/978-1-60327-353-4_17, © Springer Science+Business Media, LLC 2013

none of these methods can provide definitive diagnosis of leptospirosis in a timely manner to aid in patient management (1).

Real-time polymerase chain reaction (qPCR) is a methodology that is revolutionizing the diagnosis of infectious diseases in clinical laboratories in a timely manner (3). There are a limited number of qPCR assays for detecting pathogenic *Leptospira* (4–9). Two of the assays target genes that are universally present in bacteria, the 16S rRNA (16S) gene (5, 6) and the gene encoding DNA gyrase subunit b (*gyrB*) (9). Other assays target genes such as *lipL32* (4, 8) and leptospiral immunglobulin-like protein (*ligA* and *B*) (7) that are considered to be genes restricted to pathogenic *Leptospira* species. The most common qPCR assays for *Leptospira* are based on SYBR green technology (4, 6, 9), TaqMan probes (5, 7), or, more recently, Light Upon eXtension (LUX) technology (8).

The *lipL32* gene of pathogenic *Leptospira* spp. encodes an outer membrane lipoprotein that is not present in nonpathogenic species and may play a role in virulence (10–12). A qPCR assay using a TaqMan probe that detects *lipL32* is a rapid method that can be used on pure culture of pathogenic *Leptospira* spp. or to detect *Leptospira* in clinical specimens such as whole blood, sera, and urine, although sera is not recommended for testing due to false negative results (13). The time of sample collection after the onset of symptoms of disease determines what is the best sample to test since leptospires are present in the blood for about the first week after the onset of symptoms but they can be found in urine for several weeks after that (1). Another important consideration for qPCR with clinical samples is the presence of inhibitors in the sample that may decrease efficiency or completely inhibit the reaction. An additional qPCR reaction which targets a gene present in humans (*rnaseP*) helps to determine if sample is a true negative or if it is due to inhibition.

2. Materials

2.1. Sample Requirements

2.1.1. Blood Samples Stored at 4°C Temporarily or at −20°C or −80°C for Long-Term Storage (see Note 1)

1. Whole blood is collected in tube with EDTA or sodium citrate (see Note 2).
2. Plasma separated from whole blood collected in EDTA (see Note 2).
3. Sera.

2.1.2. Urine Processed as Soon as Possible and Stored at 4°C Temporarily or at −20°C or −80°C for Long-Term Storage

1. Urine collected in a sterile cup (see Note 3).
2. 15 or 50 ml conical tube, depending on urine volume collected.
3. Phosphate-Buffered Saline (PBS) 0.01 M, pH 7.2–7.4.

2.2. DNA Extraction

1. Extraction kit:
   (a) For cultures: QIAamp DNA minikit (QIAGEN, Valencia, CA).
   (b) For clinical samples: QIAamp DNA blood minikit (QIAGEN).
2. Ethyl alcohol (97–100%).
3. Water bath or heating block to 56°C.
4. 200 μl of culture or clinical sample.
5. Negative and positive extraction control samples (see Note 4).

2.3. Real-Time PCR

2.3.1. For Use on the Lightcycler® 1.2 Platform

1. Lightcycler® 1.2 (Roche Diagnostics, Indianapolis, IN).
2. LC Carousel Centrifuge (Roche Diagnostics).
3. External computer with Lightcycler® software.
4. Lightcycler® FastStart DNA Master HybProbe (Order #12239272001, Roche Diagnostics).
5. Lightcycler® Capillaries (20 μl) (Part #1190933900, Roche Diagnostics).
6. Oligonucleotide primers and probe for the leptospire-specific (*lipL32*) and the internal control RNase P gene (*rnaseP*) stored at −20°C (Table 1) (see Note 5).
7. Positive and negative qPCR controls (see Note 6).
8. Extract DNA from culture or clinical sample.

2.3.2. For Use on the ABI 7500 Platform

1. ABI 7500 Real-Time PCR System (Applied Biosystems, Foster City, CA).
2. External computer with ABI 7500 software.

Table 1
Oligonucelotide primers and probes for qPCR of pathogenic Leptospira (lipL32) and a positive internal control for clinical samples (RNAse P)

Target Gene	Name	Nucleotide sequence
lipL32	Forward primer (45F)	5′-AAG CAT TAC CGC TTG TGG TG-3′
	Reverse primer (286R)	5′-GAA CTC CCA TTT CAG CGA TT-3′
	Probe (189p)	FAM-5′-AA AGC CAG GAC AAG CGC CG-3′-BHQ1
rnaseP	Forward primer	5′-CCA AGT GTG AGG GCT GAA AAG-3′
	Reverse primer	5′-TGT TGT GGC TGA TGA ACT ATA AAA GG-3′
	Probe	FAM-5′-CC CCA GTC TCT GTC AGC ACT CCC TTC-3′-BHQ1

3. Platinum® Quantitative PCR SuperMix-UDG (Part #11730-025, Invitrogen, Carlsbad, CA).
4. Microamp™ Optical 96-Well Reaction Plate (Part #N801-0560, Applied Biosystems).
5. Microamp™ Optical Caps (Part #4323032, Applied Biosystems) or Optical Adhesive Film (Part #4311971, Applied Biosystems).
6. Oligonucleotide primers and probe for the leptospire-specific (*lipL32*) and the internal control RNase P gene (*rnaseP*) stored at −20°C (Table 1) (see Note 5).
7. Positive and negative qPCR controls (see Note 6).
8. Extract DNA from culture or clinical sample.

3. Methods

Real-time PCR is more sensitive than conventional PCR and therefore, proper facilities and technique are a necessity for preventing contamination between samples. It is recommended that separate rooms are used for DNA extraction, master mix preparation, and DNA addition to the master mix. These rooms should have their own pipets, tips, lab coats, and gloves, with no transfer between the rooms. Lab workers are recommended not to enter the master mix room after working in the DNA addition room. It is also recommended that DNA from culture is extracted and added to the master mix in a different location than clinical samples since there is a high potential for contamination. Consistent use of negative controls throughout the process will help to determine if contamination has occurred at any step during the process.

3.1. Sample Preparation

3.1.1. Whole Blood/Serum Samples

1. Blood samples should be processed as soon as possible after collection.
2. Sera should be separated from the clot and placed in a new tube.
3. If samples will be processed within 1 or 2 weeks of sampling they can be stored at 4°C and shipped on cold packs. If samples will not be processed within 2 weeks they should be frozen at −20°C or −80°C and shipped on dry ice.

3.1.2. Urine

1. Transfer the urine sample from the sterile collection cup to the appropriate sized conical tube.
2. Centrifuge the urine at 3,000 × *g* for 15 min.
3. Discard the supernatant and resuspend the pellet in an equivalent volume of PBS 0.01 M, pH 7.2–7.4.
4. Centrifuge at 3,000 × *g* for 15 min.

5. Repeat the wash step one time.
6. Discard the supernatant and resuspend the pellet in 500 μl PBS.
7. If samples will be processed within 1 or 2 weeks of sampling they can be stored at 4°C and shipped on cold packs. If samples will not be processed within 2 weeks they should be frozen at −20°C or −80°C and shipped on dry ice.

3.2. DNA Extraction

1. 200 μl of culture, whole blood, plasma, serum, or urine is transferred to a 1.5 ml PCR-grade microcentrifuge tube.
2. The QIAamp DNA minikit is used for DNA extraction from culture samples and QIAamp DNA blood minikit is used for DNA extraction from clinical samples.
3. DNA is extracted as per the manufacturer's "Blood and Body Fluid Spin Protocol" with the following adjustments:
   (a) Optional step 9a is not performed.
   (b) For the elution step, 100 μl of Buffer AE or distilled PCR-grade water is added to the QIAamp Spin Column and incubated for 5 min before centrifuging. This step is not repeated (the total elution volume is 100 μl).

3.3. Real-Time PCR Assays

3.3.1. Targeting Pathogenic Leptospira spp. (lipL32) Assay

1. Thaw components needed for making the qPCR master mix (see Note 7 and Table 2).
2. Prepare the master mix for the number of reactions based on the required volumes/concentrations for the qPCR kit and platform as presented in Table 2 (see Note 8). Samples should be run in duplicate. Total master mix volume should also account for positive and negative controls (see Note 9).
3. Aliquot the appropriate volume of master mix into each capillary/well.
4. Add negative control to the appropriate capillary/well and cover with cap.
5. Transfer capillaries to the DNA addition room for use as soon as possible (see Note 10). If DNA is not going to be added to the master mix immediately, store at 4°C away from light.
6. Add the appropriate volume of DNA template to each well making sure to add PCR-grade water (negative control) after every fourth well. If using capillaries, cap each sample after the DNA is added. If using optical caps, place the strips after DNA has been added to the entire row.
7. If using the Lightcycler® carousel system, capillaries will have to be transferred from the plastic container to the carousel. Capillaries must be gently pushed down to the neck of the capillary; take care not to break the capillary.

Table 2
Master mix components and concentrations or volumes for qPCR for pathogenic Leptospira

Component	Final concentration or volume/sample	
	Lightcycler® master mix	Invitrogen master mix
Forward primer (45F)	400 nM	500 nM
Reverse primer (286R)	400 nM	500 nM
Probe (189p)	132.5 nM	100 nM
$MgCl_2$	3.0 mM	–
Lightcycler® fast-start enzyme	2.0 μl	–
Platinum® Quantitative PCR SuperMix-UDG	–	12.5 μl
DNA	1.0 μl (culture sample) or 5.0 μl (clinical sample)	1.0 μl (culture sample) or 5.0 μl (clinical sample)
PCR-grade water	As needed	As needed
Total volume	20.0 μl	25.0 μl

8. Centrifuge the samples as per the manufacturer's instructions.
9. Place the plate/carousel into the appropriate qPCR platform and run the assay as indicated in Table 4 with FAM and a single acquisition mode occurring during the second step during amplification (see Note 11). Sample names can also be input at this time.
10. Analyze the data once the 45 cycles have completed (see Note 12):
    (a) For the Lightcycler® 1.2 click on the quantification tab, select fit points, and then click proportional. Click on the noise band and select all of the negative controls. Manually set the threshold baseline above the background from negative control samples. The upper limit for the threshold baseline is a fluorescence of 0.2.
    (b) For the ABI 7500 click on the results tab and select the amplification plot. The manual C_t choice should be selected and then drag the threshold setting bar until it is above baseline (see Note 13). The analyze button will need to be clicked on if any changes are made.
11. A rough guideline for interpretation of C_t results is that a value <40 would be considered positive and >40 would be considered suspect, but every lab should evaluate this based on its own experience. Interpretation criteria should also be considered on a case-by-case basis incorporating patient and test results.

Table 3
Master mix components and concentrations/volumes for qPCR internal control targeting the *rnaseP* gene

Component	Final concentration or volume/sample	
	Lightcycler® master mix	Invitrogen master mix
Forward primer	400 nM	400 nM
Reverse primer	400 nM	400 nM
Probe	100 nM	132.5 nM
$MgCl_2$	4.0 mM	–
Lightcycler® fast-start enzyme	2.0 μl	–
Platinum® Quantitative PCR SuperMix-UDG	–	12.5 μl
Extract clinical sample DNA	5.0 μl	5.0 μl
PCR-grade water	As needed	As needed
Total volume	20.0 μl	25.0 μl

Table 4
Temperature–time profiles for qPCR of pathogenic Leptospira (lipL32) and a positive internal control for clinical samples (RNAse P)

Step	Lightcycler® 1.2		ABI 7500	
	lipL32	*rnaseP*	*lipL32*	*rnaseP*
Pre-incubation	95°C for 8 min	95°C for 8 min	50°C for 2 min	50°C for 2 min
Denature	–	–	95°C for 10 min	95°C for 10 min
Amplification (acquisition at the end of cycle)	95°C for 3 s	95°C for 10 s	95°C for 15 s	95°C for 15 s
	60°C for 15 s	60°C for 30 s	60°C for 60 s	60°C for 60 s
Number of amplification cycles	45	45	45	45
Cool	45°C for 90 s	45°C for 5 min	–	–

3.3.2. Internal Control Assay RNase P Gene (rnaseP)

1. Follow steps 1 through 10 above using Table 3 for components needed for making the qPCR master mix and Table 4 for assay conditions on the qPCR instruments.
2. To be considered positive for *rnaseP*, a specimen has to have a C_t value <40 (see Note 14).

4. Notes

1. Sera is not the blood sample of choice; based on experimental evidence, leptospires are better detected in whole blood or plasma (13).
2. EDTA or sodium citrate should be used as the anticoagulant for whole blood since heparin can be an inhibitor of qPCR.
3. It is recommended that at least 10 ml of urine is used for testing.
4. Non-spiked and specimens spiked with a pathogenic *Leptospira* spp. should be used as controls for the DNA extraction process (this is in addition to positive and negative controls for the actual real-time PCR).
5. Repeated freeze–thaw cycles of the primers and probe can affect results. It is best to aliquot the primers and probe into appropriate volumes for one full run. Exposure of the probe to light should be limited.
6. PCR-grade water can be used as a negative control. The positive control should be DNA from a pathogenic *Leptospira* spp. that is at a concentration that is near the limit of detection of the assay, but still gives a positive signal every time (3). DNA was quantified using the Nanodrop instrument (Nanodrop Technologies, Willmington, DE, USA). A genome size of 4.659 Mb was used to determine the genomic equivalents (GE) per μl of the purified DNA (14, 15). Serial dilutions of genomic DNA were made starting at 1×10^6 GE/μl down to 1×10^0 GE/μl.
7. Thaw $MgCl_2$ and water at room temperature and primers, probe, enzyme 1a, and enzyme 1b at 4°C. After the reagents have thawed, vortex/spin water, $MgCl_2$, and primers. Combine enzymes 1a and 1b by putting the contents of 1b into 1a. Flick/spin enzymes mixture and probe. Combine the reagents in the same order as listed in Table 2. Mix the master mix by pipetting up and down, and then spinning. Do not vortex master mix.
8. Since culture and positive control DNA only requires 1.0 μl but clinical sample DNA requires the use of 5.0 μl, the master mix calculations can be based on the use of 5.0 μl of DNA. Once the master mix has been aliquoted, 4.0 μl of PCR-grade water can be added to samples that only require 1.0 μl DNA.
9. It is recommended that a negative control (PCR-grade water) is used in the master mix room and also in the DNA addition room to help determine where potential contamination issues arise. In the master mix room only one tube/capillary is needed

as a negative control but in the DNA addition room negative control should be added after every fourth tube/capillary to monitor for contamination (3).

10. When using the Lightcycler® carousel platform samples can be left in the plastic manufacturer's container that the capillaries come in. Capillaries are less likely to break when capped in the plastic container instead of the carousel. In addition, plastic containers can be thoroughly bleached, returned to the master mix room, and used again.
11. ROX should be turned off for the ABI 7500 real-time PCR machine.
12. If any of the negative control samples give a positive signal or the positive control does not give a result with the normal range, the experiment will have to be repeated.
13. It is a good idea to set the same threshold baseline value for each run.
14. If a C_t value was greater than 40 or negative for *rnaseP*, the specimen should be diluted 1:5 or 1:10 in water and the qPCR should be repeated for *rnaseP* and *lipL32*.

Acknowledgments

The author would like to thank Duy Bui and Dr. Alex Hoffmaster for suggestions on manuscript content.

"The findings and conclusions in this report are those of the author(s) and do not necessarily represent the official position of the Centers for Disease Control and Prevention."

References

1. Levett PN (2001) Leptospirosis. Clin Microbiol Rev 14:296–326
2. Levett PN, Morey RE, Galloway RL, Steigerwalt AG (2006) *Leptospira broomii* sp. nov., isolated from humans with leptospirosis. Int J Syst Evol Microbiol 56:671–673
3. Espy MJ, Uhl JR, Sloan LM, Buckwalter SP, Jones MF, Vetter EA, Smith TF et al (2006) Real-time PCR in clinical microbiology: applications for routine laboratory testing. Clin Microbiol Rev 19:165–256
4. Levett PN, Morey RE, Galloway RL, Turner DE, Steigerwalt AG, Mayer LW (2005) Detection of pathogenic leptospires by real-time quantitative PCR. J Med Microbiol 54:45–49
5. Smythe LD, Smith IL, Smith GA, Dohnt MF, Symonds ML, Barnett LJ et al (2002) A quantitative PCR (TaqMan) assay for pathogenic *Leptospira* spp. BMC Infect Dis 2:13
6. Merien F, Portnoi D, Bourhy P, Charavay F, Berlioz-Arthaud A, Baranton G (2005) A rapid and quantitative method for the detection of *Leptospira* species in human leptospirosis. FEMS Microbiol Lett 249:139–147
7. Palaniappan RU, Chang YF, Chang CF, Pan MJ, Yang CW, Harpending P et al (2005) Evaluation of lig-based conventional and real time PCR for the detection of pathogenic leptospires. Mol Cell Probes 19:111–117
8. Roczek A, Forster C, Raschel H, Hormansdorfer S, Bogner KH, Hafner-Marx A et al (2008)

Severe course of rat bite-associated Weil's disease in a patient diagnosed with a new *Leptospira*-specific real-time quantitative LUX-PCR. J Med Microbiol 57:658–663

9. Slack AT, Symonds ML, Dohnt MF, Smythe LD (2006) Identification of pathogenic *Leptospira* species by conventional or real-time PCR and sequencing of the DNA gyrase subunit B encoding gene. BMC Microbiol 6:95
10. Picardeau M, Bulach DM, Bouchier C, Zuerner RL, Zidane N, Wilson PJ et al (2008) Genome sequence of the saprophyte *Leptospira biflexa* provides insights into the evolution of *Leptospira* and the pathogenesis of leptospirosis. PLoS One 3:e1607
11. Haake DA, Chao G, Zuerner RL, Barnett JK, Barnett D, Mazel M et al (2000) The leptospiral major outer membrane protein LipL32 is a lipoprotein expressed during mammalian infection. Infect Immun 68:2276–2285
12. Nally JE, Whitelegge JP, Bassilian S, Blanco DR, Lovett MA (2007) Characterization of the outer membrane proteome of *Leptospira interrogans* expressed during acute lethal infection. Infect Immun 75:766–773
13. Stoddard RA, Gee JE, Wilkins PP, McCaustland K, Hoffmaster AR (2009) Detection of pathogenic *Leptospira* spp. through TaqMan polymerase chain reaction targeting the LipL32 gene. Diagn Microbiol Infect Dis 64:247–255
14. Ren SX, Fu G, Jiang XG, Zeng R, Miao YG, Xu H et al (2003) Unique physiological and pathogenic features of *Leptospira interrogans* revealed by whole-genome sequencing. Nature 422:888–893
15. Nascimento AL, Ko AI, Martins EA, Monteiro-Vitorello CB, Ho PL, Haake DA et al (2004) Comparative genomics of two *Leptospira interrogans* serovars reveals novel insights into physiology and pathogenesis. J Bacteriol 186:2164–2172

Chapter 18

Sensitive and Rapid Detection of *Campylobacter jejuni* and *Campylobacter coli* Using Loop-Mediated Isothermal Amplification

Wataru Yamazaki

Abstract

Loop-mediated isothermal amplification (LAMP) is an established nucleic acid amplification method offering rapid, accurate, and cost-effective diagnosis of infectious diseases. From the beginning of DNA extraction to final detection *of Campylobacter jejuni* and *Campylobacter coli*, the assay requires less than 50 and 90 min from a colony on selective media, and human feces, respectively. For chicken meat samples, the assay requires approximately 24–48 h from the beginning of the enrichment culture to final detection. The sensitivity of the LAMP assay is tenfold higher than that of the equivalent PCR assay. LAMP amplification can be judged by both turbidimeter analysis and visual assessment with the unaided eye. The LAMP assay is a powerful tool for rapid, simple, and sensitive detection of *C. jejuni* and *C. coli*, which may facilitate the investigation of *C. jejuni* and *C. coli* contamination in chicken, as well as the early diagnosis of *C. jejuni* and *C. coli* infection in humans.

Key words: Loop-mediated isothermal amplification, LAMP, *Campylobacter jejuni*, *Campylobacter coli*, Rapid, sensitive, and cost-effective detection, Human feces, Chicken meat sample.

1. Introduction

Campylobacter is widely acknowledged as one of the most frequent causes of acute bacterial gastroenteritis in humans worldwide. *Campylobacter jejuni* and *Campylobacter coli* are the predominant cause of *Campylobacter* gastroenteritis. Identification of *C. jejuni* and *C. coli* using conventional culture- and biochemical-based assays is time-consuming and laborious, requiring more than 4 days. Although PCR assays provide more rapid identification of *C. jejuni* and *C. coli* than conventional assays, they require the use of electrophoresis to

Mark Wilks (ed.), *PCR Detection of Microbial Pathogens: Second Edition*, Methods in Molecular Biology, vol. 943,
DOI 10.1007/978-1-60327-353-4_18, © Springer Science+Business Media, LLC 2013

detect amplified products, which is time-consuming and tedious (1). Real-time PCR assays have been devised but are not routinely used due to their requirement for an expensive thermal cycler with a fluorescence detector (2).

Loop-mediated isothermal amplification (LAMP) is an established nucleic acid amplification method offering rapid, accurate, and cost-effective diagnosis of infectious diseases (3). LAMP is based on the principle of autocycling strand displacement DNA synthesis performed by the *Bst* DNA polymerase large fragment for the detection of a specific DNA sequence with specific characteristics (3). This offers a number of advantages, first, all reactions can be carried out under isothermal conditions ranging from 60 to 65°C. Second, its use of multiple primers (four or more) recognizing six or more distinct regions on the target nucleotides means that specificity is extremely high. Third, detection is simplified by visual assessment using the unaided eye without electrophoresis. The LAMP assay is faster and easier to perform than conventional PCR assays, as well as being more specific (3–5) and requires only a simple heating-block or a water bath providing a constant temperature (4).

For amplification, the LAMP assay takes 15–25 min with a single colony on selective agar from *C. jejuni* and *C. coli* strains and 60 min with human feces (6) and enrichment broth cultures from chicken meat samples (7). From the beginning of DNA extraction to final detection *of C. jejuni* and *C. coli*, the assay requires less than 50 and 90 min from a colony on selective media and human feces, respectively. For chicken meat samples, the assay requires approximately 24–48 h from the beginning of the enrichment culture to final detection. The sensitivities for *C. jejuni* and *C. coli* are found to be 7.9 and 3.8 CFU per LAMP reaction tube (Fig. 1a, b), respectively. The sensitivity of the LAMP assay is tenfold higher than that of the PCR assay. LAMP amplification can be judged by both turbidimeter analysis and visual assessment with the unaided eye (Fig. 1c).

2. Materials

2.1. Bacterial Culture

1. Selective media, namely Butzler, Skirrow, and modified mCCDA agars (Oxoid, Ltd., Hampshire, UK), as well as blood agar (Oxoid).
2. Enrichment broths, namely Preston and Bolton (Oxoid).

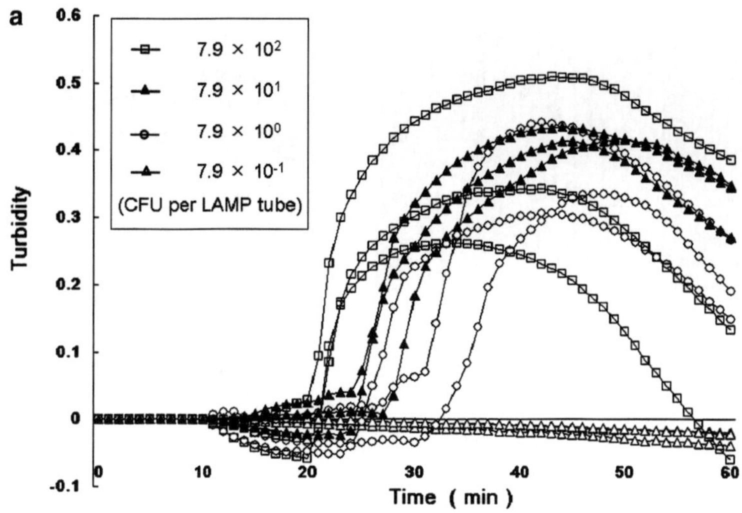

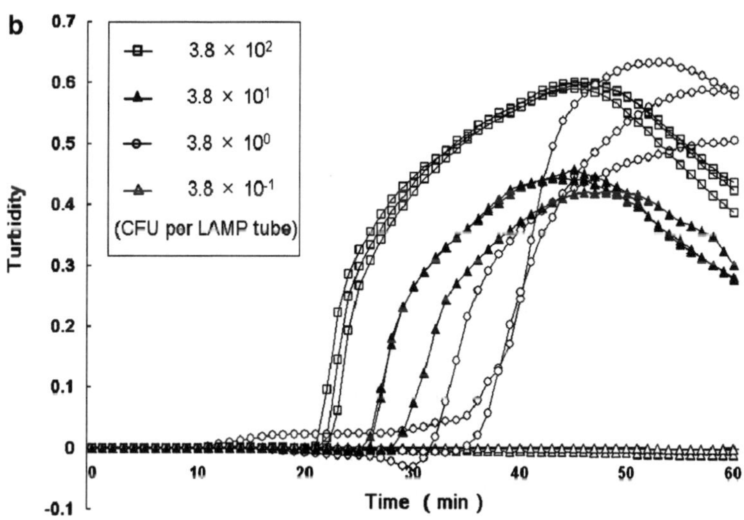

Fig. 1. Sensitivity test to detect *C. jejuni* LMG8841^T and *C. coli* JCM2529^T by using a real-time turbidimeter and visual assessment. The curves from *left* to *right* in (**a**) and (**b**) show increasing dilutions of bacteria (790 to 0.79 CFU per test tube in (**a**) and 380 to 0.38 CFU per test tube in (**b**)). (**a**) Detection of *C. jejuni*; (**b**) detection of *C. coli*; (**c**) visual detection of *C. jejuni* and *C. coli* by observation of turbidity. Tube 1, third sample containing 790 CFU for *C. jejuni*; tube 2, third sample containing 7.9 CFU for *C. jejuni*; tube 3, third sample containing 0.8 CFU for *C. jejuni*; tube 4, negative control; tube 5, third sample containing 380 CFU for *C. coli*; tube 6, third sample containing 3.8 CFU for *C. coli*; tube 7, third sample containing 0.4 CFU for *C. coli*; tube 8, negative control (reproduced from see ref. 7 with permission from American Society for Microbiology).

Fig. 1. (continued)

2.2. Human Fecal Samples

1. Fresh fecal samples obtained from clinical patients with suspected *C. jejuni/coli* infection.
2. Sterile plastic tubes (10 ml).
3. Phosphate-buffered saline (PBS).

2.3. Chicken Meat Samples

1. Fresh chicken meat samples with suspected contamination by *C. jejuni/coli*.
2. Enrichment broths, namely Preston or Bolton (Oxoid), which are used to make a tenfold (w/v) dilution of chicken meat sample, stored at 4°C until use (see Note 1).
3. Stomacher (ELMEX, Tokyo, Japan).
4. Sterilized plastic stomacher bag.

2.4. DNA Template

1. Heating-block (for 95–100°C use).
2. Centrifuge for microcentrifuge tubes (900–20,000 × *g*).
3. Centrifuge for eight connected tubes.
4. Sterilized 0.5-ml and 1.5-ml microcentrifuge tube.
5. NaOH (sodium hydroxide; 1 M): stored at room temperature. Diluted 40-fold in sterile distilled water and adjusted to 25 mM, followed by storage at −20°C until use.

Table 1
LAMP primers used

Target gene	Primer	Sequence (59′ to 3′)	Gene location (bp)
cj0414[a]	CJ-FIP	ACAGCACCGCCACCTATAGT – AGAAGCTTTTTTAAACTAGGGC (F1c-F2)	95–76 (F1c), 25–46 (F2)
	CJ-BIP	AGGCAGCAGAACTTACGCATT – GAGTTTGAAAAAACATTCTACCTCT (B1-B2c)	101–121 (B1), 181–157 (B2c)
	CJ-F3	GCAAGACAATATTATTGATCGC (F3)	3–24
	CJ-B3	CTTTCACAGGCTGCACTT (B3c)	218–201
	CJ-LF	CTAGCTGCTACTACAGAACCAC (LFc)	74–53
	CJ-LB	CATCAAGCTTCACAAGGAAA (LB)	124–143
CCO0367[b]	CC-FIP	AAGAGATAAACACCATGATCCCAG – TCATGAATGAGCTTACTTTAGC (F1c-F2)	730–707 (F1c), 665–686 (F2)
	CC-BIP	GCGGCAAAGACTTATGATAAAGC – TACCGCCATTCCTAAAACAAG (B1-B2c)	748–770 (B1), 810–790 (B2c)
	CC-F3	TGGGAGCGTTTTTGATCT (F3)	641–658
	CC-B3	AATCAAACTCACCGCCAT (B3c)	828–811
	CC-LF	CCACTACAGCAAAGGTGATG (LFc)	706–687
	CC-LB	CCACGATAGCCTTTATGGA (LB)	771–789

Primer FIP consisted of the F1 complementary sequence and the F2 sequence. Primer BIP consisted of the B1 sequence and the B2 complementary sequence. Primer B3 and LF consisted of the B3 and LF complementary sequences, respectively.

[a]Presumed to encode an oxidoreductase in the accession number AL111168 sequence, which was submitted to GenBank by Parkhill et al. (12)

[b]Presumed to be a *gufA* gene in the accession number AAFL01000003 sequence, which was submitted to GenBank by Fouts et al. (13)

6. Tris–HCl buffer (pH 7.5, 1 M), stored at room temperature or at −20°C until use.
7. Sterilized disposable 1-μl loop.
8. Vortex mixer.

2.5. Primer Design and Preparation of Primer Mixture

Sequences and locations of each primer are shown in Table 1 (see Note 2). The following primer volumes are required for each reaction (see Note 3).

0.4 μl of FIP (100 μmol l^{-1}).
0.4 μl of BIP (100 μmol l^{-1}).
0.2 μl of LF (100 μmol l^{-1}).
0.2 μl of LB (100 μmol l^{-1}).
0.05 μl of F3 (100 μmol l^{-1}).

0.05 μl of B3 (100 μmol l^{-1}).
(Total 1.3 μl).

The primer mixture was prepared in either 1.5 or 0.5-ml microcentrifuge tubes, followed by storage at −20°C until use.

2.6. LAMP Assay (See Note 4)

1. Loopamp DNA amplification kit (Eiken Chemical Co Ltd, Tokyo, Japan), which includes *Bst* DNA Polymerase, 2 × Reaction Mix, and distilled water.
2. Heating-block or water bath for end-point detection, or Loopamp turbidimeter (Eiken Chemical Co Ltd) for both end-point and real-time detection (to be used at 65°C).
3. Loopamp LAMP reaction tube (Eiken Chemical Co Ltd).

3. Methods

Although the LAMP assay is less affected by the inhibitory effects of clinical sample components than the PCR assay (8), removal of inhibitory factors and concentration of the small number of target bacterial cells in clinical and food samples are essential for sensitive and reliable detection by LAMP. Although commercially available kits such as the QIAamp DNA Mini Kit (Qiagen, Hilden, Germany) offer a sophisticated approach to DNA extraction from clinical samples, these kits are time-consuming, laborious, and costly. To remove larger debris from fecal samples, chicken samples, and components of enrichment broths which contain DNA amplification inhibitors, as well as to concentrate the small number of *Campylobacter* cells, a simple, rapid, and cost-effective DNA extraction protocol was described using a combination of NaOH-heat treatment and three-step centrifugation procedures (7).
During the DNA polymerization by *Bst* polymerase, a pyrophosphate ion is released from dNTP as a by-product. Production of large amounts of pyrophosphate ions lead to a reaction with magnesium ions from the LAMP reaction solution, which in turn produces magnesium pyrophosphate as simple turbidity. The increased turbidity in the reaction mixture caused by the production of insoluble white precipitate correlates with the amount of synthesized DNA (4). The white precipitate can be observed with the unaided eye, as well as measured for turbidity using a Loopamp turbidimeter.

3.1. DNA Extraction from Culture

1. Using a disposable loop (for 1 μl inoculation), inoculate a single loopful of fresh culture from selective media or blood agar in a 1.5-ml microcentrifuge tube containing 50 μl of NaOH (25 mM) (see Note 5).
2. Heat the cell mixture at 95–100°C for 10 min.

3. Add 4 μl of Tris–HCl buffer (1 M, pH 7.5) to neutralize the solution.
4. Centrifuge cell debris at 20,000 × *g* 4°C for 5 min.
5. Use 2 μl of the supernatant as template DNA for the LAMP assay.

3.2. DNA Extraction from Human Fecal Sample

1. In a sterilized plastic tube (10 ml), prepare human fecal homogenates by adjusting the concentration to 10% using PBS.
2. Mix the homogenate using a vortex mixer, transfer 1 ml of the homogenate into a 1.5-ml microcentrifuge tube, followed by centrifugation at 900 × *g* (3,300 rpm) for 1 min (see Note 6).
3. Transfer supernatant into a new 1.5-ml microcentrifuge tube.
4. Centrifuge for 5 min at 10,000 × *g* (10,800 rpm) and remove the supernatant (see Note 7).
5. Resuspend the pellets in 100 μl of NaOH (25 mM) (see Note 5).
6. Mix the mixture using a vortex mixer, then heat at 95–100°C for 10 min.
7. Add 8 μl of Tris–HCl buffer (1 M pH 7.5) to neutralize the solution.
8. Centrifuge cell debris at 20,000 × *g*, 4°C for 5 min.
9. Use 2 μl of the supernatant as template DNA for the LAMP assay (see Note 8).

3.3. DNA Extraction from Chicken Meat Sample

1. In a plastic stomacher bag or sterilized glass tube, weigh 2–25 g of chicken meat sample and adjust the concentration to 10% w/v using Preston or Bolton enrichment broth (see Note 1).
2. Treat the chicken meat/enrichment broth mixture from plastic stomacher bags by light hand massaging or by homogenization using a stomacher for 15–30 s. Mix the chicken meat/enrichment broth mixture from sterilized glass tubes two to three times using a vortex mixer for 2–3 s (see Note 9).
3. Incubate the chicken meat/enrichment broth mixture at 37–42°C for 20–48 h.
4. Transfer 1 ml of the cultivated enrichment broth into a 1.5-ml microcentrifuge tube and centrifuge at 900 × *g* for 1 min (see Note 6).
5. Transfer supernatant into a new 1.5-ml microcentrifuge tube.
6. Centrifuge for 5 min at 10,000 × *g* and remove the supernatant (see Note 7). Using a sterile cotton swab, remove any fatty components derived from chicken meat, if seen on the microcentrifuge tube wall.
7. Resuspend pellets in 50 μl of NaOH (25 mmol/l) (see Note 5).

8. Mix the mixture using a vortex mixer, then heat at 95–100°C for 10 min.
9. Add 4 μl of Tris–HCl buffer (1 M, pH 7.5) to neutralize the solution.
10. Centrifuge cell debris at 20,000 × *g*, 4°C for 5 min.
11. Use 2 μl of the supernatant as template DNA for the LAMP assay.

3.4. Preparation of LAMP Reagents

1. Thaw LAMP reagents to room temperature, then keep on ice.
2. Prepare the master mix in a 1.5- or 0.5-ml microcentrifuge tube kept on ice. The following component amounts are required for each reaction.

   12.5 μl of 2× reaction mix.

   1.3 μl of primer mixture for *C. jejuni* detection.

   1.3 μl of primer mixture for *C. coli* detection.

   1 μl of *Bst* DNA Polymerase.

   6.9 μl of sterilized distilled water.
3. After dispensing, gently tap the tubes approximately two to three times.
4. Centrifuge the tubes for 2–3 s. The resulting mixture can be used as the master mix for the LAMP reaction.

3.5. Operating Procedure (Perform all Reactions on Ice)

1. Dispense 23 μl of the master mix into Loopamp reaction tubes.
2. Add 2 μl of template DNA to the master mix.
3. Mix the mixture by pipetting or tapping (see Note 10).
4. Close the tube cap, then centrifuge the tubes for 2–3 s (see Note 10).

3.6. LAMP Reaction

1. Incubate the mixture at 65°C for 60 min using a Loopamp turbidimeter (end-point or real-time detection), heating-block, or water bath.
2. To terminate the reaction, inactivate the polymerase at 80°C for 5 min or 95°C for 2 min (see Note 11).
3. The reaction is considered positive when the turbidity reaches 0.1 within 60 min in a Loopamp turbidimeter, or when a white precipitate is visible to the unaided eye in the LAMP reaction tube using a Loopamp turbidimeter, heating-block, and water bath (see Note 12).
4. If necessary, differentiate *C. jejuni*- from *C. coli*-positive samples by conducting separate LAMP assays using the primer sets used to identify *C. jejuni* and *C. coli* and at the LAMP conditions described above. Add 1.3 μl of sterilized distilled water to substitute for removal of the nontarget primer.

4. Notes

1. For routine use in our laboratory, we store 1 L of enrichment broth in sterile glass bottles at 4°C. First, we place approximately 25-g of chicken sample in a stomacher bag, weigh the bag and contents and then aseptically pour in nine times the volume of broth to give a tenfold (w/v) dilution. For example if the weight of chicken is 25 g, then 225 ml of broth are added. When the available sample is limited, the use of smaller portions down to 2 g is acceptable.
2. LAMP primers, which target two sequences presumed to contain an oxidoreductase gene and a *gufA* gene in *C. jejuni* and *C. coli*, respectively, were designed with Primer ExplorerV4 software (http://primerexplorer.jp/elamp4.0.0/index.html; Fujitsu System Solutions, Tokyo, Japan) using sequence data submitted to GenBank (see Table 1).
3. Use the highly purified LAMP primers for rapid and stably reproducible gene amplification. HPLC-grade purification is recommended for the production of FIP and BIP primers, whereas HPLC- or sequence-grade purification is required for the production of the other LAMP primers (LF, LB, F3, and B3). Sequence-grade is important for effective LAMP amplification. HPLC grade is best, but it is expensive. LF, LB, F3, and B3 primers are enable to be substituted by lower-grade (sequence-grade) primers. FIP and BIP primers are, however, unable to be substituted by lower-grade primers. I do not know the detail of the reason. Presumably, FIP and BIP primers play the most important role among the six primers for LAMP amplification.
4. The mechanism of the LAMP assay is complex and difficult to describe using simple diagrams. The Eiken Genome website clearly explains the details of the LAMP assay principle using a number of diagrams and animations (http://loopamp.eiken.co.jp/e/index.html). Given the high sensitivity of the LAMP assay in synthesizing large amounts of DNA, the initial presence of the slightest amount of tainted product in the reaction may yield false-positive results. This type of contamination can be avoided by carrying out sample and reagent preparations on different clean benches. Amplification detection should be conducted using a turbidimeter, heating-block or water bath from which both reaction and detection can be accomplished while keeping the tube cap closed.
5. Using templates boiled using distilled water for DNA amplification often yields false-negative results (9), which can be avoided by NaOH-heat treatment, which potentially lyses and inactivates one or more unidentified inhibitory factors in bacterial cells during the LAMP reaction.

6. The first centrifugation for 1 min at 900 × *g* is carried out to remove larger debris in the samples and enrichment broths, which in turn decreases the influence of inhibitory factors of fecal, chicken and enrichment broth components.
7. The second centrifugation for 5 min at 10,000 × *g* is carried out to concentrate the small number of bacterial cells in the samples. A total of 1 ml of 10% fecal homogenate and enrichment broth cultures are then concentrated to a volume of roughly 100 and 50 μl by centrifugation, respectively.
8. Given that clinical patients excrete large amounts of *C. jejuni* and C. *coli* in their feces, enrichment procedures to isolate these bacteria in fecal samples are not usually necessary. If required, add the fecal sample in a sterilized glass tube containing a ninefold volume of enrichment broth, and follow the protocol for the detection of chicken meat sample, described in Subheading 3.3.
9. Food components, such as organic and phenolic compounds, glycogen, fats, and calcium ions, have been previously reported to inhibit DNA polymerase activity (10). Prolonged stomaching procedures appear to increase the release of inhibitory factors from chicken meat samples, stomaching procedures should therefore be performed for no longer than 30 s. Compared to the stomaching procedure, light hand massaging is preferable due to the low release of inhibitory factors (11). Although in our preliminary tests, liver samples have occasionally inhibited LAMP amplification possibly due to one or more unidentified inhibitory factors, inhibition may be avoided using a commercial DNA extraction kit.
10. Given that bubbles in the solution interfere with turbidity measurements and may cause false results, avoid creating bubbles when mixing the master mix and sample solutions. If bubbles are present, remove them by tapping the tubes approximately two to three times, and spin down the solution.
11. If the inactivation step at 80°C for 5 min or at 95°C for 2 min is omitted, visual assessment with the unaided eye should be performed quickly after the 60-min amplification. Failure to quickly assess the solution may lead to an observed nonspecific positive reaction in the negative sample due to residual heat in the LAMP reaction tube.
12. A turbidity decline occurs by the sedimentation of the white precipitation at the later phase of the LAMP reaction. At that time, a great amount of white precipitation is observed at the bottom of a reaction tube. When the turbidity reach 0.1 within 60 min, the sample is judged positive. And therefore, the result correlates with the naked eye measurement. To obtain clear image photographs, when we took the photographs of reaction

tubes, we lightly tapped the tubes two to three times by hand to make the sedimentation diffuse in the tube. And therefore, the sedimentation of the white precipitation is not observed in the photograph. Caps from used LAMP reaction tubes should not be opened. Contamination of amplified LAMP products from other samples may lead to false interpretation of test results, as well as contaminate the testing area. Keep the caps of used tubes completely closed and dispose of tubes by incineration or after double bagging with a sealable vinyl bag. To prevent dispersion of the amplified LAMP products, do not autoclave LAMP products before disposal.

References

1. Yamazaki-Matsune W, Taguchi M, Seto K, Kawahara R, Kawatsu K, Kumeda Y, Kitazato M, Nukina M, Misawa N, Tsukamoto T (2007) Development of a multiplex PCR assay for identification of *Campylobacter coli*, *Campylobacter fetus*, *Campylobacter hyointestinalis* subsp. *hyointestinalis*, *Campylobacter jejuni*, *Campylobacter lari* and *Campylobacter upsaliensis*. J Med Microbiol 56:1467–1473
2. Logan JM, Edwards KJ, Saunders NA, Stanley J (2001) Rapid identification of *Campylobacter* spp. by melting peak analysis of biprobes in real-time PCR. J Clin Microbiol 39:2227–2232
3. Notomi T, Okayama H, Masubuchi H, Yonekawa T, Watanabe K, Amino N, Hase T (2000) Loop-mediated isothermal amplification of DNA. Nucleic Acids Res 28:e63
4. Mori Y, Nagamine K, Tomita N, Notomi T (2001) Detection of loop-mediated isothermal amplification reaction by turbidity derived from magnesium pyrophosphate formation. Biochem Biophys Res Commun 289:150–154
5. Nagamine K, Hase T, Notomi T (2002) Accelerated reaction by loop-mediated isothermal amplification using loop primers. Mol Cell Probes 16:223–229
6. Yamazaki W, Taguchi M, Ishibashi M, Kitazato M, Nukina M, Misawa N, Inoue K (2008) Development and evaluation of a loop-mediated isothermal amplification assay for rapid and simple detection of *Campylobacter jejuni* and *Campylobacter coli*. J Med Microbiol 57:444–451
7. Yamazaki W, Taguchi M, Kawai T, Kawatsu K, Sakata J, Inoue K, Misawa N (2009) Comparison of loop-mediated isothermal amplification assay and conventional culture methods for detection of *Campylobacter jejuni* and *Campylobacter coli* in naturally contaminated chicken meat samples. Appl Environ Microbiol 75:1597–1603
8. Kaneko H, Kawana T, Fukushima E, Suzutani T (2007) Tolerance of loop-mediated isothermal amplification to a culture medium and biological substances. J Biochem Biophys Methods 70:499–501
9. Mohran ZS, Arthur RR, Oyofo BA, Peruski LF, Wasfy MO, Ismail TF, Murphy JR (1998) Differentiation of *Campylobacter* isolates on the basis of sensitivity to boiling in water as measured by PCR-detectable DNA. Appl Environ Microbiol 64:363–365
10. Wilson IG (1997) Inhibition and facilitation of nucleic acid amplification. Appl Environ Microbiol 63:3741–3751
11. Kanki M, Sakata J, Taguchi M, Kumeda Y, Ishibashi M, Kawai T, Kawatsu K, Yamasaki W, Inoue K, Miyahara M (2009) Effect of sample preparation and bacterial concentration on *Salmonella enterica* detection in poultry meat using culture methods and PCR assaying of preenrichment broths. Food Microbiol 26:1–3
12. Parkhill J, Wren BW, Mungall K, Ketley JM, Churcher C, Basham D, Chillingworth T, Davies RM, Feltwell T, Holroyd S, Jagels K, Karlyshev AV, Moule S, Pallen MJ, Penn CW, Quail MA, Rajandream MA, Rutherford KM, van Vliet AH, Whitehead S, Barrell BG (2000) The genome sequence of the food-borne pathogen *Campylobacter jejuni* reveals hypervariable sequences. Nature 403:665–668
13. Fouts DE, Mongodin EF, Mandrell RE, Miller WG, Rasko DA, Ravel J, Brinkac LM, DeBoy RT, Parker CT, Daugherty SC, Dodson RJ, Durkin AS, Madupu R, Sullivan SA, Shetty JU, Ayodeji MA, Shvartsbeyn A, Schatz MC, Badger JH, Fraser CM, Nelson KE (2005) Major structural differences and novel potential virulence mechanisms from the genomes of multiple *Campylobacter* species. PLoS Biol 3:e15

Chapter 19

PCR Detection of *Helicobacter pylori* in Clinical Samples

Emiko Rimbara, Masanori Sasatsu, and David Y. Graham

Abstract

Helicobacter pylori is an important pathogen whose primary niche is the human stomach. *H. pylori* is etiologically associated with gastric inflammation (gastritis), peptic ulcer disease, and gastric cancer. Both noninvasive (e.g., urea breath and stool antigen tests) and invasive (gastric biopsy for histology, culture, or PCR) tests are used for diagnosis. PCR detection of *H. pylori* has been reported using a variety of clinical samples including gastric biopsy, gastric juice, saliva, dental plaque, and stools as well as environmental samples. Whenever possibly, noninvasive tests are preferred over invasive tests. *H. pylori* are excreted in the stool. Culture from stool is variable whereas stool antigen testing is widely used. Stool consists of a complicated mixture of commensal bacteria and chemicals and often includes inhibitors of PCR. Nevertheless, simple extraction methods are available to efficiently extract DNA from human stools and nested-PCR targeting the 23S rRNA gene have proven to be highly sensitive for the detection of *H. pylori*. Detection of clarithromycin susceptibility/resistance is important clinically and the mutation of the 23S rRNA gene responsible for resistance can also be detected using stool. This described method can be modified for other clinical samples such as gastric juice or biopsy material.

Key words: *Helicobacter pylori*, Stool, Feces, Gastric juice, Nested-PCR, 23 S rRNA

1. Introduction

Helicobacter pylori is a Gram-negative spiral bacterial pathogen whose primary niche is the human stomach. *H. pylori* is etiologically related to gastritis and peptic ulcer disease and gastric cancer. Many diagnostic methods have been developed including the urea breath tests, rapid urease tests, and measurement of anti-*H. pylori* antibody from serum and urine, special histologic staining and immunostaining, and stool antigen testing. Many PCR methods targeting putative *H. pylori* specific genes have also been reported (1–4). For clinical studies, a positive diagnosis requires a positive culture or two positive indirect tests (i.e., histology and rapid

Mark Wilks (ed.), *PCR Detection of Microbial Pathogens: Second Edition*, Methods in Molecular Biology, vol. 943,
DOI 10.1007/978-1-60327-353-4_19,

urease test). PCR targeting *H. pylori*-specific gene can be one of the choices. Most commonly DNA extracted from gastric biopsies or gastric juice is used for the detection of *H. pylori* by PCR. A number of different commercial kits (e.g., from Qiagen or Promega) are available as well as well-described laboratory methods using proteinase K, CTAB, and phenol–chloroform extraction can be used for extraction of DNA from gastric biopsy material.

PCR-targeting *H. pylori* specific gene from gastric biopsy is a relatively simple procedure. In contrast, stools are complex mixtures that often contain PCR inhibitors and other commensal bacteria such that false-negative and false-positive results are a larger problem. Besides detection of the presence of the infection, PCR may also provide information regarding presence of *H. pylori* virulence factors as well as information regarding antimicrobial susceptibility/resistance via genotyping PCR and DNA sequencing of PCR products. Because drug-resistance among *H. pylori* has been increasing worldwide and is a major cause of treatment failure, there is an increasing need for pretreatment antimicrobial susceptibility testing (i.e., tailored therapy). PCR-based methods targeting genes related to antibacterial resistance are suitable to both diagnose the presence of the infection and provide information regarding antimicrobial resistance (i.e., detection of clarithromycin resistance caused by the mutation of the position 2,142 and 2,143 of the 23 S rRNA gene) (5, 6).

Here, we describe nested-PCR methods targeting the 23 S rRNA using DNA extracted from stool for detection of *H. pylori* from feces as well as for detection of mutations of the 23 S rRNA gene. The method is based on a published method developed at the Tokyo University of Pharmacy and Life Sciences, Tokyo (7, 8).

2. Materials

2.1. DNA Extraction from Stool (see Note 1*)*

1. Beads to disrupt cells. Add beads to a 2 mL screw cap tube by adding 250 mg of silica powder (63–210 μm), 32.5 mg of ceramic beads (1–2 mm), and 75 mg of glass beads. The tube is then placed into the cell crusher (see Note 2).
2. Cell crusher: FastPrep FP120 instrument (Qbiogene, Carlsbad, CA, USA) (see Note 2).
3. 0.1 M phosphate buffer (pH 7.0): Mix 168 mL of 0.2 M Na_2HPO_4 and 32 mL of KH_2PO_4, and the adjust the volume to 200 mL.
4. 7.5 M guanidine solution containing 5% sarcosine (see Note 3): Dissolve 35.82 g guanidine hydrochloride (MW 95.53) in 20 mL sterile ultra pure water, adjust to 50 mL, and add 2.5 g *N*-lauroyl-sarcosine.

5. 3.5 M sodium acetate (pH 5.2): Dissolve 47.6 g NaOAc·$3H_2O$ (MW 136.08) in 80 mL ultra pure water, adjust to pH 5.2 by glacial acetic acid, and adjust volume to 100 mL.
6. Wizard SV Gel and PCR Clean-up systems (Promega, Madison, WI, USA).

2.2. PCR and Electrophoresis

1. DNA polymerase (see Note 4): Ex Taq polymerase (Takara Biomedicals, Kyoto, Japan) and GoTaq® Green Master Mix is used for first PCR second PCR, respectively.
2. dNTP Mixture (Takara Biomedicals, Kyoto, Japan).
3. 10× Ex Taq Buffer 20 mM Mg^{2+} plus (Takara Biomedicals, Kyoto, Japan).
4. Bovine Serum Albumin solution (Takara Biomedicals, Kyoto, Japan).
5. Primer: The primers Hp23S 1835F (5′-GGTCTCAGCAAAG AGTCCCT-3′) and Hp23S 2327R (5′-CCCACCAAGCATT GTCCT-3′) are used for first PCR, and the primers Hp23S 1942F AGGATGCGTCAGTCGCAAGAT and Hp23S 2308R CCTGTGGATAACACAGGCCAGT are used for second PCR (see Note 5).
6. Electrophoresis buffer (50× TAE buffer): Dissolve 242 g Tris-HCl in 500 mL H_2O, add 100 mL 0.5 M Na_2 EDTA (pH 8.0) and 57.1 mL glacial acetic acid, and adjust volume to 1 L with H_2O.
7. DNA ladder marker (range between 100 and 1,000 bp).

3. Methods

To detect *H. pylori* DNA from stool by PCR, the extraction step is a critical step to eliminate PCR inhibitors in stool. Moreover, because the number of *H. pylori* cells is less than for many commercial bacteria present in stool, an efficient method to extract DNA from stool is required. There are a number of commercial kits available for DNA extraction from stool, however, we found the method using beads to crush the cells to be efficient and simple and the method requires a smaller amount of sample (50–100 mg) compared to commercial kits which often require more than 250 mg.

To avoid false-positive results a negative control using water as a template should be performed. Duplicate or triplicate reactions are also suggested for increased reliability. To prevent the false negatives, see Notes 7 and 8.

3.1. DNA Extraction from Feces (Fig. 1)

1. 50–100 mg of feces is added to the tube containing the beads (see Note 6).
2. Add 980 mL phosphate buffer and 180 mL of the 7.5 M guanidine solution containing 5% sarcosine to tube and mix gently using yellow pipette tip to break up any lumps of stool.
3. Place the tube in the FastPrep FP120 instrument and run it 20 s at 5.5 speed.
4. Centrifuge the tube at 16,000 × *g* for 30 s.
5. Transfer the supernatant (Approximately 750 mL) to new 1.5 mL eppendorf tube and add 250 mL of 3.5 M sodium acetate (pH 5.2). Mix the tube gently.
6. Centrifuge 5 min at 16,000 × *g* and transfer 700 mL of the supernatant to new eppendorf 1.5 mL tubes.
7. Add 700 mL of Membrane binding solution (Promega) and mix by pipetting.
8. Add 700 mL of mixture to SV Minicolumn in a Collection Tube (Promega) and incubate at room temperature for 1 min.

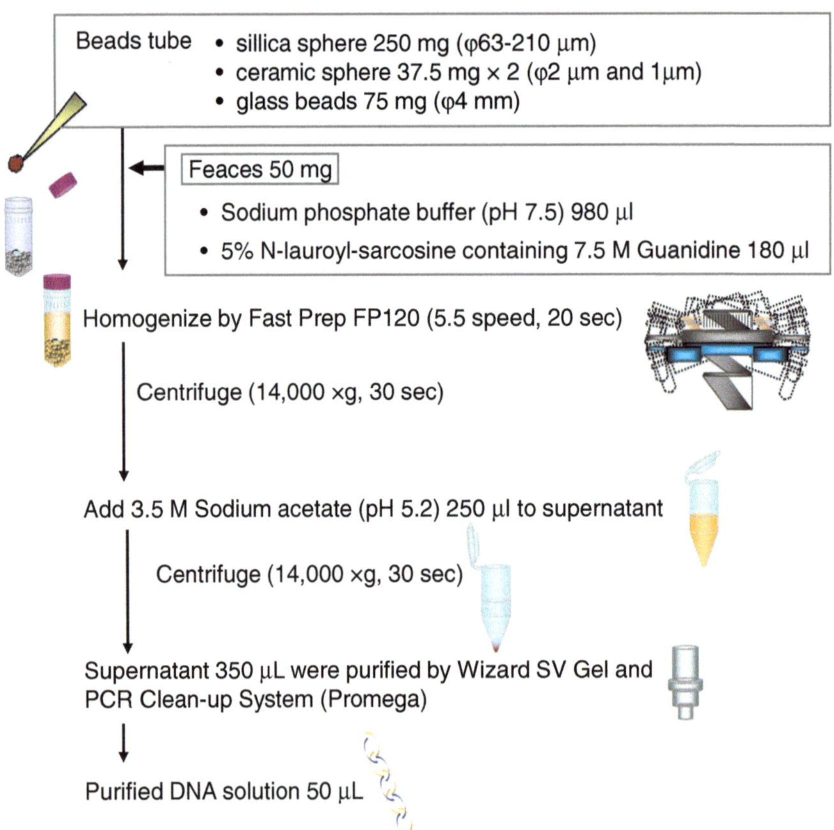

Fig. 1. Method for the isolation of DNA from feces.

9. Centrifuge at 16,000 × *g* for 1 min, discard flow through and reinsert Minicolumn into the Collection Tube.
10. Add the remaining 700 mL of the mixture (step 7) to the SV Minicolumn in the Collection Tube and repeat steps 8 and 9.
11. Add 700 μL of Membrane Wash Solution (ethanol added as per the instructions) to the column and centrifuge at 16,000 × *g* for 1 min. Discard flowthrough and reinsert the Minicolumn into the Collection Tube.
12. Repeat Step 11 with 500 μL of the Membrane Wash Solution. Centrifuge at 16,000 × *g* for 5 min.
13. Empty the Collection Tube and centrifuge the column assembly for 1 min with the microcentrifuge lid open to allow evaporation of any residual ethanol.
14. Transfer the column to a 1.5 mL eppendorf tube and add 58 mL of nuclease-free water to the column. Incubate at room temperature for 2 min and then centrifuge at 16,000 × *g* for 1 min.
15. Store the DNA solution at −20°C until using for PCR.

3.2. Nested-PCR Targeting the 23 S rRNA Gene of H. Pylori (see Note 7)

1. Prepare the first PCR reaction mixture on ice by adding the following quantities of reagents: 50 mL in the order listed: 33.75 mL of sterile ultra pure water, 5.0 mL of 10× Ex Taq Buffer, 4.0 mL of dNTP Mixture (each 2.5 mM), 5.0 mL of Bovine Serum Albumin solution, 1.0 mL of primer Hp23S 1835 F (10 pmol/mL), 1.0 mL of primer Hp23S 2327R (10 pmol/mL), and 0.25 mL TaKaRa Ex Taq (5 U/mL).
2. Aliquot 45 mL of the PCR reaction mixture to each PCR tube and add 5 mL of the DNA template. Mix well (see Note 8).
3. Using thermal cycler, perform PCR: initial denaturation at 95°C for 2 min, followed by 5 cycles: 94°C for 30 s, 57°C for 30 s, and 72°C for 30 s; then 30 cycles: 94°C for 15 s, 57°C for 15 s, and 72°C for 20 s.
4. Prepare the second PCR mixture on ice by adding: 50 μL in the order listed: 20 μL of sterile ultra pure water, 25.0 μL of Master Mix, 1.0 μL of primer Hp23S 1942 F (10 pmol/μL), and 1.0 μL of primer Hp23S 2308R (10 pmol/μL).
5. Aliquot 47 μL of the PCR reaction mixture to each PCR tube and add 3 μL of the first PCR product. Mix well.
6. Perform PCR as follows; initial denaturation at 95°C for 2 min, followed by 25 cycles: 94°C for 10 s and 63°C for 20 s.
7. Final PCR products are confirmed using electrophoresis in 2.5% agarose gels stained with ethidium bromide. The PCR product should be 367 bp (see Note 9). An example of the results is shown in Fig. 2.

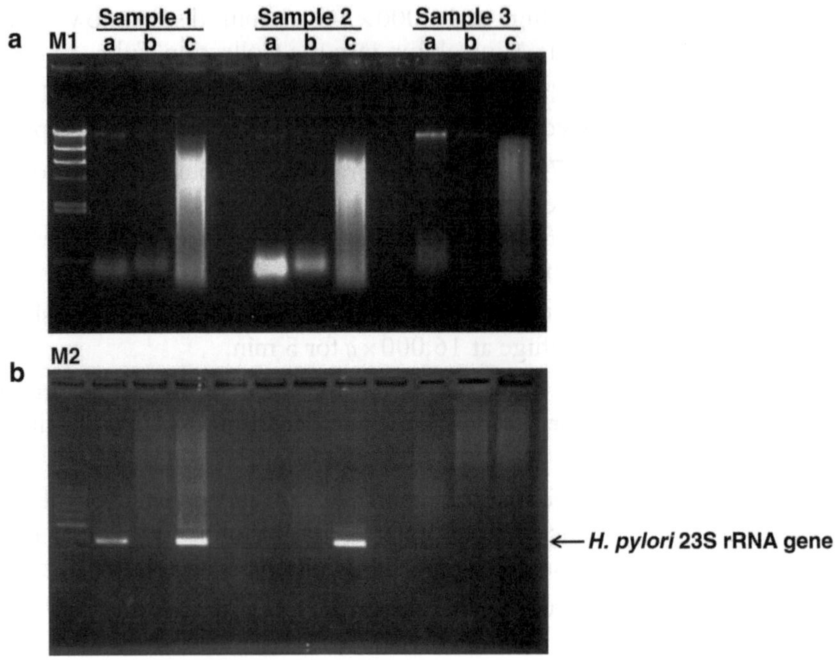

Fig. 2. Comparison of methods for the isolation of DNA from feces in *H. pylori*-positive (1 and 2) and -negative (3) samples. (**a**) Electrophoresis of DNA isolated from feces (1% agarose gel). Lane M1, *Hin*dIII-digested λ DNA; lanes (**a**) and (**b**) DNA isolated from 200 to 50 mg feces, respectively, using a QIAamp DNA Stool Mini kit; lanes (**c**) DNA isolated from 50 mg feces using the method described in this chapter. (**b**) Agarose gel (2.5%) electrophoresis of the nested PCR H. pylori 23 S rRNA gene amplification product (*arrow*) from DNA isolated as described in (**a**). Lane M2, 100 bp DNA ladder reproduced from *see* ref. (7) with permission from The Society for General Microbiology.

4. Notes

1. For DNA extraction from other clinical samples, such as biopsy, saliva, and plaque, it may be best if the combination of beads is changed. In case of gastric juice: (1) Centrifuge 300 μL of gastric juice and resuspend in 100 μL of sterile ultra pure water. (2) Boil for 3 min. (3) Mix with 100 μl of Membrane binding solution and proceed from step 8 in Subheading 3.1 (9).
2. Many kinds of cell crushers are available commercially and could be used. For example, we confirmed that the Mini-BeadBeater-1 (BioSpec Products, Inc., OK) was effective using the following combination of beads: 1,500 mg of Zirconia/silica (0.1 mm), 300 mg of Zirconia/silica (1.0 mm), and 300 mg of Zirconia/silica (2.3 mm).
3. If a precipitate appears, it can be dissolved by heating at 50°C or using a microwave.
4. Since the first PCR is more important procedure than the second PCR, a DNA polymerase with a high amplification efficiency is strongly recommended for the first PCR. ExTaq polymerase can also be used for the second PCR.

5. The primer targeting other *H. pylori* genes, such as 16S rRNA, *ureA*, or *vacA*, can be applied instead of the 23S rRNA (1–3). Primers for nested-PCR should be designed based on a region in which there is a high homology between *H. pylori* strains and a low homology with other bacteria. As new *Helicobacter sp.* are being identified rapidly, there is always the possibility of false-positive results such that PCR results along should be interpreted with caution when used for treatment decisions. Results of studies based on PCR determination of the presence of *H. pylori* in environmental samples, or clinical samples from sites other than stomach or stool, especially from patients without proven *H. pylori* infections, should be viewed with considerable caution (10). At a minimum one must show that the primer pairs used are specific (e.g., not positive in patients without *H. pylori* infections). Possibly, the use of multiple *H. pylori*-specific virulence genes (e.g., CagA or VacA) (1, 11, 12) would allow more accurate PCR detection in environmental samples.
6. Stool stored in freezer in long duration (e.g., more than 5 year) may produce false-negative results. Avoid repeating the freeze and thaw of stool. This amount is assumed when the stool is solid. When the stool contains much fiber or much water increase the amount of sample (more than 100 mg) to avoid false-negative results. When the loose stool is used, centrifuge approximately 300–600 μL of the sample at 12,000 × *g* for 5 min and use the precipitate.
7. To avoid the false-positive results, negative control (water) should perform on each PCR step. Positive controls should also be performed; however the DNA template for positive control should be added after finishing all samples and negative controls to avoid contamination.

   False-negative results are usually related to DNA extraction or the first PCR. If a false negative is suspected, the first PCR step should be performed again and changing the amount of DNA template (see Note 8). Otherwise, restart from the DNA extraction step. False-negative results caused by PCR inhibitors in the DNA samples can be excluded by adding a positive control (with serial dilution) to the DNA samples and comparing the PCR results with the simple PCR using same positive controls.

   In the case of DNA extracted from biopsies or gastric juice, the first PCR can be deleted and one can perform a single PCR using Ex Taq polymerase and primers Hp23S 1942F and Hp23S 2308R as follows; initial denaturation at 95°C for 2 min, followed by 5 cycles: 94°C for 30 s, 60°C for 30 s, and 72°C for 30 s; then 30 cycles: 94°C for 15 s, 60°C for 15 s, and 72°C for 20 s.

8. The amount of DNA template can be increased up to 10 μL or decreased to 1 μL depending on the DNA concentration and purity of the sample. Measurement of A260/280 might be useful but does not always correspond to the results probably because the proportion of the DNA attributable to *H. pylori* differs between samples. To avoid false-negative results we recommend testing 3, 5, and 10 μL of DNA template.
9. To ensure that the amplified product is *H. pylori*, DNA sequencing of PCR product is strongly recommended. This also can provide information regarding clarithromycin-resistance caused by the mutation at position 2,142 and 2,143 of the 23S rRNA gene. The mutation is usually the transition from adenine to guanine, but the transversion from adenine to cytosine has also been reported. The primers, Hp23S 1942F and Hp23S 2308R, can be used for DNA sequencing.

Acknowledgments

Dr. Rimbara and Dr. Sasatsu are supported by a grant from the High-Tech Research Centre Project for Private Universities provided by the Ministry of Education, Culture, Sports, Science and Technology and by the Matching Fund Subsidy for Private Schools of Japan. This work was supported in part by the Office of Research and Development Medical Research Service Department of Veterans Affairs. Dr. Graham is supported in part by Public Health Service grant DK56338 which funds the Texas Medical Center Digestive Diseases Center and R01 CA116845. We especially thank Drs. Norihisa Noguchi, Prof. Takashi Kawai, and Prof. Shinichi Takahashi for their expertise, advice, and encouragement.

References

1. Park CY, Kwak M, Gutierrez O, Graham DY, Yamaoka Y (2003) Comparison of genotyping H*elicobacter pylori* directly from biopsy specimens and genotyping from bacterial cultures. J Clin Microbiol 41:3336–3338
2. Smith SI, Oyedeji KS, Arigbabu AO, Cantet F, Megraud F, Ojo OO, Uwaifo AO, Otegbayo JA, Ola SO, Coker AO (2004) Comparison of three PCR methods for detection of *Helicobacter pylori* DNA and detection of *cagA* gene in gastric biopsy specimens. World J Gastroenterol 10:1958–1960
3. Mapstone NP, Lynch DA, Lewis FA, Axon AT, Tompkins DS, Dixon MF, Quirke P (1993) Identification of *Helicobacter pylori* DNA in the mouths and stomachs of patients with gastritis using PCR. J Clin Pathol 46:540–543
4. Hammar M, Tyszkiewicz T, Wadstrom T, O'Toole PW (1992) Rapid detection of *Helicobacter pylori* in gastric biopsy material by polymerase chain reaction. J Clin Microbiol 30:54–58
5. Taylor DE, Ge Z, Purych D, Lo T, Hiratsuka K (1997) Cloning and sequence analysis of two copies of a 23 S rRNA gene from *Helicobacter pylori* and association of clarithromycin resistance with 23 S rRNA mutations. Antimicrob Agents Chemother 41:2621–2628
6. Versalovic J, Shortridge D, Kibler K, Griffy MV, Beyer J, Flamm RK, Tanaka SK, Graham

DY, Go MF (1996) Mutations in 23 S rRNA are associated with clarithromycin resistance in *Helicobacter pylori*. Antimicrob Agents Chemother 40:477–480

7. Noguchi N, Rimbara E, Kato A, Tanaka A, Tokunaga K, Kawai T, Takahashi S, Sasatsu M (2007) Detection of mixed clarithromycin-resistant and -susceptible *Helicobacter pylori* using nested PCR and direct sequencing of DNA extracted from faeces. J Med Microbiol 56:1174–1180
8. Rimbara E, Noguchi N, Yamaguchi T, Narui K, Kawai T, Sasatsu M (2005) Development of a highly sensitive method for detection of clarithromycin-resistant *Helicobacter pylori* from human feces. Curr Microbiol 51:1–5
9. Rimbara E, Tamura R, Tanuma M, Noguchi N, Kawai T, Sasatsu M (2009) Evaluation of clarithromycin resistance in *Helicobacter pylori* obtained from culture isolates, gastric juice, and feces. Helicobacter 14:156–157
10. Sugimoto M, Wu JY, Abudayyeh S, Hoffman J, Brahem H, Al-Khatib K, Yamaoka Y, Graham DY (2009) Unreliability of results of PCR detection of *Helicobacter pylori* in clinical or environmental samples. J Clin Microbiol 47:738–742
11. Tiwari SK, Khan AA, Ahmed KS, Ali SM, Ahmed I, Habeeb A, Kauser F, Hussain MA, Ahmed N, Habibullah CM (2005) Polymerase chain reaction based analysis of the cytotoxin associated gene pathogenicity island of *Helicobacter pylori* from saliva: an approach for rapid molecular genotyping in relation to disease status. J Gastroenterol Hepatol 20:1560–1566
12. Sinha SK, Martin B, Gold BD, Song Q, Sargent M, Bernstein CN (2004) The incidence of *Helicobacter pylori* acquisition in children of a Canadian First Nations community and the potential for parent-to-child transmission. Helicobacter 9:59–68

Chapter 20

Rapid Detection of the *Escherichia coli* Genospecies in Water by Conventional and Real-Time PCR

Andrée F. Maheux, Luc Bissonnette, and Michel G. Bergeron

Abstract

The presence of *Escherichia coli* has long been established as the most reliable microbiological indication of fecal contamination in water. Current recommended culture-based methods for assessing water quality by the detection of *E. coli* are lengthy and lack ubiquity (ability to detect most if not all strains of a target microorganism). We describe rapid and sensitive conventional and real-time PCR assays specific to *E. coli* and Shigella, based on the nucleotide sequence of the highly conserved elongation factor Tu (*tuf*) gene enabling the detection of all members of the genospecies.

Key words: *Escherichia coli* genospecies, Shigella, Water, Molecular microbiology, PCR, Real-time PCR

1. Introduction

An important component of public health promotion applied to the microbiological safety of water is the determination of the presence of fecal contamination indicators. This strategy is used to provide indications of the risk of gastrointestinal disease caused by waterborne viral, bacterial, and parasitic pathogens released in the environment from human and animal feces (1–4). Since 1986, *Escherichia coli* has been established as the most reliable microbiological indicator of human fecal contamination to assess the risk of waterborne gastrointestinal diseases (3, 5).

Despite more than 100 years of technological refinement, testing water for the presence of fecal contamination indicators such as *E. coli*, fecal, and total coliforms is still performed by lengthy culture-based methods (6) that also lack ubiquity (7–10).

Mark Wilks (ed.), *PCR Detection of Microbial Pathogens: Second Edition*, Methods in Molecular Biology, vol. 943,
DOI 10.1007/978-1-60327-353-4_20, © Springer Science+Business Media, LLC 2013

1.1. Nucleic Acid-Based Methods for Water Molecular Microbiology

Nucleic acid-based amplification procedures are the methods of choice for the sensitive and specific detection of genomic targets, but their high level of performance is counterbalanced by the ease of cross-contamination of samples, especially by highly abundant amplicons. Rapidity, specificity, ubiquity, and sensitivity (or detection limit) are the most valuable analytical characteristics of molecular diagnostics procedures (11, 12). Specificity is the capacity of an assay to detect only a specific genetic group. On the other hand, ubiquity is the ability to detect all or most isolates of a target (group of) species from diverse genetic lineages, isolation sites, and strains of diverse geographical, or temporal origin. In the context of microbiological testing of water or for environmental monitoring of fecal contamination indicators or human pathogens, nucleic acid-based detection methods have the potential to provide a faster and more specific result than recommended culture-based methods (13, 14).

1.2. The E. coli Genospecies

The *E. coli* genospecies encompasses a wide variety of commensal and specialized pathogenic lineages, such as enterotoxigenic, enteropathogenic, enterohemorrhagic, enteroaggregative, and enteroinvasive (from now on designated diarrheagenic) *E. coli* and Shigella strains (15–21). In clinical microbiology, *E. coli* and Shigella are classified in different genera on the basis of biochemical and pathogenicity tests, despite the fact that DNA-relatedness studies have demonstrated that *E. coli* and Shigella should be considered as a single genetic species (15, 20, 22–24).

As we recently demonstrated (10), culture-based methods for detecting *E. coli* lack ubiquity for the detection of several diarrheagenic *E. coli* strains (e.g., *E. coli* O157:H7), which are apparently refractory to phenotypic detection by culture-based methods due to loss or non-expression of the β-glucuronidase (*uidA*) gene. Furthermore, it is well known that identification methods based on estimating the activity of a single enzyme lack robustness; this may lead to misinterpretations since enzymatic activity can be transient and highly regulated by environmental factors.

Recently, we also demonstrated that several PCR assays developed for the specific detection of *E. coli* lack specificity and ubiquity against *E. coli* and Shigella strains, but that the primers we developed for amplifying the *tuf* gene enable the detection of all 79 *E. coli* and 11 Shigella strains of our microbial testing panel which was composed of type strains as well as clinical and environmental isolates of various geographical origin (25). For the design of molecular diagnostics assays, Paradis et al. (20) have demonstrated that *tuf* provides better discrimination between different *Enterobacteriaceae* genospecies than the 16S rRNA gene. This chapter describes how to perform a conventional PCR assay for detecting *tuf* sequences specific to the *E. coli* genospecies, the *E. coli*-PCR assay, as well as its real-time (rtPCR) version which can also be quantitative (qPCR), when coupled to calibration procedures.

2. Materials

2.1. Solutions (see Note 1)

2.1.1. 1 M Tris pH 8.0 is made by dissolving 12.1 g of Tris base into 80 mL of reverse-osmosis (RO) purified water (Millipore) with a resistivity of 18 MΩ cm min at 25°C, the pH is adjusted to 8.0 with 6 N HCl and the volume made up to 100 mL. The stock solution is dispensed as 10 mL aliquots and stored at −20°C.

2.1.2. 500 mM EDTA pH 8.0 is made by dissolving 18.6 g EDTA·$2H_2O$ disodium salt and 2 g NaOH in 80 mL of RO water. The pH is adjusted to 8.0 with 6 N HCl or 6 N NaOH and the volume made up to 100 mL. The stock solution is dispensed as 10 mL aliquots and stored at −20°C.

2.1.3. 5× TE buffer (50 mM Tris–HCl, 5 mM EDTA, pH 8.0) is made by diluting 2.5 mL of 1 M Tris pH 8.0 and 5 mL of 500 mM EDTA pH 8.0 with 400 mL of RO water. pH is verified and adjusted with 1 N or 6 N HCl, and volume made up to 500 mL. The solution is dispensed as 1 mL aliquots and stored at −20°C.

2.1.4. 1× TE buffer (10 mM Tris–HCl, 1 mM EDTA, pH 8.0) is made by diluting 10 mL of 1 M Tris pH 8.0 and 2 mL of 500 mM EDTA pH 8.0 with 800 mL of RO water. pH is verified and adjusted with 1 N or 6 N HCl, and volume made up to 1,000 mL. The solution is dispensed as 1 mL aliquots and stored at −20°C.

2.2. PCR Reagents (see Note 1)

2.2.1. PCR-grade water is RO water that is filtered through 0.1 μm disposable units, autoclaved for 30 min at 121°C, dispensed as 1 mL aliquots, and stored at −20°C.

2.2.2. Taq 10× PCR buffer: 100 mM Tris–HCl pH 9.0, 500 mM KCl, 15 mM $MgCl_2$, and 1% Triton X-100. Dispense as single-use (100 μL) aliquots and store at −20°C.

2.2.3. 10× pre-mix PCR buffer: 100 mM Tris–HCl pH 9.1, 500 mM KCl, 25 mM $MgCl_2$, 1% Triton X-100, 33 mg/mL BSA, 2 mM dNTPs. Dispense as single-use (100 μL) aliquots and store at −20°C.

2.2.4. Amplification primers synthesized by Integrated DNA Technologies (Coralville, IA, USA) were resuspended in RO water to a final concentration of 10 μM, dispensed in single-use (40 μL) aliquots, and stored at −20°C. The nucleotide sequence and annealing temperature of the primers are shown in Table 1.

2.2.5. Dual-labeled (TaqMan) detection probes synthesized and HPLC-purified by Biosearch Technologies, Inc. (Novato,

Table 1
Composition and properties of PCR primers and dual-labeled (TaqMan) detection probes

Genetic target (reference)	Primer/probe name	Primer sequence (5′→3′)	T_{ann} [a] (°C)	Amplicon size (bp)
E. coli tuf primers (25)	TEcol553	TGGGAAGCGAAAATCCTG	58	212
	TEcol754	CAGTACAGGTAGACTTCTG		
E. coli TaqMan probe (this work)	TEco573-T1-B1	TET-AACTGGCTGGCTTCC TGG-BHQ-1	62	
B. atrophaeus subsp. *globigii atpD* primers (26)	ABgl158	5′-CACTTCATTTAGGCGAC GATACT-3′	60	211
	ABgl345a	5′-TTGTCTGTGAATCGGAT CTTTCTC-3′		
B. atrophaeus subsp. *globigii atpD* TaqMan probe (26)	ABgl-T1-A1	FAM-CGTCCCAATGTTACAT TACCAACCGGCACT-(BHQ-1)-GAAATAGG[b]	71	

TET: tetrachlorofluorescein, a chemical relative of fluorescein
BHQ-1 is a Black Hole Quencher™ dye
[a] T_{ann}: annealing temperature
[b] The BHQ-1 moiety of this dual-labeled probe is covalently linked to the T nucleotide at position 30

CA, USA) were resuspended in RO water to a final concentration of 10 μM, dispensed as single-use (10 μL) aliquots, and stored protected from light at –20°C. The nucleotide sequence of the probes are shown in Table 1.

2.2.6. 100 mM Ultrapure dNTPs set was purchased from GE Healthcare Life Sciences (Baie d'Urfé, Québec, Canada). Working solutions (4 mM) are prepared by diluting in PCR-grade water, dispensed as single-use (25 μL) aliquots, and stored at –20°C.

2.2.7. Bovine serum albumin (BSA fraction V) is made up to 66 mg/mL in autoclaved RO water, dispensed as single-use (50 μL) aliquots, and stored at –20°C.

2.2.8. 2.5 mg/mL 8-methoxypsoralen (methoxsalen) is dissolved in dimethyl sulfoxide (DMSO), dispensed as single-use (25 μL) aliquots, and stored protected from light at –20°C.

2.2.9. *Taq* DNA polymerase–TaqStart antibody complex mixture: native *Taq* DNA polymerase (3.33 U/μL; Promega, Madison, WI, USA) is mixed with TaqStart antibody (1.1 μg/μL; Clontech, Mountain View, CA, USA) in a ratio of 2:1. Dispense the *Taq* DNA polymerase–TaqStart antibody complex mixture as single-use (10 μL) aliquots and store at –20°C (see Note 2).

2.3. Positive Amplification Control

2.3.1. Isolation and purification of high molecular weight genomic DNA from *E. coli* ATCC 11775. Bacterial genomic DNA is isolated from mid-log-phase cultures by using a BioSprint 15 DNA blood kit (QIAGEN, Mississauga, Ontario, Canada) automated with a KingFisher mL instrument (Thermo Fisher Scientific, Walthman, MA, USA), or using the manual GNOME DNA kit (Qbiogene, Inc., Carlsbad, CA, USA) according to the manufacturer's instructions.

2.3.2. Quantification of high molecular weight genomic DNA preparation is done by measuring $A_{260\ \mathrm{nm}}$ whilst purity is estimated by determining the $(A_{260\ \mathrm{nm}}-A_{320\ \mathrm{nm}})/(A_{280\ \mathrm{nm}}-$ $-A_{320\ \mathrm{nm}})$ ratio (see Note 3).

2.3.3. Stock solution of genomic DNA is diluted with 5× TE to a final concentration of 10 ng/μL, dispensed as 10 μL aliquots, and stored at 4°C for up to 10 years.

2.3.4. Quantification of working DNA solution is done by measuring $A_{260\ \mathrm{nm}}$ while purity is estimated by determining the $(A_{260\ \mathrm{nm}}-A_{320\ \mathrm{nm}})/(A_{280\ \mathrm{nm}}-A_{320\ \mathrm{nm}})$ ratio (see Note 3).

2.3.5. Physical integrity of high molecular weight DNA should be evaluated by electrophoresis in a 0.8% agarose gel (see Note 4).

2.4. Internal Process Control (see Note 5)

2.4.1. Spores of *Bacillus atrophaeus* subsp. *globigii*, prepared according to Picard et al. (26). Briefly, the sporulation of *B. atrophaeus* subsp. *globigii* strain CCRI-9827 was induced by growth on a sporulation agar medium. Subsequently, spores were purified on a sodium bromide gradient, resuspended in PCR-grade water, enumerated with a Petroff-Hausser counting chamber, calibrated to 1,000 spores/μL, dispensed as single-use (25 μL) aliquots, and stored at −20°C. Typical bacterial spore preparations have a purity of at least 99.9%.

2.5. Agarose Gel Electrophoresis

2.5.1. 10× TBE gel electrophoresis buffer: 1 M Tris, boric acid 0.9 M, EDTA 0.01 M, pH 8.4.

2.5.2. 1× TBE gel electrophoresis buffer is made by diluting 10× TBE 10 times with RO water.

2.5.3. Low electroendosmosis molecular biology grade agarose was purchased from Wisent, Inc. (St-Bruno, Québec, Canada)

2.5.4. Ethidium bromide (EtBr) solution is prepared by dissolving EtBr to 5 mg/mL in autoclaved RO water. Dispense as 500 μL aliquots and store protected from light at room temperature.

2.5.5. Gel electrophoresis loading buffer is made by dissolving 20 g sucrose, 0.279 g $EDTA_2 \cdot 2H_2O$ disodium salt, and 50 mg bromophenol blue in 8 mL of autoclaved RO water. The volume is adjusted to 10 mL, and the loading buffer is dispensed as 1 mL aliquots that are stored at 4°C.

2.5.6. Molecular weight marker: 100 μL of a 100-base pair DNA ladder (GE Healthcare Life Sciences) is diluted with 1,800 μL of 1× TE and 700 μL of gel electrophoresis loading buffer, dispensed as single-use (15 μL) aliquots, and stored at –20°C.

3. Methods

This molecular amplification assay for the rapid detection of the *E. coli* genospecies is described in two versions: a conventional and a real-time assay.

Conventional PCR is the commonest version of the amplification method. It is the most economical, in terms of equipment and supplies, but only produces qualitative results. In addition, it requires post-amplification analysis steps that increase assay duration, hands-on time, as well as the risk of human error and cross-contamination of reagents, materials, and surfaces (see Note 6).

The recurrent production of amplicons from target microorganisms may lead to an increased rate of false-positive results, due to cross-contamination of laboratory personnel, reagents, and surfaces. One of the most useful solutions is to perform a closed-tube assay, the operational format of most real-time PCR instruments which enable nucleic acid amplification and concomitant specific detection of amplicons through the use of fluorescent intercalating agents or specialized labeled probes (27, 28). In theory, performing routine closed-tube rtPCR tests requires more expensive equipment and reagents but, on the other hand, may necessitate a less complex laboratory infrastructure, thereby reducing implementation costs.

3.1. The Conventional E. coli PCR Assay

3.1.1. The preparation of the PCR master mix is done in the PCR reagent preparation room (see Note 6). The unitary PCR master mix (19.0 μL) is made by combining PCR-grade water (12.8 μL), 10× Taq PCR buffer (2.0 μL), amplification primers (0.8 μL of each primer), dNTPs (1.0 μL), BSA (1 μL), 8-methoxypsoralen (0.48 μL), and *Taq* DNA polymerase–TaqStart antibody complex (0.15 μL). To make a larger master mix, multiply the volumes by $n+4$, n being the number of samples (and replicates) and four corresponding to the number of controls (three negative and one positive controls). The final reagent concentrations are 50 mM KCl, 10 mM Tris–HCl (pH 9.0), 0.1% Triton X-100, 1.5 mM $MgCl_2$, 0.4 μM of each amplification primer, 200 μM of each dNTP, 3.3 mg/mL BSA, 0.06 μg/μL 8-methoxypsoralen, and 0.025 U of *Taq* DNA polymerase–TaqStart antibody complex.

3.1.2. The PCR master mix is decontaminated in a UV cross-linker (see Note 7) according to Picard et al. (29).

3.1.3. The PCR master mix is dispensed as 19 μL aliquots in 0.2 mL thin-wall PCR microcentrifuge tubes (see Note 8).

3.1.4. PCR negative control 1 (see Note 9): 1 μL of PCR-grade water is added to a reaction tube before leaving the PCR reagent preparation room.

3.1.5. The assembly of PCR reactions is done in the sample preparation room (see Note 6).

3.1.6. PCR negative control 2 (see Note 9): 1 μL of PCR-grade water is added to a reaction tube when entering the sample preparation room, before the introduction of samples to be tested in reaction tubes.

3.1.7. 1 μL of sample (genomic DNA or nucleic acid extract) is added to its corresponding PCR reaction tube.

3.1.8. Preparation of positive amplification control (see Note 11): tenfold dilutions of the stock solution of *E. coli* genomic DNA (10 ng/μL; see Subheading 2.3) are made in 1× TE buffer to a final concentration of 0.1 ng/μL. A solution of 10^4 genome copies per μL (GC/μL) is made by diluting 50.9 μL of 0.1 ng/μL DNA solution with 49.1 μL of 1× TE. The solution is made fresh and discarded after use. Tenfold dilutions of the 10^4 GC/μL standard solution are made with 1× TE to generate solutions of 1,000 and 100 GC/μL. The 100 GC/μL solution is the positive amplification control. The solution is made fresh, stored at 4°C until needed, and discarded after use. 1 μL of 100 GC/μL solution is added to its corresponding PCR reaction tube.

3.1.9. PCR negative control 3 (see Note 9): 1 μL of PCR-grade water is added to a reaction tube before leaving the sample preparation room.

3.1.10. Mix the contents of all PCR reaction tubes by vortexing 2–3 s and briefly spin them in a microcentrifuge.

3.1.11. Thermal cycling is performed in a molecular amplification room using a PTC-200 DNA Engine thermocycler (MJ Research, now Bio-Rad Laboratories, Hercules, CA, USA) with the following conditions: 3 min at 95°C, then 40 cycles of 1 s at 95°C, 30 s at 58°C, and 30 s at 72°C, with a final extension step of 5 min at 72°C (see Note 10).

3.1.12. In a separate room for post-PCR procedures, all PCR reaction tubes are quickly spun in a microcentrifuge. 2 μL of gel electrophoresis loading buffer is added, all tubes are mixed by vortexing and recentrifuged briefly.

3.1.13. A 11 × 14 cm 2% agarose gel is made by adding 2 g of low electroendosmosis agarose to 100 mL of TBE 1× and dissolving by heating in a microwave oven. Before pouring the gel, 5 μL of 5 mg/mL EtBr is added and well mixed.

3.1.14. Gel loading: 10 μL of PCR samples are loaded on the gel, in parallel with 13 μL of 100-bp ladder solution prepared as in Item 2.5.6.

3.1.15. Electrophoresis is performed at 170 V for 30 min. A typical pattern of results is shown in Fig. 1a.

3.2. The Conventional PCR Assay for the Internal Process Control (B. atrophaeus subsp. globigii Spores)

3.2.1. The preparation of the PCR master mix is done in the PCR reagent preparation room (see Note 6). The unitary PCR master mix (19.0 μL) is made by combining PCR-grade water (11.25 μL), 10× pre-mix PCR buffer (6.0 μL), amplification primers (0.8 μL of each primer), and *Taq* DNA polymerase–TaqStart antibody complex (0.15 μL). To make a larger master mix, multiply the volumes by $n+4$, n being the number of samples (and replicates) and four corresponding to the number of controls (three negative and one positive controls). The final reagent concentrations are 50 mM KCl, 10 mM Tris–HCl (pH 9.1), 0.1% Triton X-100, 2.5 mM $MgCl_2$, 0.4 μM of each amplification primer, 200 μM of each dNTP, 3.3 mg/mL BSA, and 0.025 U of *Taq* DNA polymerase–TaqStart antibody complex.

3.2.2. The PCR master mix is dispensed as 19 μL aliquots in 0.2 mL thin-wall PCR microcentrifuge tubes (see Note 8).

3.2.3. PCR negative control 1 (see Note 9): 1 μL of PCR-grade water is added to a reaction tube before leaving the PCR reagent preparation room.

3.2.4. The assembly of PCR reactions is done in the sample preparation room (see Note 6).

3.2.5. PCR negative control 2 (see Note 9): 1 μL of PCR-grade water is added to a reaction tube when entering the sample preparation room, before the introduction of samples to be tested in reaction tubes.

3.2.6. 1 μL of sample (genomic DNA or nucleic acid extract) is added to its corresponding PCR reaction tube (see Note 6).

3.2.7. Preparation of positive amplification control. Spores of *B. atrophaeus* subsp. *globigii* (1,000 spores/μL; see Item 2.4.1) are diluted 1:10 with PCR-grade water to a final

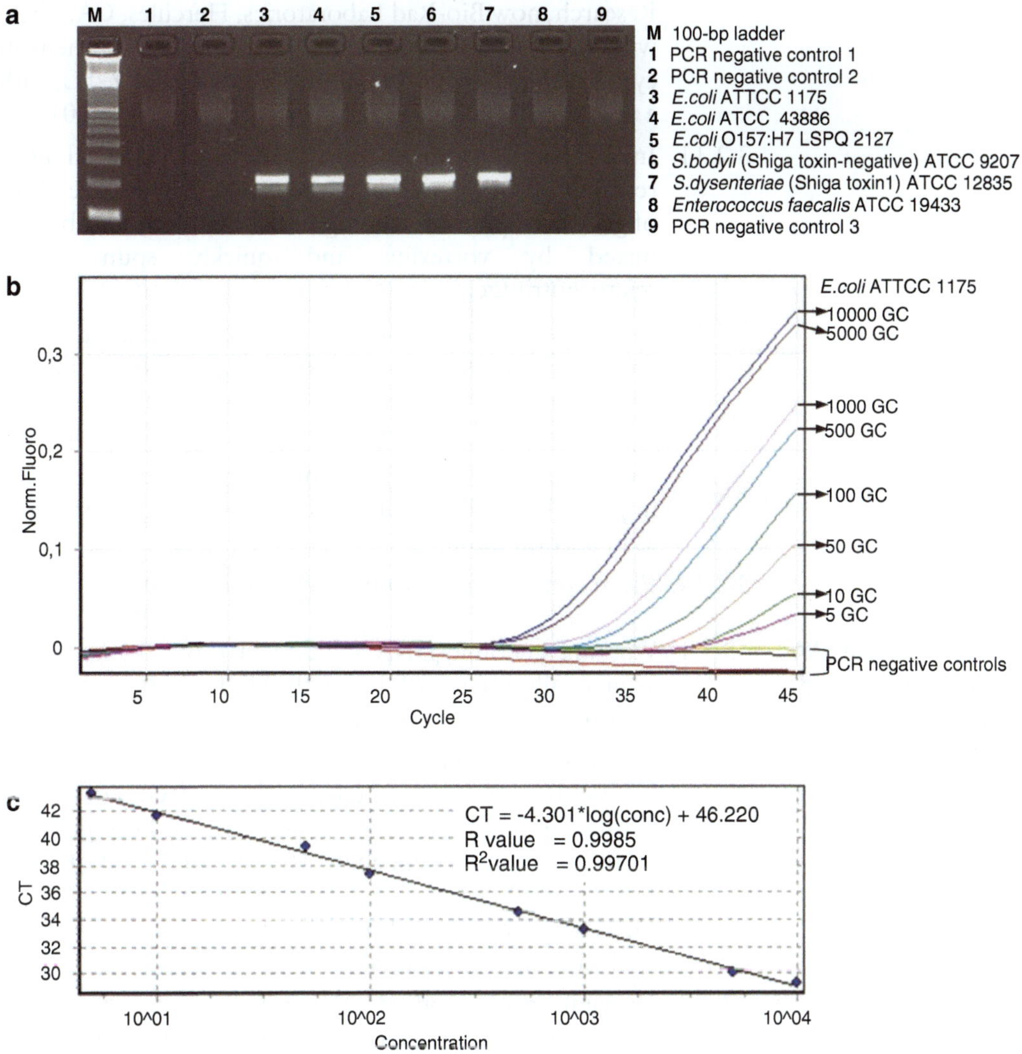

Fig. 1. PCR and rtPCR assays for the molecular detection of the *E. coli* genospecies. (**a**) Ubiquity of detection of the conventional PCR assay for *E. coli*, Shiga toxin-producing *E. coli* and Shigella, and Shiga toxin-negative Shigella; (**b**) Thermal amplification profile of calibrated genomic DNA solutions of *E. coli* ATCC 11775 with the *E. coli* genospecies rtPCR assay; (**c**) Standard curve derived from the thermal amplifications profiles of (**b**).

concentration of 100 spores/μL. The solution is made fresh and discarded after use. 1 μL of the 100 spores/μL dilution is added to its corresponding PCR reaction tube.

3.2.8. PCR negative control 3 (see Note 9): 1 μL of PCR-grade water is added to a reaction tube before leaving the sample preparation room.

3.2.9. Mix the contents of all PCR reaction tubes by vortexing 2–3 s and briefly spin in a microcentrifuge.

3.2.10. Thermal cycling is performed in a molecular amplification room using a PTC-200 DNA Engine thermocycler (MJ

Research, now Bio-Rad Laboratories, Hercules, CA, USA) with the following conditions: 3 min at 95°C, then 40 cycles of 1 s at 95°C, 30 s at 60°C, and 30 s at 72°C, with a final extension step of 5 min at 72°C (see Note 10).

3.2.11. In a separate room for post-PCR procedures, all PCR reaction tubes are quickly spun in a microcentrifuge. 2 μL of gel electrophoresis loading buffer is added, all tubes are mixed by vortexing and quickly spun in a microcentrifuge.

3.2.12. A 11 × 14 cm 2% agarose gel is made by adding 2 g of low electroendosmosis agarose to 100 mL of TBE 1× and dissolving by heating in a microwave oven. Before pouring the gel, 5 μL of 5 mg/mL EtBr is added and well mixed.

3.2.13. Gel loading: 10 μL of PCR samples are loaded on the gel, in parallel with 13 μL of 100-bp ladder solution prepared as in Section 2.5.6.

3.2.14. Electrophoresis is performed at 170 V for 30 min.

3.3. Preparation of DNA Solutions for Standard Curve Determination

3.3.1. Working genomic DNA solution (10 ng/μL) is diluted sequentially (factor of 10) with TE 1× buffer to a final concentration of 0.1 ng/μL.

3.3.2. Conversion of genomic DNA solution from concentration to number of genome copies per μL (GC/μL). The 0.1 ng/μL DNA solution is diluted with TE 1× to generate a 100 μL solution containing 10^4 GC/μL (see Note 11). For *E. coli* ATCC 11775, dilute 50.9 μL of 0.1 ng/μL DNA solution with 49.1 μL of TE 1×. The solution is made fresh and discarded after use.

3.3.3. Preparation of standard DNA solutions (see Note 12). The solution at 10^4 genome copies/μL is diluted sequentially with TE 1× to generate solutions at 5,000, 1,000, 500, 100, 50, 10, and 5 GC/μL. The solutions are made freshly for each assay and discarded after use.

3.4. The Real-Time E. coli-rtPCR Assay

3.4.1. The preparation of the rtPCR master mix is done in the PCR reagent preparation room (see Note 6). The unitary rtPCR master mix (24.0 μL) is made by combining PCR-grade water (15.7 μL), 10× Taq PCR buffer (2.5 μL), amplification primers (1.0 μL of each primer), dual-labeled (TaqMan) detection probe (0.5 μL), dNTPs (1.25 μL), BSA (1.25 μL), 8-methoxypsoralen (0.6 μL), and *Taq* DNA polymerase–TaqStart antibody complex (0.19 μL). To make a larger master mix, multiply the volumes by $n+13$, n being the number of samples (and replicates) and 13 corresponding to the number of controls (three negative and one positive controls) and the number of calibrated DNA solutions to build the standard curve. The final reagent concentrations are 10 mM Tris–HCl (pH 9.0), 50 mM KCl, 0.1% Triton X-100, 1.5 mM $MgCl_2$, 0.4 μM of each amplification primer, 0.2 μM of dual-labeled detection probe, 200 μM of each dNTP, 3.3 mg/mL BSA, 0.06 μg/μL 8-methoxypsoralen, and 0.025 U of *Taq* DNA polymerase–TaqStart antibody complex.

3.4.2. Molecular decontamination of rtPCR master mix (see Note 7) was achieved in a Spectrolinker™ model XL-1000 UV crosslinker (Spectronics Corporation), according to Picard et al. (29).

3.4.3. The rtPCR master mix is dispensed as 24 μL aliquots in 0.2 mL thin-wall PCR microcentrifuge tubes (see Note 8).

3.4.4. rtPCR negative control 1 (see Note 9): 1 μL of PCR-grade water is added to a reaction tube before leaving the PCR reagent preparation room.

3.4.5. The assembly of rtPCR reactions is done in the sample preparation room (see Note 6).

3.4.6. rtPCR negative control 2 (see Note 9): 1 μL of PCR-grade water is added to a reaction tube when entering the sample preparation room.

3.4.7. 1 μL of sample is added to its corresponding PCR reaction tube.

3.4.8. Preparation of positive amplification control (see Note 11). *E. coli* genomic DNA solution (10 ng/μL; see Subheading 2.3) is diluted sequentially (factor of 10) with 1× TE buffer to a final concentration of 0.1 ng/μL. A solution of 10^4 genome copies per μL (GC/μL) is made by diluting 50.9 μL of 0.1 ng/μL DNA solution with 49.1 μL of 1× TE. The solution is made fresh and discarded after use. The 10^4 GC/μL is diluted sequentially (factor of 10) with 1× TE to generate solutions at 1,000 and 100 GC/μL. The 100 GC/μL solution is the positive amplification control. The solution is made fresh, stored at 4°C until needed, and discarded after use. 1 μL of 100 GC/μL solution is added to its corresponding PCR reaction tube.

3.4.9. rtPCR negative control 3 (see Note 9): 1 μL of PCR-grade water is added to a reaction tube before leaving the sample preparation room.

3.4.10. All rtPCR reaction tubes are mixed by vortexing briefly, followed by a short spin in a microcentrifuge.

3.4.11. Thermal cycling is performed in a molecular amplification room using a Rotor-Gene 3000 (Corbett Life Sciences, now QIAGEN, Inc., Mississauga, Ontario, Canada) with the following conditions: 3 min at 95°C, then 45 cycles of 2 s at 95°C, 10 s at 58°C, and 20 s at 72°C (see Note 10).

3.4.12. Figure 1b shows the real-time amplification profile obtained by performing the *E. coli* rtPCR assay with the calibrated DNA solutions prepared in Subheading 3.3 and Fig. 1c is the standard curve derived by the Rotor-Gene software.

3.5. The Real-Time rtPCR Assay for the Internal Process Control (B. atrophaeus subsp. globigii Spores)

3.5.1. The preparation of the rtPCR master mix is done in the PCR reagent preparation room (see Note 6). The unitary rtPCR master mix (24.0 μL) is made by combining PCR-grade water (13.8 μL), 10× pre-mix PCR buffer (7.5 μL), amplification primers (1.0 μL of each primer), dual-labeled detection probe (0.5 μL), and *Taq* DNA polymerase–TaqStart antibody complex (0.19 μL). To make a larger master mix, multiply the volumes by $n+4$, n being the number of samples (and replicates) and four corresponding to the number of controls (three negative and one positive controls). The final reagent concentrations are 10 mM Tris–HCl (pH 9.1), 50 mM KCl, 0.1% Triton X-100, 2.5 mM $MgCl_2$, 0.4 μM of each amplification primer, 0.2 μM of dual-labeled detection probe, 200 μM of each dNTP, 3.3 mg/mL BSA, and 0.025 U of *Taq* DNA polymerase–TaqStart antibody complex.

3.5.2. The rtPCR master mix is dispensed as 24 μL aliquots in 0.2 mL thin-wall PCR microcentrifuge tubes (see Note 8).

3.5.3. rtPCR negative control 1 (see Note 9): 1 μL of PCR-grade water is added to a reaction tube before leaving the PCR reagent preparation room.

3.5.4. The assembly of rtPCR reactions is done in the sample preparation room (see Note 6).

3.5.5. rtPCR negative control 2 (see Note 9): 1 μL of PCR-grade water is added to a reaction tube when entering the sample preparation room.

3.5.6. 1 μL of sample is added to its corresponding rtPCR reaction tube.

3.5.7. Preparation of positive amplification control. Spores of *B. atrophaeus* subsp. *globigii* (1,000 spores/μL; see Item 2.4.1) are diluted 1:10 with PCR-grade water to a final concentration of 100 spores/μL. The solution is made fresh and discarded after use. 1 μL of the 100 spores/μL solution is added to its corresponding PCR reaction tube.

3.5.8. rtPCR negative control 3 (see Note 9): 1 μL of PCR-grade water is added to a reaction tube before leaving the sample preparation room.

3.5.9. All rtPCR reaction tubes are mixed by vortexing briefly, followed by a short spin in a microcentrifuge.

3.5.10. Thermal cycling is performed in a molecular amplification room using a Rotor-Gene 3000 (Corbett Life Sciences, now QIAGEN, Inc., Mississauga, Ontario, Canada) with the following conditions: 3 min at 95°C, then 45 cycles of 15 s at 95°C and 60 s at 60°C (see Note 10).

3.6. Characteristics of the E. coli Genospecies (rt)PCR Assays

In this chapter, we describe two configurations of rapid, specific, and ubiquitous PCR tests that could be used to detect the *E. coli* genospecies in water. Figure 1a demonstrates that the *E. coli* PCR test enables the detection of members of the *E. coli* genospecies including diarrheagenic *E. coli* and Shigella strains. The analytical sensitivity of an assay is defined by the minimum number of copies in a sample that can be measured accurately with an assay (30). Because of Poisson distribution statistics, it is unlikely that nucleic acid genetic targets from a single *E. coli* cell can be efficiently and robustly recovered and transferred to an amplification tube upon sample concentration (30). According to the parameters of the PCR assay, the theoretical limit of detection should be of three genome copies per PCR reaction.

When compared to other molecular amplification methods, real-time amplification procedures are increasingly recommended as tools for the rapid detection of nucleic acids, especially in laboratory conditions that might favor cross-contamination by amplification products. Figure 1b, c provides an example of the assay that was converted to an rtPCR format for the detection of the *E. coli* genospecies. In many instances, where it might be desirable to convert such an assay to a quantitative format (qrtPCR), it must be understood that assay development and implementation requires strict adherence to the guidelines suggesting the (1) incorporation of appropriate negative, positive, and internal (process) controls, (2) building standard and calibration curves, and (3) validation and evaluation of assay performance (30).

4. Notes

1. For the development of highly sensitive PCR assays, many problems may originate from reagents and solutions that are not adequately prepared or stored or are contaminated with microbial nucleic acids. In many instances, contamination problems may be solved by the inclusion of filtration steps and preparation of single-use aliquots, even for commercially available buffers. To prevent cross-contamination and to avoid repeated freezing and thawing of reagent and stock solutions,

these should be aliquoted into smaller working volumes and stored in the reagent preparation room.

2. The use of a *Taq* (or other bacterial or archaeal thermostable) DNA polymerase modified or engineered for "hot start" reduces the production of nonspecific products by preventing the elongation from primers annealed nonspecifically to the template at lower T_{ann}.
3. Tolerance for genomic DNA purity. The $A_{260\ nm}/A_{280\ nm}$ ratio must be between 1.6 and 2.2, while the $A_{320\ nm}$ must be less than 0.200. For a DNA concentration less than 50 ng/μL, a tolerance of ±20% is accepted. For DNA solution with a concentration of 55–99.9 ng/μL and higher than 99.9 ng/μL, tolerances of ±15% and ±10% are accepted, respectively.
4. Visual inspection of the gel enables the physical integrity of the genomic DNA preparation to be estimated. If degraded on the basis of a smear on the gel, the preparation is rejected.
5. For our water molecular microbiology application, water samples are spiked with *B. atrophaeus* subsp. *globigii* spores prior to membrane filtration. These spores serve as a general control of the efficiency of recuperation of microbial particles and the extraction of nucleic acids. This internal process control could be used with any other waterborne microbial target for PCR or rtPCR.
6. The ability of PCR to produce many copies of target DNA creates the possibility of contamination by previously amplified products leading to false-positive results. Strict precautions must be observed during material preparation and the assembly of amplification reactions. First, a laboratory performing PCR analyses on environmental samples should be divided into physically separate rooms with different mechanically controlled air pressures to limit cross-contamination of reagents (31). A molecular amplification suite should consists of dedicated rooms: (1) *PCR reagent preparation room* for storage of PCR reagents and assembly of PCR master mixes (positive air pressure), (2) *sample preparation room* for storage of nucleic acids solutions and samples, and assembly of PCR reaction mixtures (negative air pressure), (3) *molecular amplification room* (neutral air pressure), and (4) *a room for post-PCR procedures* such as gel electrophoresis (negative or neutral air pressure); rooms three and four may be combined in a single laboratory space. An alternate molecular amplification laboratory design is suggested by Mitchell and colleagues (32). Second, manipulations of PCR reagents and nucleic acid samples should be done under a biological (laminar flow) hood, beforehand decontaminated by exposure to UV light (30–60 min). Third, to minimize cross-contamination of samples

during pipetting, RNase-, DNase-, and pyrogen-free polypropylene microcentrifuge tubes and specialized tips designed for PCR analysis must be used. Fourth, micropipettors should be calibrated on a regular basis and also treated to be RNase-, DNase-, and pyrogen-free.

7. Decontamination of reagents. It is well known that many basic and molecular biology reagents can be contaminated by microbial DNA that can increase the frequency of false-positive amplification and lead to the disqualification of a test result. *E. coli* can be a major contributor of contaminating DNA since it is a host of choice for the production of a large number of recombinant enzymes and biological products (33). Even in a controlled laboratory environment, the development of a molecular assay (overproduction of amplicons) or the developers themselves (shedding microorganisms or microbial DNA) may be linked to an unacceptable rate of false-positive amplification due to cross-contamination. Therefore, confinement procedures must be applied and reagents should be tested for potential molecular contamination and decontaminated when necessary (32). We have observed that no empirical rule can be applied for determining the parameters of optimal decontamination by UV exposure since this is dependent on (1) the level of nucleic acid contamination of reagents and (2) the quality, stability, and/or intensity of the UV light source. We recommend regular reagent validation and control experiments to determine the efficiency of a molecular decontamination procedure.
8. The utilization of thin-walled tubes is highly recommended as they provide the best heat transfer performance, thereby improving specificity and reproducibility.
9. Negative controls. When assembling PCR reactions, three types of negative controls are recommended. PCR negative control 1 serves to verify that the PCR master mix was not contaminated during its preparation and dispensing, while PCR negative controls 2 and 3 serve to determine if no cross-contamination with samples or control DNA has happened during the assembly of PCR reactions.
10. Thermal cycling profiles may have to be optimized if a different brand or model of (real-time) thermal cycler is used. Ramping speeds are adjusted such that an amplification run is generally completed in approximately 1 h.
11. Conversion of genomic DNA solution from concentration to number of genome copies per μL (GC/μL). The 0.1 ng/μL DNA solution is diluted with TE 1× to generate a 100 μL solution at 10^4 GC/μL The general formula [genome size (bp) $\times 1.0959 \times 10^{-5}$] directly provides the volume of 0.1 ng/μL

DNA solution that contains 10^6 GC. For *E. coli* ATCC 11775, a genome size value of 4.64×10^6 bp is used.

12. In accordance with the recommendations of Shepley and Wolk (34), calibrated DNA solutions used for building a q(rt)PCR standard curve must be made and tested for each amplification run.

Acknowledgment

We thank Dr Maurice Boissinot for critical reading of the manuscript.

References

1. United States Environmental Protection Agency. (1993) Preventing waterborne disease - AfocusonEPA'sresearch,EPA/640/K-93/001, Office of Research and Development, United States Environmental Protection Agency, Washington, DC, 20 pages.
2. Sinton LW, Finlay RK, Hannah DJ (1998) Distinguishing human from faecal contamination in water: a review. N Z J Mar Freshwater Res 32:323–348
3. Edberg SC, Rice EW, Karlin RJ, Allen MJ (2000) *Escherichia coli*: the best biological water indicator for public health protection. J Appl Microbiol 88 suppl:S106–S116
4. Doyle MP, Erickson MC (2006) Closing the door on the fecal coliform assay. Microbe 1:162–163
5. United States Environmental Protection Agency. (1986) Ambient water quality criteria for bacteria - 1986, EPA440/5-84-002, Office of Water Regulations and Standards Criteria and Standards Division, United States Environmental Protection Agency, Washington DC, 18 pages.
6. Rompré A, Servais P, Baudart J, de-Roubin M-R, Laurent P (2002) Detection and enumeration of coliforms in drinking water: current methods and emerging approaches. J Microbiol Meth 49:31–54
7. Doyle MP, Schoeni JL (1984) Survival and growth characteristics of *Escherichia coli* associated with hemorrhagic colitis. Appl Environ Microbiol 48:855–856
8. Chang GW, Brill J, Lum R (1989) Proportion of β-D-glucuronidase-negative *Escherichia coli* in human fecal samples. Appl Environ Microbiol 55:335–339
9. Feng P, Lum R, Chang GW (1991) Identification of *uidA* gene sequences in β-D-glucuronidase-negative *Escherichia coli*. Appl Environ Microbiol 57:320–323
10. Maheux AF, Huppé V, Boissinot M, Picard FJ, Bissonnette L, Bernier J-LT, Bergeron MG (2008) Analytical limits of four β-glucuronidase and β-galactosidase-based commercial culture methods used to detect *Escherichia coli* and total coliforms. J Microbiol Meth 75:506–514
11. Boissinot M, Bergeron MG (2002) Toward rapid real-time molecular diagnostic to guide smart use of antimicrobials. Curr Opin Microbiol 5:478–482
12. Picard FJ, Bergeron MG (2002) Rapid molecular theranostics in infectious diseases, (Drug Discov) Today 7:1092–1100
13. Gilbride KA, Lee D-Y, Beaudette LA (2006) Molecular techniques in wastewater: understanding microbial communities, and real-time process control. J Microbiol Meth 66:1–20
14. Santo Domingo JW, Bambic DG, Edge TA, Wuertz S (2007) Quo vadis source tracking? Towards a strategic framework for environmental monitoring of fecal pollution. Water Res 41:3539–3552
15. Lawrence JG, Ochman H, Hartl DL (1991) Molecular and evolutionary relationships among enteric bacteria. J Gen Microbiol 137:1911–1921
16. Nataro JP, Kaper JB (1998) Diarrheagenic *Escherichia coli*. Clin Microbiol Rev 11: 142–201
17. Pupo GM, Lan R, Reeves PR (2000) Multiple independent origins of Shigella clones of *Escherichia coli* and convergent evolution of many of their characteristics. Proc Natl Acad Sci USA 97:10567–10572
18. Fukushima M, Kakinuma K, Kawaguchi R (2002) Phylogenetic analysis of *Salmonella*,

Shigella, and *Escherichia coli* strains on the basis of *gyrB* gene sequence. J Clin Microbiol 40:2779–2785

19. Lan R, Reeves PR (2002) *Escherichia coli* in disguise: molecular origins of Shigella. Microbes Infect 4:1125–1132
20. Paradis S, Boissinot M, Paquette N, Bélanger SD, Martel EA, Boudreau DK, Picard FJ, Ouellette M, Roy PH, Bergeron MG (2005) Phylogeny of the *Enterobacteriaceae* based on genes encoding elongation factor Tu and F-ATPase β-subunit. Int J Syst Evol Microbiol 55:2013–2025
21. Touchon M, Hoede C, Tenaillon O, Barbe V, Baeriswyl S, Bidet P, Bingen E, Bonacorsi S, Bouchier C, Bouvet O, Calteau A, Chiapello H, Clermont O, Cruveiller S, Danchin A, Diard M, Dossat C, Karoui ME, Frapy E, Garry L, Ghigo JM, Gilles AM, Johnson J, Le Bouguénec C, Lescat M, Mangenot S, Martinez-Jéhanne V, Matic I, Nassif X, Oztas S, Petit MA, Pichon C, Rouy Z, Ruf CS, Schneider D, Tourret J, Vacherie B, Vallenet D, Médigue C, Rocha EP, Denamur E (2009) Organised genome dynamics in the *Escherichia coli* species results in highly diverse adaptive paths. PLoS Genet 5:e1000344
22. Brenner DJ, Fanning GR, Skerman FJ, Falkow S (1972) Polynucleotide sequence divergence among strains of *Escherichia coli* and closely related organisms. J Bacteriol 109:953–965
23. Brenner DJ, Fanning GR, Miklos GV, Steigerwalt AG (1973) Polynucleotide sequence relatedness among *Shigella* species. Int J Syst Bacteriol 23:1–7
24. Brenner DG (1984) Family I Enterobacteriaceae Rhan (1937). In: Krieg NR, Holt JG (eds) Bergey's Manual of systematic bacteriology, vol 1. The Williams & Williams Co, Baltimore, pp 410–411
25. Maheux AF, Picard FJ, Boissinot M, Bissonnette L, Paradis S, Bergeron MG (2009) Analytical comparison of nine PCR primer sets designed to detect the presence of *Escherichia coli/Shigella* in water samples. Water Res 43:3019–3028
26. Picard FJ, Gagnon M, Bernier MR, Parham NJ, Bastien M, Boissinot M, Peytavi R, Bergeron MG (2009) Internal control for nucleic acid testing based on the use of purified *Bacillus atrophaeus* subsp. *globigii* spores. J Clin Microbiol 47:751–757
27. Jordan JA (2000) Real-time detection of PCR products and microbiology. New Technol Life Sci 6:61–66
28. Yang S, Rothman RE (2004) PCR-based diagnostics for infectious diseases: uses, limitations, and future applications in acute-care settings. Lancet Infect Dis 4:337–348
29. Picard FJ, Ménard C, Bastien M, Boissinot M (2005-02-17) Method for the preparation of reagents for amplification and/or detection of nucleic acids that exhibit no significant contamination by nucleic acids, United States Patent Application 20050037349.
30. Bustin SA, Benes V, Garson JA, Hellemans J, Huggett J, Kubista M, Mueller R, Nolan T, Pfaffl MW, Shipley GL, Vandesompele J, Wittwer CT (2009) The MIQE guidelines: minimum information for publication of quantitative real-time PCR experiments. Clin Chem 55:611–622
31. United States Environmental Protection Agency. (2004) Quality assurance/quality control guidance for laboratories performing PCR analyses on environmental samples, EPA 815-B-04-001. Office of Water (4607), United States Environmental Protection Agency, Cincinnati, OH, 56 pages.
32. Mitchell PS, Germer JJ, Patel R (2004) Nucleic acid amplification methods: laboratory design and operations. In: Persing DH, Tenover FC, Versalovic J, Tang Y-W, Unger ER, Relman DA, White TJ (eds) Molecular microbiology: Diagnostic principles and practice. ASM Press, Washington, DC, pp 85–93
33. Glushkov SA, Bragin AG, Dymshits GM (2009) Decontamination of polymerase chain reaction reagents using DEAE-cellulose. Anal Biochem 393:135–137
34. Shepley DP, Wolk DM (2004) Quantitative molecular methods: result standardization, interpretation, and laboratory quality control. In: Persing DH, Tenover FC, Versalovic J, Tang Y-W, Unger ER, Relman DA, White TJ (eds) Molecular microbiology: Diagnostic principles and practice. ASM Press, Washington, DC, pp 95–129

Chapter 21

Multiplex Real-Time PCR (MRT-PCR) for Diarrheagenic

Francesca Barletta, Theresa J. Ochoa, and Thomas G. Cleary

Abstract

Diarrheagenic *Escherichia coli* strains are important causes of diarrhea in children from the developing world and are now being recognized as emerging enteropathogens in the developed world. Current methods of detection are too expensive and labor-intensive for routine detection of these organisms to be practical. We developed a real-time fluorescence-based multiplex PCR for the detection of all six of the currently recognized classes of diarrheagenic *E. coli*. The primers were designed to specifically amplify eight different virulence genes in the same reaction: *aggR* for enteroaggregative *E. coli* (EAEC), *stIa/stIb* and *lt* for enterotoxigenic *E. coli* (ETEC), *eaeA* for enteropathogenic *E. coli* (EPEC), *stx*1 and *stx*2 for Shiga toxin-producing *E. coli* (STEC), *ipaH* for enteroinvasive *E. coli* (EIEC), and *daaD* for diffusely adherent *E. coli* (DAEC).

Key words: Multiplex real-time PCR, Diarrhea, Diarrheagenic *E. coli*, Enteroaggregative *E. coli*, Enterotoxigenic *E. coli*, Enteropathogenic *E. coli*, Shiga toxin-producing *E. coli*, Enteroinvasive *E. coli*, Diffusely adherent *E. coli*

1. Introduction

Escherichia coli are the predominant facultative anaerobes of the human colonic flora, and they usually remain harmlessly confined to the intestinal lumen. In the case of immunosupressed host, neonates or patients with impaired gastrointestinal barriers, systemic infections (urinary tract infection, meningitis, bacteremia) may occur due to *E. coli* possessing specific virulence genes. *E. coli* strains associated with diarrhea do not possess genes associated with these systemic infections, but rather have genes that allow them to attach, invade, damage intestinal cell or to alter their function. The diarrheagenic *E. coli* have been classified into six groups based on clinical, epidemiological, and molecular criteria (Table 1).

Mark Wilks (ed.), *PCR Detection of Microbial Pathogens: Second Edition*, Methods in Molecular Biology, vol. 943,
DOI 10.1007/978-1-60327-353-4_21,

Table 1
Diarrheagenic *E. coli* groups and associated genes

Abbreviations	Name	Genes analyzed in MRT-PCR	Other associated genes
ETEC	Enterotoxigenic *E. coli*	*st, lt*	
EPEC	Enteropathogenic *E. coli*	*eaeA*	*bfpA, astA, per*
EIEC	Enteroinvasive *E. coli*	*ipaH*	*virF, ial*
STEC/EHEC	Shiga toxin-producing *E. coli or* enterohemorragic *E. coli*	*eaeA, stx1, stx2*	*astA, hlyA, ehx, uidA, fliC*
EAEC or EAggEC	Enteroaggregative *E. coli*	*aggR*	*astA, aat*
DAEC	Diffusely adherent *E. coli*	*daaD*	*daaC*

Although these *E. coli* commonly cause gastroenteritis in children they are not routinely sought in clinical laboratories worldwide because rapid, specific, inexpensive tools have been lacking.

Even though *E. coli* of specific serogroups can be associated with specific clinical syndromes, it is not generally the serologic antigens themselves that confer virulence but the production of specific proteins such as toxins and adhesins (1). In addition, serotyping is tedious and expensive. Phenotypic assays such as the Hep-2 adherence assay (2, 3) differentiate localized (EPEC), aggregative (EAEC), and diffuse adherence (DAEC) *E. coli*; however, these methods are expensive and labor-intensive. Thus, a diagnosis of the diarrheagenic group of *E. coli* requires a molecular method focused on the identification of characteristics that determine their virulence. Several associated genes have been studied, but we chose to amplify the genes in Table 1 for various reasons specific to each category of pathogen. For ETEC, STEC, and EPEC we chose genes that have a well-established central role in pathogenesis (*lt, stIa/stIb,stx1, stx2, eaeA*). For DAEC we selected a gene that was highly conserved and probably part of the virulence mechanism but not yet proven itself to be essential for virulence (*daaD*). In the case of EAEC, a pathogenic group whose mechanism of virulence is poorly understood and changing, we selected a gene that has been epidemiologically but provisionally linked with virulence (*aggR*). For EIEC we chose a gene that is conserved but its precise role in invasion is uncertain (*ipaH*).

Traditional PCR methods require amplification in a thermocycler followed by product separation by gel electrophoresis (4) or fluorescent capillary electrophoresis (5), both of which are time-consuming and laborious processes. However, as shown here, the products of PCR can also be practically detected by using a DNA binding dye, such as SYBR Green, in a multiplex format. SYBR

Green is the most widely used double-strand DNA-specific dye reported for real-time PCR. It binds to the minor groove of the DNA double helix. In the solution, the unbound dye exhibits very little fluorescence. This fluorescence is substantially enhanced when the dye is bound to double-stranded DNA. The fluorescence of the reporter dye increases as the product accumulates with each successive cycle of amplification. By recording the amount of fluorescence emission at each cycle, it is possible to monitor the PCR reaction during exponential phase (Fig. 1). Real-time PCR offers the advantage of being a faster, more robust assay because it

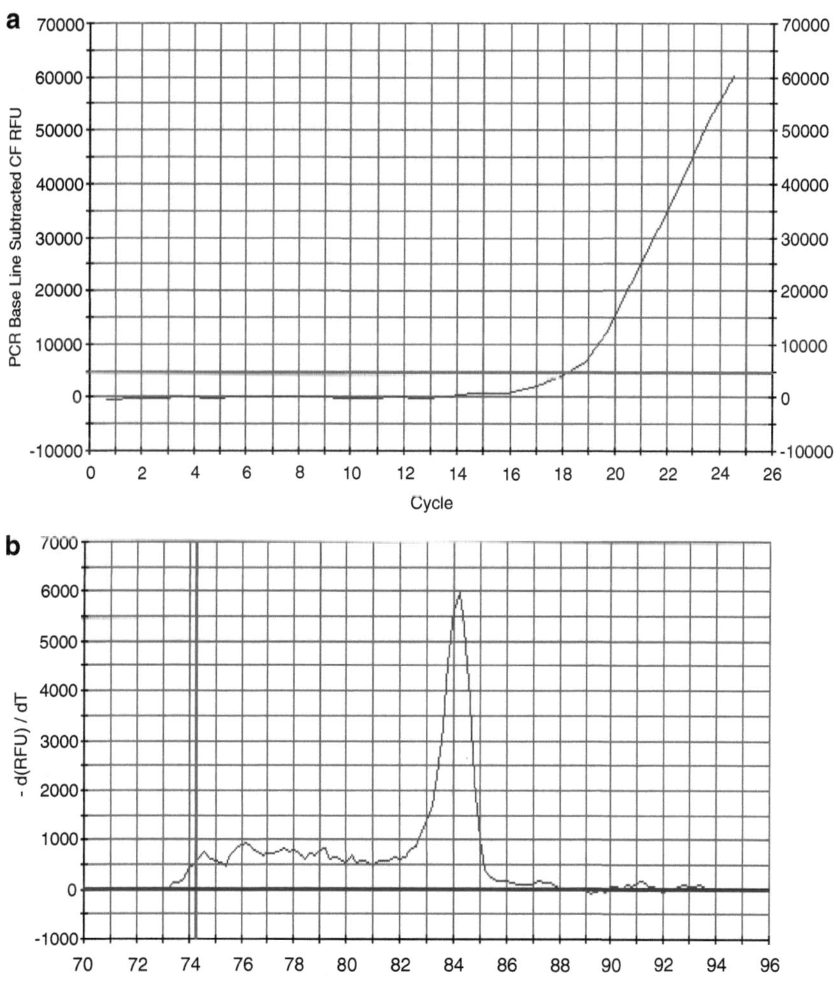

Fig. 1. Typical reading from a single PCR cycle in a real-time PCR machine. (**a**) The vertical axis represents copy number (arbitrary units) and the horizontal axis shows the PCR cycle number. The ***solid line*** is an arbitrary value, usually about 0.1 and is the "copy number" used to determine C_t. The lower a C_t value, the more copies are present in the specific sample. (**b**) Following amplification, a melting curve analysis was performed to verify the correct product by its specific melting temperature (T_m). In this example the T_m of 84°C corresponds to the ***eae***A gene of EPEC.

does not require post-PCR procedures to detect amplification products. The assay that we describe is unique in the number of different genes recognized in a single reaction (6).

The approach we used to develop this assay was to start with previously described standard PCR reactions for diagnosis of diarrheagenic *E. coli* in a multiplex format. Previously described primer pairs for each pathogenic group were tested with prototypical strains of all six pathogenic groups (7–10). Any pair that nonspecifically amplified more than one product of its target strain or amplified something in a different pathogenic group or in *E. coli* that were part of the normal flora was discarded. Extraneous uninterruptible peaks (wide peaks, small peaks, multiple peaks or peaks at a melting temperature [T_m] very different from the predicted T_m in silico) were found with some previously described primers and so these were eliminated from further consideration and new primers designed. Each set of new primers was examined by BLAST to determine its suitability for amplifying the multiple variants of each target and was analyzed by an oligonucleotide property calculator (http://www.basic.northwestern.edu/biotools/oligocalc.html) which uses the nearest-neighbor method to predict the amplicon melting temperature (T_m). Primers predicted to give a peak separate from the peaks of the primers chosen for other pathogenic groups were chosen. After individual primer pairs were selected, each pair was sequentially added to the mix to determine whether the primers themselves would amplify with no bacterial DNA present. If there was no nonspecific amplification, the primer pools were tested with representative strains of each pathogenic group and normal *E. coli*. The result of this approach was to select primers that gave specific, distinct nonoverlapping peaks that could detect several pathogenic types in the same mix (6) (Table 2).

2. Materials

2.1. MacConkey Medium

1. Master Mix

The reagents for the PCR master mix are listed in Table 3.

3. Methods

3.1. Stool Culture

Take a small stool sample (see Note 1) with a swab, streak it on a MacConkey agar plate and incubate at 37°C for 24 h. Growth of *E. coli*, which ferments lactose, appears red/pink on the agar. Select, randomly, 5 lactose-positive colonies (see Note 2) streak

Table 2
Primers for multiplex real-time PCR for diarrheagenic *E. coli*

Gene	O	Primer sequence 5′–3′	Final concentration (μM)	Amplicon size (bp)
eaeA	F	ATGCTTAGTGCTGGTTTAGG	0.56	248
	R	GCCTTCATCATTTCGCTTTC		
aggR	F	CGAAAAAGAGATTATAAAAATTAAC	0.44	100
	R	GCTTCCTTCTTTTGTGTAT		
daaD	F	TGAACGGGAGTATAAGGAAGATG	0.50	444
	R	GTCCGCCATCACATCAAAA		
ipaH	F	GTTCCTTGACCGCCTTTCCGATACCGTC	0.04	619
	R	GCCGGTCAGCCACCCTCTGAGAGTAC		
stIa	F	TTTCCCCTCTTTTAGTCAGTCAA	0.26	159
stIb	F	TGCTAAACCAGTAGAGTCTTCAAAA		138
st	R	GCAGGATTACAACACAATTCACAGCAG		
stx1	F	CTGGATTTAATGTCGCATAGTG	0.12	150
	R	AGAACGCCCACTGAGATCATC		
stx2	F	GGCACTGTCTGAAACTGCTCC	0.08	255
	R	TCGCCAGTTATCTGACATTCTG		
lt	F	TCTCTATGTGCATACGGAGC	0.36	322
	R	CCATACTGATTGCCGCAAT		

O orientation

Table 3
Master mix reagents for MRT-PCR

Reagent	Stock (μM)	Target (μM)	1 Rxn (μl)
PCR water			13.63–14.67[a]
5× buffer		1×	5.00
50 mM MgCl2		4 mM	2.00
dNTPs		200 μM	0.50
Primer fwd	25	0.04–0.56	0.04–0.56
Primer rev	25	0.04–0.56	0.04–0.56
SBRG	100×	1×	0.25
Phusion[b]	2 U/μl	0.5 U/Rxn	0.25
Fluorescein	1	10 nM	0.25
Total volume[c]			23.00

[a]The final volume of PCR water will dependent on the volume use for each set of primers
[b]Phusion High Fidelity DNA Polymerase. Finnzymes, Espoo, Finland. PCR Kit contains Buffer 5×, MgCl2, dNTPs and the polymerase
[c]Add 2 μl of crude lysate to have a final reaction volume of 25 μl

them in segments of a new MacConkey agar plate and incubate at 37°C for 24 h. To look for the occasional non-lactose fermenting pathogen, lactose negative colonies may also be sampled.

3.2. Sample Preparation

Five colonies are carefully removed from the plate by using a sterile toothpick (see Note 3). Crude lysates are prepared and used directly as a template for the PCR. DNA is extracted by boiling a single colony in 100 μl of PCR- or molecular-grade water for 5 min, followed by centrifugation at 15,000 × *g* for 10 min. Remove 50 μl of the supernatant to a new tube and store it at −20°C until the PCR analysis.

3.3. Controls

Prototypical strains (DAEC-5019, STEC-91/8626, ETEC-H10407, EPEC-2348/69, EIEC-213, and EAEC-042) are used as positives controls, and *E. coli* C600 is used as a negative control for each PCR reaction.

3.4. Primers

Primers are resuspend to 100 μM. Then 4 stocks of 25 μM are prepared and stored at −20°C. The volume for each set of primers is calculated according to final concentration wanted for each of them (Table 2).

3.5. PCR Conditions

PCR is performed using a PTC-200 thermal cycler and real-time fluorescence monitoring by a Chromo 4 optical detector (MJ Research/Bio-Rad, Hercules, CA). The amplification cycles consist of incubation at 98°C for 50 s, 60°C for 20 s, 72°C for 30 s, and 75°C for 1 s. After 25 cycles, a melting curve with a ramp speed of 2.5°C/s between 73 and 95°C is determined with a reading every 0.2°C using fluorescence of SYBR green (see Note 4). Melting peaks are automatically calculated by Opticon Monitor software (Bio-Rad, Hercules, CA), which, after subtracting background fluorescence from a set of water blanks, plot the negative derivative of fluorescence with respect to temperature ($-dF/dT$ versus T). An example result is shown in Fig. 2.

4. Interpretation

Each positive control must be analyzed independently to be sure of a correct T_m with an acceptable standard deviation for each gene (Fig. 2). Then compare each unknown sample with the positive controls. There is machine to machine variability on the T_m measurements for each gene. For example with the iCycler iQ, these were the T_m (mean + standard deviation) for each gene: *aggR* 77.1 + 0.7°C, *st* 81.4 + 0 .3°C, *eaeA* 83.9 + 0.3°C, *lt* 85.9 + 0.3°C, *stx1* 87.4 + 0.3°C, *stx2* 89.6 + 0.3°C, *ipaH* 91.5 + 0.3°C, and *daaD* 93.8 + 0.4°C.

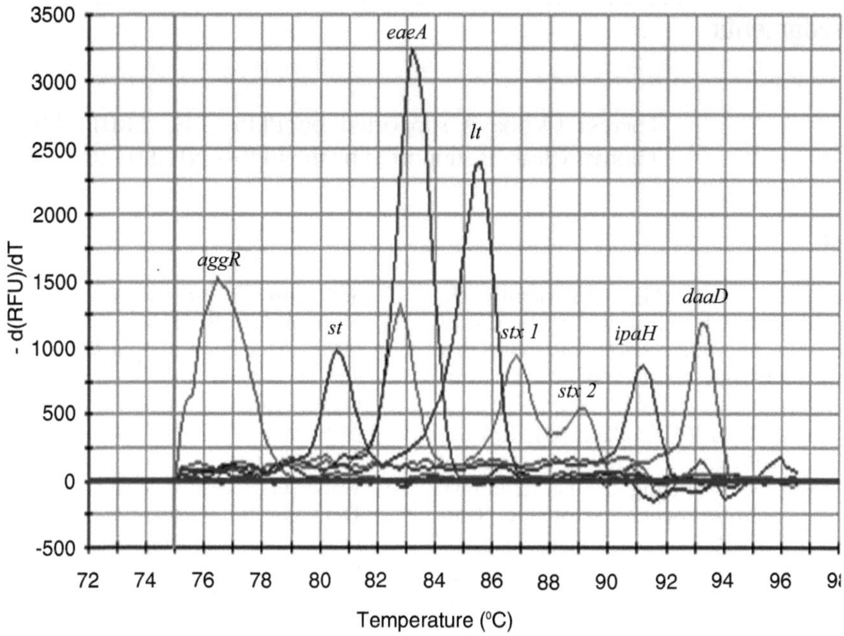

Fig. 2. Real-time PCR display showing the range of melting temperatures of the eight different diarrheagenic ***Escherichia coli*** genes. Data from individual tubes, each containing an EAEC, ETEC, EPEC, STEC, EIEC, or DAEC strain, are shown in a single figure so that the separation between individual amplicon melting curves is illustrated. Fluorescence melting peaks were obtained by plotting the negative derivative of fluorescence over temperature (–dRFU/d***T***) versus temperature.

5. Notes

1. Stool samples should be collected in a clean plastic container. Samples should be sent to the laboratory immediately or should be transferred to a transport media (Cary Blair). Do not include toilet paper or urine in the specimen.
2. The choice of five colonies for analysis represents a compromise between the cost of analysis and the need to detect infection. In addition, the analysis of a pool of 5 colonies for the detection of diarrheagenic *E. coli* by multiplex real-time PCR is a dramatically more cost effective, sensitive, and specific technique when compared to the individual colony analysis (11).
3. We recommend removing the colonies from the agar plate with a sterile toothpick to avoid agar contamination (metal loops may pick some agar from the plate), an important cause of erratic amplification.
4. Other real-time thermocyclers could be used prior standardization of the PCR conditions.

Acknowledgments

Theresa Ochoa is supported by PHS—FIC 1K01TW007405 and Thomas Cleary is supported by the PHS—NICHD R01- HD051716.

References

1. Whittam TS, Wolfe ML, Wachsmuth IK, Orskov F, Orskov I, Wilson RA (1993) Clonal relationships among *Escherichia coli* strains that cause hemorrhagic colitis and infantile diarrhea. Infect Immun 61:1619–1629
2. Cravioto A, Gross RJ, Scotland SM, Rowe B (1979) An adhesive factor found in strains of *Escherichia coli* belonging to the traditional infantile enteropathogenic serotypes. Curr Microbiol 3:95–99
3. Vial PA, Mathewson JJ, DuPont HL, Guers L, Levine MM (1990) Comparison of two assay methods for patterns of adherence to HEp-2 cells of *Escherichia coli* from patients with diarrhea. J Clin Microbiol 28:882–885
4. Nguyen TV, Le Van P, Le Huy C, Gia KN, Weintraub A (2005) Detection and characterization of diarrheagenic *Escherichia coli* from Young children in Hanoi, Vietnam. J Clin Microbiol 43:755–760
5. Brandal LT, Lindstedt S, Aas L, Stavnes T, Lassen J, Kapperud G (2007) Octaplex PCR and fluorescence-based capillary electrophoresis for identification of human diarrheagenic *Escherichia coli* and *Shigella* spp. J Microbiol Methods 68:331–341
6. Guion CE, Ochoa TJ, Walker CM, Barletta F, Cleary TG (2008) Detection of diarrheagenic Escherichia coli by use of melting-curve analysis and real-time multiplex PCR. J Clin Microbiol 46(5):1752–1757
7. Sethabutr O, Venkatesan M, Murphy GS, Eampokalap B, Hoge CW, Echeverria P (1993) Detection of *Shigellae* and enteroinvasive *Escherichia coli* by amplification of the invasion plasmid antigen H DNA sequence in patients with dysentery. J Infect Dis 167:458–461
8. Wang G, Clark CG, Rodgers FG (2002) Detection in *Escherichia coli* of the genes encoding the major virulence factors, the genes defining the 0157:H7 serotype, and components of the type 2 Shiga toxin family by multiplex PCR. J Clin Microbiol 40:3613–3619
9. Lopez-Saucedo C, Cerna JF, Villegas-Sepulveda N, Thompson R, Velazquez FR, Torres J, Tarr PI, Estrada-Garcia T (2003) Single multiplex polymerase chain reaction to detect diverse loci associated with diarrheagenic *Escherichia coli*. Emerg Infect Dis 9:127–131
10. Toma C, Lu Y, Higa N, Nakasone N, Chinen I, Baschkier A, Rivas M, Iwanaga M (2003) Multiplex PCR assay for identification of human diarrheagenic *Escherichia coli*. J Clin Microbiol 41:2669–2671
11. Barletta F, Ochoa TJ, Ecker L, Gil AI, Lanata CF, Cleary TG (2009) Validation of five-colony pool analysis using multiplex real-time PCR for detection of diarrheagenic *Escherichia coli*. J Clin Microbiol 47(6): 1915–1917

Index

Mark Wilks (ed.), *PCR Detection of Microbial Pathogens: Second Edition*, Methods in Molecular Biology, vol. 943, DOI 10.1007/978-1-60327-353-4, © Springer Science+Business Media, LLC 2013

Q

R

S

T

U

W

MIX
Papier aus verantwortungsvollen Quellen
Paper from responsible sources
FSC® C105338

If you have any concerns about our products,
you can contact us on
ProductSafety@springernature.com

In case Publisher is established outside the EU,
the EU authorized representative is:
Springer Nature Customer Service Center GmbH
Europaplatz 3, 69115 Heidelberg, Germany

Printed by Libri Plureos GmbH
in Hamburg, Germany